Handbook of Mathematical Analysis

Handbook of Mathematical Analysis

Edited by Matt Ferrier

Larsen & Keller
www.larsen-keller.com

Handbook of Mathematical Analysis
Edited by Matt Ferrier
ISBN: 979-8-88836-048-4 (Hardback)

Larsen & Keller
Published by Larsen and Keller Education,
5 Penn Plaza,
19th Floor,
New York, NY 10001, USA

Cataloging-in-Publication Data

Handbook of mathematical analysis / edited by Matt Ferrier.
p. cm.
Includes bibliographical references and index.
ISBN 979-8-88836-048-4
1. Mathematical analysis. 2. Algebra. 3. Mathematics. I. Ferrier, Matt.
QA300 .A67 2023
515--dc23

For more information regarding Larsen and Keller Education and its products, please visit the publisher's website www.larsen-keller.com

Table of Contents

Preface

Mathematical analysis is a branch of mathematics that involves the study of continuous functions, limits, and related theories, such as differentiation, integration, infinite sequences, series and analytic functions in the context of real and complex numbers, and functions. The field of mathematical analysis developed from calculus, which involves the basic concepts and techniques of analysis. The study of mathematical analysis encompasses the study of various topics such as differential equations, measure theory, numerical analysis and vector analysis. These branches of mathematical analysis are applied in various fields of study. Differential equations are used in a wide range of disciplines such as engineering, physics, economics and biology. Newton's second law, the Schrodinger equation, and the Einstein field equations are also based on differential equations. This book is a detailed explanation of the various concepts, theories, and applications of mathematical analysis. It will prove to be immensely beneficial to the students and researchers associated with this area of mathematics.

Various studies have approached the subject by analyzing it with a single perspective, but the present book provides diverse methodologies and techniques to address this field. This book contains theories and applications needed for understanding the subject from different perspectives. The aim is to keep the readers informed about the progresses in the field; therefore, the contributions were carefully examined to compile novel researches by specialists from across the globe.

Indeed, the job of the editor is the most crucial and challenging in compiling all chapters into a single book. In the end, I would extend my sincere thanks to the chapter authors for their profound work. I am also thankful for the support provided by my family and colleagues during the compilation of this book.

Editor

1

Fractional Integration and Differentiation of the Generalized Mathieu Series

Ram K. Saxena [1] and Rakesh K. Parmar [2,*]

[1] Department of Mathematics and Statistics, Jai Narain Vyas University, Jodhpur-342004, Rajasthan, India; ram.saxena@yahoo.com
[2] Department of Mathematics, Government College of Engineering and Technology, Bikaner-334004, Rajasthan, India
* Correspondence: rakeshparmar27@gmail.com

Academic Editor: Hans J. Haubold

Abstract: We aim to present some formulas for the Saigo hypergeometric fractional integral and differential operators involving the generalized Mathieu series $S_\mu(r)$, which are expressed in terms of the Hadamard product of the generalized Mathieu series $S_\mu(r)$ and the Fox–Wright function ${}_p\Psi_q(z)$. Corresponding assertions for the classical Riemann–Liouville and Erdélyi–Kober fractional integral and differential operators are deduced. Further, it is emphasized that the results presented here, which are for a seemingly complicated series, can reveal their involved properties via the series of the two known functions.

Keywords: Mathieu series; generalized Mathieuseries; fractional calculus operators

---

1. Introduction and Preliminaries

Fractional calculus, which has a long history, is an important branch of mathematical analysis (calculus) where differentiations and integrations can be of arbitrary non-integer order. During the past four decades or so, fractional calculus has been widely and extensively investigated and has gained importance and popularity due mainly to its demonstrated applications in numerous and diverse fields of science and engineering such as turbulence and fluid dynamics, stochastic dynamical system, plasma physics and controlled thermonuclear fusion, nonlinear control theory, image processing, nonlinear biological systems, and astrophysics (see, for detail, [1–5]).

We recall Saigo fractional integral and differential operators involving Gauss's hypergeometric function ${}_2F_1$ as a kernel. Let $\alpha,\ \beta,\ \eta \in \mathbb{C}$, $\Re(\alpha) > 0$ and $x > 0$, then Saigo's fractional integral and differential operators $\left(I_{0+}^{\alpha,\beta,\eta} f\right)(x)$, $\left(I_{-}^{\alpha,\beta,\eta} f\right)(x)$ and $\left(D_{0+}^{\alpha,\beta,\eta} f\right)(x)$, $\left(D_{-}^{\alpha,\beta,\eta} f\right)(x)$ are defined as (see, for example, [1,2,4–6]):

$$\left(I_{0+}^{\alpha,\beta,\eta} f\right)(x) = \frac{x^{-\alpha-\beta}}{\Gamma(\alpha)} \int_0^x (x-t)^{\alpha-1}\, {}_2F_1\left(\alpha+\beta, -\eta; \alpha; 1-\frac{t}{x}\right) f(t)\, \mathrm{d}t, \tag{1}$$

$$\left(I_{-}^{\alpha,\beta,\eta} f\right)(x) = \frac{1}{\Gamma(\alpha)} \int_x^\infty (t-x)^{\alpha-1} t^{-\alpha-\beta}\, {}_2F_1\left(\alpha+\beta, -\eta; \alpha; 1-\frac{x}{t}\right) f(t)\, \mathrm{d}t, \tag{2}$$

and

$$\begin{aligned}\left(D_{0+}^{\alpha,\beta,\eta} f\right)(x) &= \left(I_{0+}^{-\alpha,-\beta,\alpha+\eta} f\right)(x) \\ &= \left(\frac{d}{dx}\right)^n \left(I_{0+}^{-\alpha+n,-\beta-n,\alpha+\eta-n} f\right)(x) \quad (n = [\Re(\alpha)]+1),\end{aligned} \tag{3}$$

$$\begin{aligned}\left(D_{-}^{\alpha,\beta,\eta} f\right)(x) &= \left(I_{-}^{-\alpha,-\beta,\alpha+\eta} f\right)(x)\\ &= (-1)^n \left(\frac{d}{dx}\right)^n \left(I_{-}^{-\alpha+n,-\beta-n,\alpha+\eta} f\right)(x) \quad (n = [\Re(\alpha)] + 1).\end{aligned} \tag{4}$$

Here and in what follows, $[x]$ denotes the greatest integer less than or equal to the real number x. When $\beta = -\alpha$, the operators in (1)–(4) coincide with the classical Riemann–Liouville fractional integrals and derivatives of order $\alpha \in \mathbb{C}$ with $\Re(\alpha) > 0$ and $x > 0$ (see, e.g., [1,4]):

$$\left(I_{0+}^{\alpha,-\alpha,\eta} f\right)(x) = \left(I_{0+}^{\alpha} f\right)(x) \equiv \frac{1}{\Gamma(\alpha)} \int_0^x (x-t)^{\alpha-1} f(t)\, dt, \tag{5}$$

$$\left(I_{-}^{\alpha,-\alpha,\eta} f\right)(x) = \left(I_{-}^{\alpha} f\right)(x) = \frac{1}{\Gamma(\alpha)} \int_x^{\infty} (t-x)^{\alpha-1} f(t)\, dt, \tag{6}$$

and

$$\begin{aligned}\left(D_{0+}^{\alpha,-\alpha,\eta} f\right)(x) &= \left(D_{0+}^{\alpha} f\right)(x) = \left(\frac{d}{dx}\right)^n \frac{1}{\Gamma(n-\alpha)} \int_0^x (x-t)^{n-\alpha-1} f(t)\, dt\\ &= \left(\frac{d}{dx}\right)^n \left(I_{0+}^{n-\alpha} f\right)(x) \qquad (n = [\Re(\alpha)] + 1),\end{aligned} \tag{7}$$

$$\begin{aligned}\left(D_{-}^{\alpha,-\alpha,\eta} f\right)(x) &= \left(D_{-}^{\alpha} f\right)(x) = (-1)^n \left(\frac{d}{dx}\right)^n \frac{1}{\Gamma(n-\alpha)} \int_x^{\infty} (t-y)^{n-\alpha-1} f(t)\, dt\\ &= (-1)^n \left(\frac{d}{dx}\right)^n \left(I_{-}^{n-\alpha} f\right)(x) \qquad (n = [\Re(\alpha)] + 1).\end{aligned} \tag{8}$$

Here and in the following, let $\mathbb{C}$, $\mathbb{R}^+$, and $\mathbb{N}$ be the sets of complex numbers, positive real numbers, and positive integers, respectively, and let $\mathbb{N}_0 := \mathbb{N} \cup \{0\}$.

If $\beta = 0$ in (1)–(4) yields the so-called Erdélyi–Kober fractional integrals and derivatives of order $\alpha \in \mathbb{C}$ with $\Re(\alpha) > 0$ and $x > 0$ (see, e.g., [1,4]):

$$\left(I_{0+}^{\alpha,0,\eta} f\right)(x) = \left(I_{\eta,\alpha}^{+} f\right)(x) = \frac{x^{-\alpha-\eta}}{\Gamma(\alpha)} \int_0^x (x-t)^{\alpha-1} t^{\eta} f(t)\, dt, \tag{9}$$

$$\left(I_{-}^{\alpha,0,\eta} f\right)(x) = \left(K_{\eta,\alpha}^{-} f\right)(x) \equiv \frac{x^{\eta}}{\Gamma(\alpha)} \int_x^{\infty} (t-x)^{\alpha-1} t^{-\alpha-\eta} f(t)\, dt, \tag{10}$$

and

$$\begin{aligned}\left(D_{0+}^{\alpha,0,\eta} f\right)(x) &= \left(D_{\eta,\alpha}^{+} f\right)(x)\\ &= \left(\frac{d}{dx}\right)^n \left(I_{0+}^{-\alpha+n,-\alpha,\alpha+\eta-n} f\right)(x) \quad (n = [\Re(\alpha)] + 1),\end{aligned} \tag{11}$$

$$\begin{aligned}\left(D_{-}^{\alpha,0,\eta} f\right)(x) &= \left(D_{\eta,\alpha}^{-} f\right)(x)\\ &= (-1)^n \left(\frac{d}{dx}\right)^n \left(I_{-}^{-\alpha+n,-\alpha,\alpha+\eta} f\right)(x) \quad (n = [\Re(\alpha)] + 1),\end{aligned} \tag{12}$$

$$\left(D_{\eta,\alpha}^{+} f\right)(x) = x^{-\eta} \left(\frac{d}{dx}\right)^n \frac{1}{\Gamma(n-\alpha)} \int_0^x t^{\alpha+\eta} (x-t)^{n-\alpha-1} f(t)\, dt \quad (n = [\Re(\alpha)] + 1), \tag{13}$$

$$\left(D_{\eta,\alpha}^{-}f\right)(x) = x^{\eta+\alpha}\left(\frac{d}{dx}\right)^{n}\frac{1}{\Gamma(n-\alpha)}\int_{x}^{\infty} t^{-\eta}(t-x)^{n-\alpha-1}f(t)\,\mathrm{d}t \quad (n = [\Re(\alpha)]+1). \tag{14}$$

A detailed account of such operators along with their properties and applications has been considered by several authors (see, for details, [1–5]).

The following familiar infinite series of the form

$$S(r) = \sum_{n\geq 1}\frac{2n}{(n^2+r^2)^2}, \qquad r > 0 \tag{15}$$

is known in literature as the Mathieu series. Émile Leonard Mathieu was the first to investigate such a series in 1890 in his book elasticity of solid bodies [7]. An alternative version of (15)

$$\widetilde{S}(r) = \sum_{n\geq 1}(-1)^{n-1}\frac{2n}{(n^2+r^2)^2}, \qquad r > 0 \tag{16}$$

was introduced by Pogány et al. [8]. Closed form integral representations for $S(r)$ and $\widetilde{S}(r)$ are given by (see e.g., [8,9])

$$S(r) = \frac{1}{r}\int_0^\infty \frac{x\sin(rx)}{\mathrm{e}^x-1}\,\mathrm{d}x \tag{17}$$

and

$$\widetilde{S}(r) = \frac{1}{r}\int_0^\infty \frac{x\sin(rx)}{\mathrm{e}^x+1}\,\mathrm{d}x, \tag{18}$$

respectively. Several interesting problems and solutions deal with integral representations and bounds for the following mild generalization of the Mathieu series and its alternative version with a fractional power defined by ([10], p. 2, Equation (16)) (see also, [11], p. 181)

$$S_\mu(r) = \sum_{n\geq 1}\frac{2n}{(n^2+r^2)^{\mu+1}} \qquad (r > 0,\ \mu > 0) \tag{19}$$

and

$$\widetilde{S}_\mu(r) = \sum_{n\geq 1}(-1)^{n-1}\frac{2n}{(n^2+r^2)^{\mu+1}} \qquad (r > 0,\ \mu > 0), \tag{20}$$

respectively. Such a series has been widely considered in mathematical literature (see, e.g., papers of Cerone and Lenard [10], Diananda [12] and Pogány et al. [8]). Various applications of the familiar Mathieu series and its generalizations in probability theory with other variants such as trigonometric Mathieu series, harmonic Mathieu series, Fourier–Mathieu series and some other particular forms of the Mathieu series can be found in a recent paper [13].

Recently, Tomovski and Pogány [14] studied the several integral representations of the generalized fractional order Mathieu-type power series

$$S_\mu(r;z) = \sum_{n\geq 1}\frac{2n\,z^n}{(n^2+r^2)^{\mu+1}}, \qquad (\mu > 0,\ r\in\mathbb{R},\ |z| < 1). \tag{21}$$

Obviously, we have

$$S_\mu(r;1) = S_\mu(r) \quad \text{and} \quad S_\mu(r;-1) = -\widetilde{S}_\mu(r).$$

Various other investigations and generalizations of the Mathieu series with its alternative variants can also be found in [11,14–24], and the references cited therein.

The concept of the Hadamard product (or the convolution) of two analytic functions is useful in our present investigation. It can help us to decompose a newly emerged function into two known functions. If, in particular, one of the power series defines an entire function, then the Hadamard product series defines an entire function, too. Let

$$f(z) := \sum_{n=0}^{\infty} a_n z^n \ (|z| < R_f) \ \text{and} \ g(z) := \sum_{n=0}^{\infty} b_n z^n \ (|z| < R_g)$$

be two power series whose radii of convergence are denoted by R_f and R_g, respectively. Then, their Hadamard product is the power series defined by

$$(f * g)(z) := \sum_{n=0}^{\infty} a_n\, b_n z^n = (g * f)(z) \ (|z| < R) \tag{22}$$

where

$$R = \lim_{n\to\infty} \left| \frac{a_n\, b_n}{a_{n+1}\, b_{n+1}} \right| = \left(\lim_{n\to\infty} \left| \frac{a_n}{a_{n+1}} \right| \right) \cdot \left(\lim_{n\to\infty} \left| \frac{b_n}{b_{n+1}} \right| \right) = R_f \cdot R_g,$$

therefore, in general, we have $R \geq R_f \cdot R_g$ [25,26]. For various other investigations involving the Hadamard product (or the convolution), the interested reader may be referred to several recent papers on the subject (see, for example, [27,28] and the references cited in each of these papers).

In this paper, our aim is to study the compositions of the generalized fractional integration and differentiation operators (1)–(4) with the generalized Mathieu series (21) in terms of the Hadamard product (22) of the generalized Mathieu series and the Fox–Wright function. Further, corresponding assertions for the classical Riemann–Liouville and Erdélyi–Kober fractional integral and differential operators are deduced. The results presented in Theorems together with Corollaries are sure to be new and potentially useful, mainly because they are expressed in terms of the Hadamard product with two known functions. At least, a seemingly complicated resulting series expressed in terms of two known functions *means* that certain properties involved in the complicated resulting series can be revealed via the series of the known functions.

2. Fractional Integration of the Mathieu Series

We first recall the Fox–Wright function ${}_p\Psi_q(z)$ $(p,\, q \in \mathbb{N}_0)$ with p numerator and q denominator parameters defined for $\alpha_1, \ldots, \alpha_p \in \mathbb{C}$ and $\beta_1, \ldots, \beta_q \in \mathbb{C} \setminus \mathbb{Z}_0^-$ by (see, for details, [1,3]; see also [4,29]):

$${}_p\Psi_q \left[\begin{matrix} (\alpha_1, A_1), \cdots, (\alpha_p, A_p); \\ (\beta_1, B_1), \cdots, (\beta_q, B_q); \end{matrix} \; z \right] = \sum_{n=0}^{\infty} \frac{\Gamma(\alpha_1 + A_1 n) \cdots \Gamma(\alpha_p + A_p n)}{\Gamma(\beta_1 + B_1 n) \cdots \Gamma(\beta_q + B_q n)} \frac{z^n}{n!} \tag{23}$$

$$\left(A_j \in \mathbb{R}^+ \ (j = 1, \ldots, p);\ B_j \in \mathbb{R}^+ \ (j = 1, \ldots, q);\ 1 + \sum_{j=1}^{q} B_j - \sum_{j=1}^{p} A_j \geq 0 \right),$$

where the equality in the convergence condition holds true for

$$|z| < \nabla := \left(\prod_{j=1}^{p} A_j^{-A_j} \right) \cdot \left(\prod_{j=1}^{q} B_j^{B_j} \right).$$

In particular, when $A_j = B_k = 1$ $(j = 1, \ldots, p;\ \ k = 1, \ldots, q)$, (23) reduces immediately to the generalized hypergeometric function ${}_pF_q$ $(p,\, q \in \mathbb{N}_0)$ (see, e.g., [29]):

$${}_pF_q \left[\begin{matrix} \alpha_1, \ldots, \alpha_p; \\ \beta_1, \ldots, \beta_q; \end{matrix} \; z \right] = \frac{\Gamma(\beta_1) \cdots \Gamma(\beta_q)}{\Gamma(\alpha_1) \cdots \Gamma(\alpha_p)} \, {}_p\Psi_q \left[\begin{matrix} (\alpha_1, 1), \cdots, (\alpha_p, 1); \\ (\beta_1, 1), \cdots, (\beta_q, 1); \end{matrix} \; z \right]. \tag{24}$$

Lemma 1. *Let α, β, $\eta \in \mathbb{C}$. Then, there exists the relation*

(a) *If $\Re(\alpha) > 0$ and $\Re(\sigma) > \max[0, \Re(\beta-\eta)]$, then*

$$(I_{0+}^{\alpha,\beta,\eta} t^{\sigma-1})(x) = \frac{\Gamma(\sigma)\Gamma(\sigma+\eta-\beta)}{\Gamma(\sigma-\beta)\Gamma(\sigma+\alpha+\eta)} x^{\sigma-\beta-1} \tag{25}$$

In particular, for $x > 0$, we have

$$(I_{0+}^{\alpha} t^{\sigma-1})(x) = \frac{\Gamma(\sigma)}{\Gamma(\sigma+\alpha)} x^{\sigma+\alpha-1} \qquad (\Re(\alpha) > 0,\ \Re(\sigma) > 0), \tag{26}$$

$$(I_{\eta,\alpha}^{+} t^{\sigma-1})(x) = \frac{\Gamma(\sigma+\eta)}{\Gamma(\sigma+\alpha+\eta)} x^{\sigma-1} \qquad (\Re(\alpha) > 0,\ \Re(\sigma) > -\Re(\eta)). \tag{27}$$

(b) *If $\Re(\alpha) > 0$ and $\Re(\sigma) < 1 + \min[\Re(\beta), \Re(\eta)]$, then*

$$(I_{-}^{\alpha,\beta,\eta} t^{\sigma-1})(x) = \frac{\Gamma(1-\sigma+\beta)\Gamma(1-\sigma+\eta)}{\Gamma(1-\sigma)\Gamma(1-\sigma+\alpha+\beta+\eta)} x^{\sigma-\beta-1}. \tag{28}$$

In particular, for $x > 0$, we have

$$(I_{-}^{\alpha} t^{\sigma-1})(x) = \frac{\Gamma(1-\alpha-\sigma)}{\Gamma(1-\sigma)} x^{\sigma+\alpha-1} \qquad (0 < \Re(\alpha) < 1 - \Re(\sigma)), \tag{29}$$

$$(K_{\eta,\alpha}^{-} t^{\sigma-1})(x) = \frac{\Gamma(1-\sigma+\eta)}{\Gamma(1-\sigma+\alpha+\eta)} x^{\sigma-1} \qquad (\Re(\sigma) < 1 + \Re(\sigma)). \tag{30}$$

We begin the exposition of the main results by presenting the composition formulas of generalized fractional integrals, (1) and (2), involving the generalized Mathieu series in terms of the Hadamard product (22) of the generalized Mathieu series (21) and the Fox–Wright function (23). It is emphasized that the results presented here, which are for a seemingly complicated series, can reveal their involved properties via the series of the two known functions.

Theorem 1. *Let α, β, η, $\sigma \in \mathbb{C}$ and $\rho > 0$, $\mu > 0$, $r \in \mathbb{R}$ be such that $\Re(\alpha) > 0$ and $\Re(\sigma) > \max[0, \Re(\beta-\eta)]$. Then, the following Saigo hypergeometric fractional integral $I_{0+}^{\alpha,\beta,\eta}$ of $S_\mu(r, t^\rho)$ holds true:*

$$\begin{aligned}&\left(I_{0+}^{\alpha,\beta,\eta}\left\{t^{\sigma-1} S_\mu(r,t^\rho)\right\}\right)(x)\\ &\quad= x^{\sigma-\beta+\rho-1} S_\mu(r,x^\rho) * {}_3\Psi_2\left[\begin{array}{c}(1,1),(\sigma+\rho,\rho),(\sigma+\eta-\beta+\rho,\rho);\\ (\sigma-\beta+\rho,\rho),(\sigma+\alpha+\eta+\rho,\rho);\end{array} x^\rho\right].\end{aligned} \tag{31}$$

Proof. Using the definitions (1) and (21), by changing the order of integration and applying the relation (25), we find that $x > 0$

$$\begin{aligned}&\left(I_{0+}^{\alpha,\beta,\eta}\left\{t^{\sigma-1} S_\mu(r,t^\rho)\right\}\right)(x) = \sum_{k=1}^{\infty} \frac{2k}{(k^2+r^2)^{\mu+1}} \left(I_{0+}^{\alpha,\beta,\eta} t^{\sigma+\rho k-1}\right)(x)\\ &= x^{\sigma-\beta-1} \sum_{k=1}^{\infty} \frac{2k}{(k^2+r^2)^{\mu+1}} \frac{\Gamma(\sigma+\rho k)\Gamma(\sigma+\eta-\beta+\rho k)}{\Gamma(\sigma-\beta+\rho k)\Gamma(\sigma+\alpha+\eta+\rho k)} x^{\rho k}.\end{aligned} \tag{32}$$

by applying the Hadamard product (22) in (32), which in the view of (21) and (23), yields the desired formula (31). □

Theorem 2. *Let α, β, η, $\sigma \in \mathbb{C}$ and $\rho > 0$, $\mu > 0$, $r \in \mathbb{R}$ be such that $\Re(\alpha) > 0$ and $\Re(\sigma) < 1 + \min[\Re(\beta), \Re(\eta)]$. Then, the following Saigo hypergeometric fractional integral $I_{-}^{\alpha,\beta,\eta}$ of $S_\mu\left(r, \frac{1}{t^\rho}\right)$ holds true:*

$$\begin{aligned}&\left(I_{-}^{\alpha,\beta,\eta}\left\{t^{\sigma-1}\, S_\mu\left(r, \frac{1}{t^\rho}\right)\right\}\right)(x) \\ &\quad = x^{\sigma-\rho-\beta-1}\, S_\mu\left(r, \frac{1}{x^\rho}\right) * {}_3\Psi_2\left[\begin{array}{l} (1,1), (1-\sigma+\beta+\rho,\rho), (1-\sigma+\eta+\rho,\rho); \\ (1-\sigma+\rho,\rho), (1-\sigma+\alpha+\beta+\eta+\rho,\rho); \end{array} \frac{1}{x^\rho}\right]. \end{aligned} \tag{33}$$

Proof. Using the definitions (2) and (21), by changing the order of integration and applying the relation (28)

$$\begin{aligned}\left(I_{-}^{\alpha,\beta,\eta}\left\{t^{\sigma-1} S_\mu\left(r, \frac{1}{t^\rho}\right)\right\}\right)(x) &= \sum_{k=1}^{\infty} \frac{2k}{(k^2+r^2)^{\mu+1}}\left(I_{-}^{\alpha,\beta,\eta}\, t^{\sigma-\rho k-1}\right)(x) \\ = x^{\sigma-\beta-1} \sum_{k=1}^{\infty} \frac{2k}{(k^2+r^2)^{\mu+1}} &\frac{\Gamma(1-\sigma+\beta+\rho k)\Gamma(1-\sigma+\eta+\rho k)}{\Gamma(1-\sigma+\rho k)\Gamma(1-\sigma+\alpha+\beta+\eta+\rho k)} x^{-\rho k}. \end{aligned} \tag{34}$$

by applying the Hadamard product (22) in (34), which in the view of (21) and (23), yields the desired formula (33). □

Further, we deduce the fractional integral formulas for the classical Riemann–Liouville and Erdélyi–Kober fractional integral and differential operators by letting $\beta = -\alpha$ and $\beta = 0$ respectively, which are asserted by Corollaries 1–4 below.

Corollary 1. *Let α, $\sigma \in \mathbb{C}$ and $\rho > 0$, $\mu > 0$, $r \in \mathbb{R}$ be such that $\Re(\alpha) > 0$ and $\Re(\sigma) > 0$. Then, the following Riemann–Liouville fractional integral I_{0+}^{α} of $S_\mu(r, t^\rho)$ holds true:*

$$\left(I_{0+}^{\alpha}\left\{t^{\sigma-1}\, S_\mu(r, t^\rho)\right\}\right)(x) = x^{\sigma+\rho+\alpha-1}\, S_\mu(r, x^\rho) * {}_2\Psi_1\left[\begin{array}{l} (1,1), (\sigma+\rho,\rho); \\ (\sigma+\alpha+\rho,\rho); \end{array} x^\rho\right]. \tag{35}$$

Corollary 2. *Let α, η, $\sigma \in \mathbb{C}$ and $\rho > 0$, $\mu > 0$, $r \in \mathbb{R}$ be such that $\Re(\alpha) > 0$ and $\Re(\sigma) > -\Re(\eta)$. Then, the following Erdélyi–Kober fractional integral $I_{\eta,\alpha}^{+}$ of $S_\mu(r, t^\rho)$ holds true:*

$$\left(I_{\eta,\alpha}^{+}\left\{t^{\sigma-1}\, S_\mu(r, t^\rho)\right\}\right)(x) = x^{\sigma+\rho-1}\, S_\mu(r, x^\rho) * {}_2\Psi_1\left[\begin{array}{l} (1,1), (\sigma+\eta+\rho,\rho); \\ (\sigma+\alpha+\eta+\rho,\rho); \end{array} x^\rho\right]. \tag{36}$$

Corollary 3. *Let α, $\sigma \in \mathbb{C}$ and $\rho > 0$, $\mu > 0$, $r \in \mathbb{R}$ be such that $0 < \Re(\alpha) < 1 - \Re(\sigma)$. Then, the following Riemann–Liouville fractional integral I_{-}^{α} of $S_\mu\left(r, \frac{1}{t^\rho}\right)$ holds true:*

$$\left(I_{-}^{\alpha}\left\{t^{\sigma-1}\, S_\mu\left(r, \frac{1}{t^\rho}\right)\right\}\right)(x) = x^{\sigma+\alpha-\rho-1}\, S_\mu\left(r, \frac{1}{x^\rho}\right) * {}_2\Psi_1\left[\begin{array}{l} (1,1), (1-\sigma-\alpha+\rho,\rho); \\ (1-\sigma+\rho,\rho); \end{array} \frac{1}{x^\rho}\right]. \tag{37}$$

Corollary 4. *Let α, η, $\sigma \in \mathbb{C}$ and $\rho > 0$, $\mu > 0$, $r \in \mathbb{R}$ be such that $\Re(\alpha) > 0$ and $\Re(\sigma) < 1 + \Re(\eta)$. Then, the following Erdélyi–Kober fractional integral $K_{\eta,\alpha}^{-}$ of $S_\mu\left(r, \frac{1}{t^\rho}\right)$ holds true:*

$$\left(K_{\eta,\alpha}^{-}\left\{t^{\sigma-1}\, S_\mu\left(r, \frac{1}{t^\rho}\right)\right\}\right)(x) = x^{\sigma-\rho-1}\, S_\mu\left(r, \frac{1}{x^\rho}\right) * {}_2\Psi_1\left[\begin{array}{l} (1,1), (1-\sigma+\eta+\rho,\rho); \\ (1-\sigma+\alpha+\eta+\rho,\rho); \end{array} \frac{1}{x^\rho}\right]. \tag{38}$$

The results obtained in this section can be presented in terms of Gauss's hypergeometric functions by taking $\rho = 1$. Here, we present results for the classical Riemann–Liouville fractional integral operators.

Corollary 5. *Let the conditions of Corollary 1 be satisfied, and let* $\Re(\sigma) > 0$ *and* $\Re(\sigma+\alpha+1) > 0$*. Then, for* $x > 0$*, there holds the relation*

$$\left(I_{0+}^{\alpha}\left\{t^{\sigma-1}\,S_{\mu}(r,t)\right\}\right)(x) = x^{\sigma+\alpha}\frac{\Gamma(\sigma+1)}{\Gamma(\sigma+\alpha+1)}\,S_{\mu}(r,x) * {}_2F_1\left[\begin{array}{c}1,\sigma+1;\\ \sigma+\alpha+1;\end{array} x\right]. \tag{39}$$

Corollary 6. *Let the conditions of Corollary 3 be satisfied, and let* $\Re(1-\sigma) > 0$ *and* $\Re(2-\sigma-\alpha) > 0$*. Then, for* $x > 0$*, there holds the relation*

$$\left(I_{-}^{\alpha}\left\{t^{\sigma-1}\,S_{\mu}\left(r,\frac{1}{t}\right)\right\}\right)(x) = x^{\sigma+\alpha-2}\frac{\Gamma(2-\sigma-\alpha)}{\Gamma(2-\sigma)}\,S_{\mu}\left(r,\frac{1}{x}\right) * {}_2F_1\left[\begin{array}{c}1,2-\sigma-\alpha;\\ 2-\sigma;\end{array} \frac{1}{x}\right]. \tag{40}$$

3. Fractional Differentiation of the Mathieu Series

In this section, we present the composition formulas of generalized fractional derivatives, (3) and (4), involving the generalized Mathieu series in terms of the Hadamard product (22) of the generalized Mathieu series (21) and the Fox–Wright function (23).

Lemma 2. *Let* α, β, $\eta \in \mathbb{C}$*. Then, there exists the relations*

(a) *If* $\Re(\alpha) > 0$ *and* $\Re(\sigma) > -\min[0, \Re(\alpha+\beta+\eta)]$*, then*

$$(D_{0+}^{\alpha,\beta,\eta}\,t^{\sigma-1})(x) = \frac{\Gamma(\sigma)\Gamma(\sigma+\alpha+\beta+\eta)}{\Gamma(\sigma+\beta)\Gamma(\sigma+\eta)}x^{\sigma+\beta-1} \tag{41}$$

In particular, for $x > 0$*, we have*

$$(D_{0+}^{\alpha}t^{\sigma-1})(x) = \frac{\Gamma(\sigma)}{\Gamma(\sigma-\alpha)}x^{\sigma-\alpha-1} \qquad (\Re(\alpha) > 0,\ \Re(\sigma) > 0), \tag{42}$$

$$(D_{\eta,\alpha}^{+}t^{\sigma-1})(x) = \frac{\Gamma(\sigma+\alpha+\eta)}{\Gamma(\sigma+\eta)}x^{\sigma-1} \qquad (\Re(\alpha) > 0,\ \Re(\sigma) > -\Re(\alpha+\eta)). \tag{43}$$

(b) *If* $\Re(\alpha) > 0, \Re(\sigma) < 1+\min[\Re(-\beta-n), \Re(\alpha+\eta)]$ *and* $n = [\Re(\alpha)]+1$*, then*

$$(D_{-}^{\alpha,\beta,\eta}\,t^{\sigma-1})(x) = \frac{\Gamma(1-\sigma-\beta)\Gamma(1-\sigma+\alpha+\eta)}{\Gamma(1-\sigma)\Gamma(1-\sigma+\eta-\beta)}x^{\sigma+\beta-1}. \tag{44}$$

In particular, for $x > 0$*, we have*

$$(D_{-}^{\alpha}\,t^{\sigma-1})(x) = \frac{\Gamma(1-\sigma+\alpha)}{\Gamma(1-\sigma)}x^{\sigma-\alpha-1} \qquad (\Re(\alpha) > 0, \Re(\sigma) < 1+\Re(\alpha)-n), \tag{45}$$

$$(D_{\eta,\alpha}^{-}t^{\sigma-1})(x) = \frac{\Gamma(1-\sigma+\alpha+\eta)}{\Gamma(1-\sigma-\eta)}x^{\sigma-1} \qquad (\Re(\alpha) > 0, \Re(\sigma) < 1+\Re(\alpha+\eta)-n). \tag{46}$$

Theorem 3. *Let* α, β, η, $\sigma \in \mathbb{C}$ *and* $\rho > 0$, $\mu > 0$, $r \in \mathbb{R}$ *be such that* $\Re(\alpha) \geq 0$ *and* $\Re(\sigma) > -\min[0, \Re(\alpha+\beta+\eta)]$*. Then, the following Saigo hypergeometric fractional derivative* $D_{0+}^{\alpha,\beta,\eta}$ *of* $S_{\mu}(r,t^{\rho})$ *holds true:*

$$\begin{aligned}&\left(D_{0+}^{\alpha,\beta,\eta}\left\{t^{\sigma-1}\,S_{\mu}(r,t^{\rho})\right\}\right)(x)\\ &\quad = x^{\sigma+\beta+\rho-1}\,S_{\mu}(r,x^{\rho}) * {}_3\Psi_2\left[\begin{array}{c}(1,1),(\sigma+\rho,\rho),(\sigma+\alpha+\beta+\eta+\rho,\rho);\\ (\sigma+\beta+\rho,\rho),(\sigma+\eta+\rho,\rho);\end{array} x^{\rho}\right].\end{aligned} \tag{47}$$

Proof. Using the definitions (3) and (21), by changing the order of integration and applying the relation (41), we find for $x > 0$

$$\left(D_{0+}^{\alpha,\beta,\eta}\left\{t^{\sigma-1}S_{\mu}(r,t^{\rho})\right\}\right)(x)=\sum_{k=1}^{\infty}\frac{2k}{(k^2+r^2)^{\mu+1}}\left(D_{0+}^{\alpha,\beta,\eta}t^{\sigma+\rho k-1}\right)(x)$$
$$=x^{\sigma+\beta-1}\sum_{k=1}^{\infty}\frac{2k}{(k^2+r^2)^{\mu+1}}\frac{\Gamma(\sigma+\rho k)\Gamma(\sigma+\alpha+\beta+\eta+\rho k)}{\Gamma(\sigma+\beta+\rho k)\Gamma(\sigma+\eta+\rho k)}x^{\rho k}. \tag{48}$$

by applying the Hadamard product (22) in (48), which in the view of (21) and (23), yields the desired formula (47). □

Theorem 4. *Let $\alpha,\ \beta,\ \eta,\ \sigma \in \mathbb{C}$ and $\rho > 0,\ \mu > 0,\ r \in \mathbb{R}$ be such that $\Re(\alpha) \geq 0$ and $\Re(\sigma) < 1 + \min[\Re(-\beta-n), \Re(\alpha+\eta)]$, $n = [\Re(\alpha)] + 1$. Then, the following Saigo hypergeometric fractional derivative $D_{-}^{\alpha,\beta,\eta}$ of $S_{\mu}\left(r, \frac{1}{t^{\rho}}\right)$ holds true:*

$$\left(D_{-}^{\alpha,\beta,\eta}\left\{t^{\sigma-1}S_{\mu}\left(r,\frac{1}{t^{\rho}}\right)\right\}\right)(x)$$
$$=x^{\sigma-\rho+\beta-1}S_{\mu}\left(r,\frac{1}{x^{\rho}}\right)*{}_3\Psi_2\left[\begin{matrix}(1,1),(1-\sigma-\beta+\rho,\rho),(1-\sigma+\alpha+\eta+\rho,\rho); \\ (1-\sigma+\rho,\rho),(1-\sigma+\eta-\beta+\rho,\rho);\end{matrix}\ \frac{1}{x^{\rho}}\right]. \tag{49}$$

Proof. Using the definitions (4) and (21), by changing the order of integration and applying the relation (44), we find for $x > 0$

$$\left(D_{-}^{\alpha,\beta,\eta}\left\{t^{\sigma-1}S_{\mu}\left(r,\frac{1}{t^{\rho}}\right)\right\}\right)(x)=\sum_{k=1}^{\infty}\frac{2k}{(k^2+r^2)^{\mu+1}}\left(D_{-}^{\alpha,\beta,\eta}t^{\sigma-\rho k-1}\right)(x)$$
$$=x^{\sigma+\rho+\beta-1}\sum_{k=1}^{\infty}\frac{2k}{(k^2+r^2)^{\mu+1}}\frac{\Gamma(1-\sigma-\beta+\rho k)\Gamma(1-\sigma+\alpha+\eta+\rho k)}{\Gamma(1-\sigma+\rho k)\Gamma(1-\sigma+\eta-\beta+\rho k)}x^{-\rho k}. \tag{50}$$

by applying the Hadamard product (22) in (50), which in the view of (21) and (23), yields the desired formula (49). □

Now, we deduce fractional derivative formulas for the classical Riemann–Liouville and Erdélyi–Kober fractional integral and differential operators by letting $\beta = -\alpha$ and $\beta = 0$ respectively, which are asserted by Corollaries 7–10 below.

Corollary 7. *Let $\alpha,\ \sigma \in \mathbb{C}$ and $\rho > 0,\ \mu > 0,\ r \in \mathbb{R}$ be such that $\Re(\alpha) \geq 0$ and $\Re(\sigma) > 0$. Then, the following Riemann–Liouville fractional differentiation D_{0+}^{α} of $S_{\mu}(r, t^{\rho})$ holds true:*

$$\left(D_{0+}^{\alpha}\left\{t^{\sigma-1}S_{\mu}(r,t^{\rho})\right\}\right)(x)=x^{\sigma+\rho-\alpha-1}S_{\mu}(r,x^{\rho})*{}_2\Psi_1\left[\begin{matrix}(1,1),(\sigma+\rho,\rho); \\ (\sigma-\alpha+\rho,\rho);\end{matrix}\ x^{\rho}\right]. \tag{51}$$

Corollary 8. *Let $\alpha,\ \eta,\ \sigma \in \mathbb{C}$ and $\rho > 0,\ \mu > 0,\ r \in \mathbb{R}$ be such that $\Re(\alpha) \geq 0$ and $\Re(\sigma) > -\Re(\alpha+\eta)$. Then, the following Erdélyi–Kober fractional derivative $D_{\eta,\alpha}^{+}$ of $S_{\mu}(r, t^{\rho})$ holds true:*

$$\left(D_{\eta,\alpha}^{+}\left\{t^{\sigma-1}S_{\mu}(r,t^{\rho})\right\}\right)(x)=x^{\sigma+\rho-1}S_{\mu}(r,x^{\rho})*{}_2\Psi_1\left[\begin{matrix}(1,1),(\sigma+\alpha+\eta+\rho,\rho); \\ (\sigma+\eta+\rho,\rho);\end{matrix}\ x^{\rho}\right]. \tag{52}$$

Corollary 9. *Let $\alpha, \sigma \in \mathbb{C}$ and $\rho > 0, \mu > 0, r \in \mathbb{R}$ be such that $\Re(\alpha) \geq 0$ and $\Re(\sigma) < \Re(\alpha) - [\Re(\alpha)]$. Then, the following Riemann–Liouville fractional differentiation D_{-}^{α} of $S_\mu\left(r, \frac{1}{t^\rho}\right)$ holds true:*

$$\left(D_{-}^{\alpha}\left\{t^{\sigma-1} S_\mu\left(r, \frac{1}{t^\rho}\right)\right\}\right)(x) = x^{\sigma-\rho-\alpha-1} S_\mu\left(r, \frac{1}{x^\rho}\right) * {}_2\Psi_1\left[\begin{array}{c}(1,1),(1-\sigma+\alpha+\rho,\rho);\\(1-\sigma+\rho,\rho);\end{array} \frac{1}{x^\rho}\right]. \quad (53)$$

Corollary 10. *Let $\alpha, \eta, \sigma \in \mathbb{C}$ and $\rho > 0, \mu > 0, r \in \mathbb{R}$ be such that $\Re(\alpha) \geq 0$ and $\Re(\sigma) < \Re(\alpha+\eta) - [\Re(\alpha)]$. Then, the following Erdélyi–Kober fractional differentiation $D_{\eta,\alpha}^{-}$ of $S_\mu\left(r, \frac{1}{t^\rho}\right)$ holds true:*

$$\left(D_{\eta,\alpha}^{-}\left\{t^{\sigma-1} S_\mu\left(r, \frac{1}{t^\rho}\right)\right\}\right)(x) = x^{\sigma-\rho-1} S_\mu\left(r, \frac{1}{x^\rho}\right) * {}_2\Psi_1\left[\begin{array}{c}(1,1),(1-\sigma+\alpha+\eta+\rho,\rho);\\(1-\sigma+\eta+\rho,\rho);\end{array} \frac{1}{x^\rho}\right]. \quad (54)$$

The results obtained in this section can be presented in terms of Gauss's hypergeometric functions by taking $\rho = 1$. Here, we present results for the classical Riemann–Liouville fractional derivative operators.

Corollary 11. *Let the conditions of Corollary 7 be satisfied, and let $\Re(\sigma+1) > 0$ and $\Re(\sigma-\alpha+1) > 0$. Then, for $x > 0$, there holds the relation*

$$\left(D_{0+}^{\alpha}\left\{t^{\sigma-1} S_\mu(r,t)\right\}\right)(x) = x^{\sigma-\alpha-2}\frac{\Gamma(\sigma+1)}{\Gamma(\sigma-\alpha+1)} S_\mu(r,x) * {}_2F_1\left[\begin{array}{c}1,\sigma+1;\\\sigma-\alpha+1;\end{array} x\right]. \quad (55)$$

Corollary 12. *Let the conditions of Corollary 9 be satisfied, and let $\Re(2-\sigma) > 0$ and $\Re(2-\sigma+\alpha) > 0$. Then, for $x > 0$, there holds the relation*

$$\left(D_{-}^{\alpha}\left\{t^{\sigma-1} S_\mu\left(r, \frac{1}{t}\right)\right\}\right)(x) = x^{\sigma-\alpha-2}\frac{\Gamma(2-\sigma+\alpha)}{\Gamma(2-\sigma)} S_\mu\left(r, \frac{1}{x}\right) * {}_2F_1\left[\begin{array}{c}1,2-\sigma+\alpha;\\2-\sigma;\end{array} \frac{1}{x}\right]. \quad (56)$$

4. Concluding Remarks and Observations

In our present investigation, with the help of the concept of the Hadamard product (or the convolution) of two analytic functions, we have obtained the composition formulas of the generalized fractional integrals, (1) and (2), involving the generalized Mathieu series in terms of the Hadamard product (22) of the generalized Mathieu series (21) and the Fox–Wright function (23). Further, we have also deduced the fractional integral formulas for the classical Riemann–Liouville and the Erdélyi–Kober fractional integral and differential operators by letting $\beta = -\alpha$ and $\beta = 0$, respectively. The results presented here, which are for a seemingly complicated series, can reveal their involved properties via the series of the two known functions.

Acknowledgments: The authors thanks to the anonymous referees for valuable comments and useful suggestions, improving the quality of the article.

Author Contributions: Ram K. Saxena and Rakesh K. Parmar contributed equally to this work.

References

1. Kilbas, A.A.; Srivastava, H.M.; Trujillo, J.J. *Theory and Applications of Fractional Differential Equations, Volume 204;* North-Holland Mathematical Studies; Elsevier (North-Holland) Science Publishers: New York, NY, USA, 2006.
2. Kiryakova, V.S. *Generalized Fractional Calculus and Applications;* Pitman Research Notes in Mathematics Series; Longman Scientific and Technical: Harlow, UK; John Wiley and Sons, Inc.: New York, NY, USA, 1993.
3. Mathai, A.M.; Saxena, R.K.; Haubold, H.J. *The H-Functions: Theory and Applications;* Springer: New York, NY, USA, 2010.
4. Samko, S.G.; Kilbas, A.A.; Marichev, O.I. *Fractional Integrals and Derivatives: Theory and Applications;*

Translated from the Russian: Integrals and Derivatives of Fractional Order and Some of Their Applications ("Nauka i Tekhnika", Minsk, 1987); Gordon and Breach Science Publishers: Reading, UK, 1993.
5. Srivastava, H.M.; Saxena, R.K. Operators of fractional integration and their applications. *Appl. Math. Comput.* 2001, 118, 1–52.
6. Saigo, M. A remark on integral operators involvig the Gauss hypergeometric functions. *Math. Rep. Kyushu Univ.* **1978**, *11*, 135–143.
7. Mathieu, E.L. *Traité de Physique Mathématique, VI-VII: Théorie de L'élasticité des Corps Solides*; Gauthier-Villars: Paris, France, 1890.
8. Pogány, T.K.; Srivastava, H.M.; Tomovski, Ž. Some families of Mathieu a-series and alternating Mathieu a-series. *Appl. Math. Comput.* **2006**, *173*, 69–108.
9. Emersleben, O. Über die Reihe $\sum_{k=1}^{\infty} \frac{k}{(k^2+r^2)^2}$. *Math. Ann.* **1952**, *125*, 165–171.
10. Cerone, P.; Lenard, C.T. On integral forms of generalized Mathieu series. *JIPAM J. Inequal. Pure Appl. Math.* **2003**, *4*, 1–11.
11. Milovanović, G.V.; Pogány, T.K. New integral forms of generalized Mathieu series and related applications. *Appl. Anal. Discret. Math.* **2013**, *7*, 180–192.
12. Diananda, P.H. Some inequalities related to an inequality of Mathieu. *Math. Ann.* **1980**, *250*, 95–98.
13. Srivastava, H.M.; Tomovski, Ž.; Leškovski, D. Some families of Mathieu type series and Hurwitz-Lerch zeta functions and associated probability distributions. *Appl. Comput. Math.* **2015**, *14*, 349–380.
14. Tomovski, Ž.; Pogány, T.K. Integral expressions for Mathieu-type power series and for the Butzer-Flocke-Hauss Ω-function. *Fract. Calc. Appl. Anal.* **2011**, *14*, 623–634.
15. Baricz, Á.; Butzer, P.L.; Pogány, T.K. Alternating Mathieu series, Hilbert—Eisenstein series and their generalized Omega functions. In *Analytic Number Theory, Approximation Theory, and Special Functions*; Rassias, T., Milovanović, G.V., Eds.; In Honor of Hari M. Srivastava; Springer: New York, NY, USA, 2014; pp. 775–808.
16. Choi, J.; Parmar, R.K.; Pogány, T.K. Mathieu-type series built by (p,q)—Extended Gaussian hypergeometric function. *Bull. Korean Math. Soc.* **2017**, *54*, 789–797.
17. Choi, J.; Srivastava, H.M. Mathieu series and associated sums involving the Zeta functions. *Comput. Math. Appl.* **2010**, *59*, 861–867.
18. Elezović, N.; Srivastava, H.M.; Tomovski, Ž. Integral representations and integral transforms of some families of Mathieu type series. *Integral Transforms Spec. Funct.* **2008**, *19*, 481–495.
19. Pogány, T.K.; Parmar, R.K. On p-extended Mathieu series; submitted manuscript. *arXiv* **2016**, arXiv:1606.08369.
20. Pogány T.K.; Tomovski, Ž. Bounds improvement for alternating Mathieu type series. *J. Math. Inequal.* **2010**, *4*, 315–324.
21. Saxena, R.K.; Pogány, T.K.; Saxena, R. Integral transforms of the generalized Mathieu series. *J. Appl. Math. Stat. Inform.* **2010**, *6*, 5–16.
22. Tomovski, Ž. Integral representations of generalized Mathieu series via Mittag-Leffler type functions. *Fract. Calc. Appl. Anal.* **2007**, *10*, 127–138.
23. Tomovski, Ž. New integral and series representations of the generalized Mathieu series. *Appl. Anal. Discret. Math.* **2008**, *2*, 205–212.
24. Tomovski, Ž.; Pogány, T.K. New upper bounds for Mathieu-type series. *Banach J. Math. Anal.* **2009**, *3*, 9–15.
25. Kiryakova, V. On two Saigo's fractional integral operators in the class of univalent functions. *Fract. Calc. Appl. Anal.* **2006**, *9*, 159–176.
26. Pohlen, T. The Hadamard Product and Universal Power Series. Ph.D. Thesis, Universität Trier, Trier, Germany, 2009.
27. Srivastava, H.M.; Agarwal, R.; Jain, S. Integral transform and fractional derivative formulas involving the extended generalized hypergeometric functions and probability distributions. *Math. Method Appl. Sci.* **2017**, *40*, 255–273.
28. Srivastava, R.; Agarwal, R.; Jain, S. A family of the incomplete hypergeometric functions and associated integral transform and fractional derivative formulas. *Filomat* **2017**, *31*, 125–140.
29. Srivastava, H.M.; Karlsson, P.W. *Multiple Gaussian Hypergeometric Series*; Ellis Horwood Ltd.: Chichester, UK, 1985.

2

A Note on Generalized q-Difference Equations and their Applications Involving q-Hypergeometric Functions

Hari M. Srivastava [1,2,3,*], **Jian Cao** [4] and **Sama Arjika** [5,6]

1 Department of Mathematics and Statistics, University of Victoria, Victoria, BC V8W 3R4, Canada
2 Department of Medical Research, China Medical University Hospital, China Medical University, Taichung 40402, Taiwan
3 Department of Mathematics and Informatics, Azerbaijan University, 71 Jeyhun Hajibeyli Street, Baku AZ1007, Azerbaijan
4 Department of Mathematics, Hangzhou Normal University, Hangzhou City 311121, China; 21caojian@hznu.edu.cn
5 Department of Mathematics and Informatics, University of Agadez, Post Office Box 199, Agadez 8000, Niger; rjksama2008@gmail.com
6 International Chair of Mathematical Physics and Applications (ICMPA-UNESCO Chair), University of Abomey-Calavi, Post Office Box 072, Cotonou 50, Benin
* Correspondence: harimsri@math.uvic.ca

Abstract: Our investigation is motivated essentially by the demonstrated applications of the basic (or q-) series and basic (or q-) polynomials, especially the basic (or q-) hypergeometric functions and basic (or q-) hypergeometric polynomials, in many diverse areas. Here, in this paper, we use two q-operators $\mathbb{T}(a,b,c,d,e,yD_x)$ and $\mathbb{E}(a,b,c,d,e,y\theta_x)$ to derive two potentially useful generalizations of the q-binomial theorem, a set of two extensions of the q-Chu-Vandermonde summation formula and two new generalizations of the Andrews-Askey integral by means of the q-difference equations. We also briefly describe relevant connections of various special cases and consequences of our main results with a number of known results.

Keywords: q-difference operator; q-binomial theorem; q-hypergeometric functions; q-Chu-Vandermonde summation formula; Andrews-Askey integral; q-series and q-integral identities; q-difference equations; Sears transformation

MSC: Primary 05A30; 11B65; 33D15; 33D45; Secondary 33D60; 39A13; 39B32

---

1. Introduction, Definitions and Preliminaries

Throughout this paper, we refer to [1] for definitions and notations. We also suppose that $0 < q < 1$. For complex numbers a, the q-shifted factorials are defined by

$$(a;q)_0 := 1, \tag{1}$$

$$(a;q)_n := \prod_{k=0}^{n-1}(1-aq^k) = (1-a)(1-aq)(1-aq^2)\cdots(1-aq^{n-1}) \qquad (n \in \mathbb{N} := \{1,2,3,\cdots\}) \tag{2}$$

and

$$(a;q)_\infty := \prod_{k=0}^{\infty}(1-aq^k), \tag{3}$$

where (see, for example, [1,2])

$$(a;q)_n = \frac{(a;q)_\infty}{(aq^n;q)_\infty} \qquad \text{and} \qquad (a;q)_{n+m} = (a;q)_n (aq^n;q)_m$$

and

$$\left(\frac{q}{a};q\right)_n = (-a)^{-n}\, q^{\binom{n+1}{2}} \frac{(aq^{-n};q)_\infty}{(a;q)_\infty}.$$

We adopt the following notation:

$$(a_1, a_2, \cdots, a_r; q)_m = (a_1;q)_m (a_2;q)_m \cdots (a_r;q)_m \qquad (m \in \mathbb{N} := \{1,2,3,\cdots\}).$$

Also, for m large, we have

$$(a_1, a_2, \cdots, a_r; q)_\infty = (a_1;q)_\infty (a_2;q)_\infty \cdots (a_r;q)_\infty.$$

The q-binomial coefficient is defined by

$$\begin{bmatrix} n \\ k \end{bmatrix}_q := \frac{(q;q)_n}{(q;q)_k (q;q)_{n-k}}. \tag{4}$$

The basic (or q-) hypergeometric function of the variable z and with $\mathfrak{r}$ numerator and $\mathfrak{s}$ denominator parameters is defined as follows (see, for details, the monographs by Slater ([2], Chapter 3) and by Srivastava and Karlsson ([3], p. 347, equation (272)); see also [4–6]):

$${}_{\mathfrak{r}}\Phi_{\mathfrak{s}} \left[\begin{array}{l} a_1, a_2, \cdots, a_{\mathfrak{r}}; \\ \\ b_1, b_2, \cdots, b_{\mathfrak{s}}; \end{array} q; z \right] := \sum_{n=0}^{\infty} \left[(-1)^n\, q^{\binom{n}{2}} \right]^{1+\mathfrak{s}-\mathfrak{r}} \frac{(a_1, a_2, \cdots, a_{\mathfrak{r}}; q)_n}{(b_1, b_2, \cdots, b_{\mathfrak{s}}; q)_n} \frac{z^n}{(q;q)_n},$$

where $q \neq 0$ when $\mathfrak{r} > \mathfrak{s} + 1$. We also note that

$${}_{\mathfrak{r}+1}\Phi_{\mathfrak{r}} \left[\begin{array}{l} a_1, a_2, \cdots, a_{\mathfrak{r}+1} \\ \\ b_1, b_2, \cdots, b_{\mathfrak{r}}; \end{array} q; z \right] = \sum_{n=0}^{\infty} \frac{(a_1, a_2, \cdots, a_{\mathfrak{r}+1}; q)_n}{(b_1, b_2, \cdots, b_{\mathfrak{r}}; q)_n} \frac{z^n}{(q;q)_n}.$$

We remark in passing that, in a recently-published survey-cum-expository review article, the so-called (p,q)-calculus was exposed to be a rather trivial and inconsequential variation of the classical q-calculus, the additional parameter p being redundant or superfluous (see, for details, ([7], p. 340)).

Basic (or q-) series and basic (or q-) polynomials, especially the basic (or q-) hypergeometric functions and basic (or q-) hypergeometric polynomials, are applicable particularly in several areas of Number Theory such as the Theory of Partitions and are useful also in a wide variety of fields including, for example, Combinatorial Analysis, Finite Vector Spaces, Lie Theory, Particle Physics, Non-Linear Electric Circuit Theory (see [8,9]), Mechanical Engineering (see [10]), Theory of Heat Conduction, Quantum Mechanics, Cosmology, Computation of Fractional-Order Derivatives (see [11]) and Statistics [see also ([3], pp. 350–351), and the references cited therein]. Here, in our present investigation, we are mainly concerned with the Cauchy polynomials $p_n(x,y)$ as given below (see [1,12]):

$$p_n(x,y) := (x-y)(x-qy)\cdots(x-q^{n-1}y) = \left(\frac{y}{x};q\right)_n x^n, \tag{5}$$

together with the following Srivastava-Agarwal type generating function (see also [13]):

$$\sum_{n=0}^{\infty} p_n(x,y) \frac{(\lambda;q)_n\, t^n}{(q;q)_n} = {}_2\Phi_1 \left[\begin{array}{c} \lambda, \frac{y}{x}; \\ \\ 0; \end{array} q; xt \right]. \tag{6}$$

In particular, for $\lambda = 0$ in (6), we get the following simpler generating function [12]:

$$\sum_{n=0}^{\infty} p_n(x,y)\,\frac{t^n}{(q;q)_n} = \frac{(yt;q)_\infty}{(xt;q)_\infty}. \tag{7}$$

The generating function (7) is also the homogeneous version of the Cauchy identity or the following q-binomial theorem (see, for example, [1–3,14]):

$$\sum_{k=0}^{\infty} \frac{(a;q)_k}{(q;q)_k}\,z^k = {}_1\Phi_0\left[\begin{matrix} a; \\ -; \end{matrix}\; q;z\right] = \frac{(az;q)_\infty}{(z;q)_\infty} \qquad (|z|<1). \tag{8}$$

Upon further setting $a = 0$, this last relation (8) becomes Euler's identity (see, for example, [1]):

$$\sum_{k=0}^{\infty} \frac{z^k}{(q;q)_k} = \frac{1}{(z;q)_\infty} \qquad (|z|<1) \tag{9}$$

and its inverse relation given below [1]:

$$\sum_{k=0}^{\infty} \frac{(-1)^k}{(q;q)_k}\,q^{\binom{k}{2}}\,z^k = (z;q)_\infty. \tag{10}$$

Based upon the q-binomial theorem (8) and Heine's transformations, Srivastava et al. [15] established a set of two presumably new theta-function identities (see, for details, [15]).

The following usual q-difference operators are defined by [16–18]

$$D_a\{f(a)\} := \frac{f(a)-f(qa)}{a}, \quad \theta_a = \{f(a)\} := \frac{f(q^{-1}a)-f(a)}{q^{-1}a}, \tag{11}$$

and their Leibniz rules are given by (see [19])

$$D_a^n\{f(a)g(a)\} = \sum_{k=0}^{n} \begin{bmatrix} n \\ k \end{bmatrix}_q q^{k(k-n)} D_a^k\{f(a)\}\, D_a^{n-k}\left\{g(q^k a)\right\} \tag{12}$$

and

$$\theta_a^n\{f(a)g(a)\} = \sum_{k=0}^{n} \begin{bmatrix} n \\ k \end{bmatrix}_q \theta_a^k\{f(a)\}\,\theta_a^{n-k}\left\{g(q^{-k}a)\right\}, \tag{13}$$

respectively. Here, and in what follows, D_a^0 and θ_a^0 are understood as the identity operators.

Recently, Chen and Liu [16,20] constructed the following pair of augmentation operators, which is of great significance for deriving identities by applying its various special cases:

$$\mathbb{T}(bD_x) = \sum_{n=0}^{\infty} \frac{(bD_x)^n}{(q;q)_n} \quad \text{and} \quad \mathbb{E}(b\theta_x) = \sum_{n=0}^{\infty} \frac{(b\theta_x)^n}{(q;q)_n}. \tag{14}$$

Subsequently, Chen and Gu [21] defined the Cauchy augmentation operators as follows:

$$\mathbb{T}(a,bD_x) = \sum_{n=0}^{\infty} \frac{(a;q)_n}{(q;q)_n}\,(bD_x)^n \tag{15}$$

and

$$\mathbb{E}(a,b\theta_x) = \sum_{n=0}^{\infty} \frac{(b;q)_n}{(q;q)_n}\,(-b\theta_x)^n. \tag{16}$$

On the other hand, Fang [22] and Zhang and Wang [23] considered the following finite generalized q-exponential operators with two parameters:

$$\mathbb{T}\left[\begin{matrix} q^{-N}, w \\ v \end{matrix} \middle| q; tD_x\right] = \sum_{n=0}^{N} \frac{(q^{-N}, w; q)_n}{(v, q; q)_n} (tD_x)^n \tag{17}$$

and

$$\mathbb{E}\left[\begin{matrix} q^{-N}, w \\ v \end{matrix} \middle| q; t\theta_x\right] = \sum_{n=0}^{N} \frac{(q^{-N}, w; q)_n}{(v, q; q)_n} (t\theta_x)^n. \tag{18}$$

Moreover, Li and Tan [24] constructed two generalized q-exponential operators with three parameters as follows:

$$\mathbb{T}\left[\begin{matrix} u, v \\ w \end{matrix} \middle| q; tD_x\right] = \sum_{n=0}^{\infty} \frac{(u, v; q)_n}{(w, q; q)_n} (tD_x)^n \tag{19}$$

and

$$\mathbb{E}\left[\begin{matrix} u, v \\ w \end{matrix} \middle| q; t\theta_x\right] = \sum_{n=0}^{\infty} \frac{(u, v; q)_n}{(w, q; q)_n} (t\theta_x)^n. \tag{20}$$

Finally, we recall that Cao et al. [25] constructed the following q-operators:

$$\mathbb{T}(a, b, c, d, e, yD_x) = \sum_{n=0}^{\infty} \frac{(a, b, c; q)_n}{(q, d, e; q)_n} (yD_x)^n \tag{21}$$

and

$$\mathbb{E}(a, b, c, d, e, y\theta_x) = \sum_{n=0}^{\infty} \frac{(-1)^n q^{\binom{n}{2}} (a, b, c; q)_n}{(q, d, e; q)_n} (y\theta_x)^n \tag{22}$$

and thereby generalized Arjika's results in [26] by using the q-difference equations (see, for details, [25]).

We remark that the q-operator (21) is a particular case of the homogeneous q-difference operator $\mathbb{T}(\mathbf{a}, \mathbf{b}, cD_x)$ (see [27]) by taking

$$\mathbf{a} = (a, b, c), \quad \mathbf{b} = (d, e) \quad \text{and} \quad c = y.$$

Furthermore, for $b = c = d = e = 0$, the q-operator (22) reduces to the operator $\widetilde{L}(a, y; \theta_x)$ which was investigated by Srivastava et al. [28].

Cao et al. [25] used the q-operators (21) and (22) and gave the following results.

Proposition 1. *(see ([25], theorem 3)) Let $f(a, b, c, d, e, x, y)$ be a seven-variable analytic function in a neighborhood of $(a, b, c, d, e, x, y) = (0, 0, 0, 0, 0, 0, 0) \in \mathbb{C}^7$.*

(I) If $f(a, b, c, d, e, x, y)$ satisfies the following difference equation:

$$\begin{aligned} & x\{f(a,b,c,d,e,x,y) - f(a,b,c,d,e,x,yq) \\ & \quad - (d+e)q^{-1}\left[f(a,b,c,d,e,x,yq) - f(a,b,c,d,e,x,yq^2)\right] \\ & \quad + deq^{-2}\left[f(a,b,c,d,e,x,yq^2) - f(a,b,c,d,e,x,yq^3)\right]\} \\ & = y\{[f(a,b,c,d,e,x,y) - f(a,b,c,d,e,xq,y)] \\ & \quad - (a+b+c)[f(a,b,c,d,e,x,yq) - f(a,b,c,d,e,xq,yq)] \\ & \quad + (ab+ac+bc)[f(a,b,c,d,e,x,yq^2) - f(a,b,c,d,e,xq,yq^2)] \\ & \quad - abc[f(a,b,c,d,e,x,yq^3) - f(a,b,c,d,e,xq,yq^3)]\}, \end{aligned} \tag{23}$$

then

$$f(a,b,c,d,e,x,y) = \mathbb{T}(a,b,c,d,e,yD_x)\{f(a,b,c,d,e,x,0)\}. \tag{24}$$

(II) *If $f(a,b,c,d,e,x,y)$ satisfies the following difference equation:*

$$\begin{aligned} x\{f(a,b,c,d,e,xq,y) &- f(a,b,c,d,e,xq,yq) \\ &- (d+e)q^{-1}\left[f(a,b,c,d,e,xq,yq) - f(a,b,c,d,e,xq,yq^2)\right] \\ &+ deq^{-2}\left[f(a,b,c,d,e,xq,yq^2) - f(a,b,c,d,e,xq,yq^3)\right]\} \\ = y\{&[f(a,b,c,d,e,xq,yq) - f(a,b,c,d,e,x,yq)] \\ &- (a+b+c)[f(a,b,c,d,e,xq,yq^2) - f(a,b,c,d,e,x,yq^2)] \\ &+ (ab+ac+bc)[f(a,b,c,d,e,xq,yq^3) - f(a,b,c,d,e,x,yq^3)] \\ &- abc[f(a,b,c,d,e,xq,yq^4) - f(a,b,c,d,e,x,yq^4)]\}, \end{aligned} \tag{25}$$

then

$$f(a,b,c,d,e,x,y) = \mathbb{E}(a,b,c,d,e,y\theta_x)\{f(a,b,c,d,e,x,0)\}. \tag{26}$$

Liu [29,30] initiated the method based upon q-difference equations and deduced several results involving Bailey's ${}_6\psi_6$, q-Mehler formulas for the Rogers-Szegö polynomials and q-integral version of the Sears transformation.

Lemma 1. *Each of the following q-identities holds true:*

$$D_a^k\left\{\frac{1}{(as;q)_\infty}\right\} = \frac{s^k}{(as;q)_\infty}, \tag{27}$$

$$\theta_a^k\left\{\frac{1}{(as;q)_\infty}\right\} = \frac{s^k q^{-\binom{k}{2}}}{(asq^{-k};q)_\infty}, \tag{28}$$

$$D_a^k\left\{(as;q)_\infty\right\} = (-s)^k\, q^{\binom{k}{2}}\,(asq^k;q)_\infty, \tag{29}$$

$$\theta_a^k\left\{(as;q)_\infty\right\} = (-s)^k\,(as;q)_\infty, \tag{30}$$

$$D_a^n\left\{\frac{(as;q)_\infty}{(a\omega;q)_\infty}\right\} = \omega^n\,\frac{\left(\frac{s}{\omega};q\right)_n}{(as;q)_n}\,\frac{(as;q)_\infty}{(a\omega;q)_\infty} \tag{31}$$

and

$$\theta_a^n\left\{\frac{(as;q)_\infty}{(a\omega;q)_\infty}\right\} = \left(-\frac{q}{a}\right)^n\frac{\left(\frac{s}{\omega};q\right)_n}{\left(\frac{q}{a\omega};q\right)_n}\,\frac{(as;q)_\infty}{(a\omega;q)_\infty}. \tag{32}$$

We now state and prove the q-difference formulas as Theorem 1 below.

Theorem 1. *Each of the following assertions holds true:*

$$\begin{aligned} &\mathbb{T}(r,f,g,v,w,uD_a)\left\{\frac{(as;q)_\infty}{(az,at;q)_\infty}\right\} \\ &= \frac{(as;q)_\infty}{(az,at;q)_\infty}\sum_{k=0}^{\infty}\frac{\left(r,f,g,\frac{s}{z},at;q\right)_k (zu)^k}{(v,w,as,q;q)_k}\,{}_3\Phi_2\left[\begin{matrix} rq^k, fq^k, gq^k; \\ vq^k, wq^k; \end{matrix}\; q; ut\right] \end{aligned} \tag{33}$$

and

$$\mathbb{E}(r,f,g,v,w,u\theta_a)\left\{\frac{(az,at;q)_\infty}{(as;q)_\infty}\right\}$$
$$= \frac{(az,at;q)_\infty}{(as;q)_\infty}\sum_{k=0}^{\infty}\frac{(r,f,g,\frac{z}{s},\frac{q}{at};q)_k\,(-ut)^k}{(v,w,\frac{q}{as},q;q)_k}\,{}_3\Phi_3\left[\begin{matrix} rq^k,fq^k,gq^k; \\ vq^k,wq^k,0; \end{matrix}\; q;-ut\right], \tag{34}$$

provided that $\max\{|az|,|as|,|at|,|ut|\}<1$.

Proof. By means of the definition (21) of the operator $\mathbb{T}(r,f,g,v,w,uD_a)$ and the Leibniz rule (12), we observe that

$$\mathbb{T}(r,f,g,v,w,uD_a)\left\{\frac{(as;q)_\infty}{(az,at;q)_\infty}\right\}$$
$$= \sum_{n=0}^{\infty}\frac{(r,f,g;q)_n u^n}{(v,w,q;q)_n}D_a^n\left\{\frac{(as;q)_\infty}{(az,at;q)_\infty}\right\}$$
$$= \sum_{n=0}^{\infty}\frac{(r,f,g;q)_n u^n}{(v,w,q;q)_n}\sum_{k=0}^{n}\begin{bmatrix} n \\ k \end{bmatrix}_q q^{k(k-n)}\,D_a^k\left\{\frac{(as;q)_\infty}{(az;q)_\infty}\right\}$$
$$\cdot\, D_a^{n-k}\left\{\frac{1}{(atq^k;q)_\infty}\right\}. \tag{35}$$

Now, using the q-identities (27) and (31), we find that

$$\mathbb{T}(r,f,g,v,w,uD_a)\left\{\frac{(as;q)_\infty}{(az,at;q)_\infty}\right\}$$
$$= \sum_{n=0}^{\infty}\frac{(r,f,g;q)_n u^n}{(v,w,q;q)_n}\sum_{k=0}^{n}\begin{bmatrix} n \\ k \end{bmatrix}_q q^{k(k-n)}\,\frac{(\frac{s}{z};q)_k\,z^k}{(as;q)_k}\,\frac{(as;q)_\infty}{(az;q)_\infty}\,\frac{(tq^k)^{n-k}}{(atq^k;q)_\infty}$$
$$= \frac{(as;q)_\infty}{(az,at;q)_\infty}\sum_{k=0}^{\infty}\frac{(\frac{s}{z},at;q)_k\,z^k}{(as,q;q)_k}\sum_{n=k}^{\infty}\frac{(r,f,g;q)_n\,u^n\,t^{n-k}}{(v,w;q)_n(q;q)_{n-k}}. \tag{36}$$

Upon setting $n-k=n$, the right-hand side of (36) takes the following form:

$$\frac{(as;q)_\infty}{(az,at;q)_\infty}\sum_{k=0}^{\infty}\frac{(r,f,g,\frac{s}{z},at;q)_k\,(uz)^k}{(v,w,as,q;q)_k}\sum_{n=0}^{\infty}\frac{(rq^k,fq^k,gq^k;q)_n(ut)^n}{(vq^k,wq^k,q;q)_n}. \tag{37}$$

Summarizing the above observations (35) to (37), we get the assertion (33) of Theorem 1.

Similarly, by means of the definition (22) of the operator $\mathbb{E}(r,f,g,v,w,u\theta_a)$ and the Leibniz rule (13), we get

$$\mathbb{E}(r,f,g,v,w,u\theta_a)\left\{\frac{(az,at;q)_\infty}{(as;q)_\infty}\right\}$$
$$= \sum_{n=0}^{\infty}\frac{(-1)^n\,q^{\binom{n}{2}}(r,f,g;q)_n u^n}{(v,w,q;q)_n}\theta_a^n\left\{\frac{(az,at;q)_\infty}{(as;q)_\infty}\right\}$$
$$= \sum_{n=0}^{\infty}\frac{(-1)^n\,q^{\binom{n}{2}}(r,f,g;q)_n u^n}{(v,w,q;q)_n}\sum_{k=0}^{n}\begin{bmatrix} n \\ k \end{bmatrix}_q \theta_a^k\left\{\frac{(az;q)_\infty}{(as;q)_\infty}\right\}\theta_a^{n-k}\left\{(atq^{-k};q)_\infty\right\}, \tag{38}$$

By using the q-identities (30) and (32), we obtain

$$\begin{aligned}
&\mathbb{E}(r,f,g,v,w,u\theta_a)\left\{\frac{(az,at;q)_\infty}{(as;q)_\infty}\right\} \\
&= \sum_{n=0}^{\infty}\frac{(-1)^n q^{\binom{n}{2}}(r,f,g;q)_n u^n}{(v,w,q;q)_n}\sum_{k=0}^{n}\begin{bmatrix} n \\ k \end{bmatrix}_q \frac{\left(-\frac{q}{a}\right)^k\left(\frac{z}{s};q\right)_k}{\left(\frac{q}{as};q\right)_k} \\
&\quad\cdot\frac{(az;q)_\infty}{(as;q)_\infty}(atq^{-k};q)_\infty\,(-tq^{-k})^{n-k} \\
&= \frac{(az,at;q)_\infty}{(as;q)_\infty}\sum_{k=0}^{\infty}\frac{(-1)^k q^{-\binom{k}{2}}\left(\frac{z}{s},\frac{q}{at};q\right)_k t^k}{\left(\frac{q}{as},q;q\right)_k} \\
&\quad\cdot\sum_{n=k}^{\infty}\frac{q^{\binom{n}{2}-k(n-k)}(r,f,g;q)_n\, u^n\, t^{n-k}}{(v,w;q)_n(q;q)_{n-k}}.
\end{aligned} \tag{39}$$

Now, if we set $n-k=n$, the right-hand side of (39) becomes

$$\frac{(az,at;q)_\infty}{(as;q)_\infty}\sum_{k=0}^{\infty}\frac{\left(r,f,g,\frac{z}{s},\frac{q}{at};q\right)_k(-ut)^k}{\left(v,w,\frac{q}{as},q;q\right)_k}\sum_{n=0}^{\infty}\frac{q^{\binom{n}{2}}(rq^k,fq^k,gq^k;q)_n\,(ut)^n}{(vq^k,wq^k,q;q)_n}. \tag{40}$$

Finally, by combining the above observations (38) to (40), we arrive at the assertion (34) of Theorem 1. This evidently completes the proof of Theorem 1. □

We remark that, when $g=w=0$, Theorem 1 reduces to the concluding result of Li and Tan [24].

Corollary 1. *It is asserted that*

$$\mathbb{T}(r,f,g,v,w,uD_s)\left\{\frac{1}{(xs;q)_\infty}\right\} = \frac{1}{(xs;q)_\infty}\,{}_3\Phi_2\left[\begin{matrix} r,f,g; \\ v,w; \end{matrix}\; q;xu\right] \tag{41}$$

and

$$\mathbb{E}(r,f,g,v,w,-u\theta_s)\{(xs;q)_\infty\} = (xs;q)_\infty\,{}_3\Phi_3\left[\begin{matrix} r,f,g; \\ v,w,0; \end{matrix}\; q;xu\right], \tag{42}$$

provided that $\max\{|xs|,|xu|\}<1$.

The goal in this paper is to give potentially useful generalizations of a number q-series and q-integral identities such as the q-binomial theorem or the q-Gauss sum, the q-Chu-Vandermonde summation formula and the Andrews-Askey integral.

Our paper is organized as follows. In Section 2, we give two formal generalizations of the q-binomial theorem or the q-Gauss sum by applying the q-difference equations. In Section 3, we derive a set of two extensions q-Chu-Vandermonde summation formulas by making use of the q-difference equations. Next, in Section 4, we derive two new generalizations of the Andrews-Askey integral by means of the q-difference equations. Finally, in our last section (Section 5), we present a number of concluding remarks and observations concerning the various results which we have considered in this investigation.

2. A Set of Formal Generalizations of the q-Binomial Theorem

We begin this section by recalling the following q-binomial theorem (see, for example, [1–3,14]):

$$\sum_{n=0}^{\infty}\frac{(a;q)_n\,x^n}{(q;q)_n} = \frac{(ax;q)_\infty}{(x;q)_\infty} \qquad (|x|<1). \tag{43}$$

In Theorem 2 below, we give two generalizations of the q-binomial theorem (43) by applying the q-difference equations.

Theorem 2. *Each of the following assertions holds true:*

$$\sum_{n=0}^{\infty}\frac{(a;q)_n\,a^{-n}}{(q;q)_n}\sum_{k=0}^{\infty}\frac{(q^{-n},ax;q)_k\,q^k}{(q;q)_k}\sum_{j,i\geqq 0}\frac{(r,f,g;q)_{j+i}\,u^{j+i}}{(q;q)_i\,(v,w;q)_{j+i}}\frac{\left(\frac{c}{b},axq^k;q\right)_j}{(cx,q;q)_j}(aq^k)^i\,b^j$$
$$=\frac{(ax;q)_\infty}{(x;q)_\infty}\sum_{j,i\geqq 0}\frac{(r,f,g;q)_{j+i}\,u^{j+i}}{(q;q)_i\,(v,w;q)_{j+i}}\frac{\left(\frac{c}{b},x;q\right)_j}{(cx,q;q)_j}b^j\qquad(|x|<1)\tag{44}$$

and

$$\sum_{n=0}^{\infty}\frac{(a;q)_n\,a^{-n}}{(q;q)_n}\sum_{k=0}^{\infty}\frac{(q^{-n},ax;q)_k\,q^k}{(q;q)_k}\sum_{i,j\geqq 0}\frac{(-1)^{j+i}\,q^{\binom{i}{2}}(r,f,g;q)_{j+i}}{(q;q)_i\,(v,w;q)_{j+i}}$$
$$\cdot\frac{\left(\frac{bq^{1-k}}{a},\frac{q}{cx};q\right)_j}{\left(\frac{q^{1-k}}{ax},q;q\right)_j}(uc)^{j+i}$$
$$=\frac{(ax;q)_\infty}{(x;q)_\infty}\sum_{i,j=0}^{\infty}\frac{(-1)^{j+i}\,q^{\binom{i}{2}}\,(r,f,g;q)_{j+i}}{(q;q)_i(v,w;q)_{j+i}}\frac{\left(\frac{q}{bx};q\right)_j}{\left(\frac{q}{x},q;q\right)_j}(bu)^{j+i},\tag{45}$$

provided that both sides of (44) *and* (45) *exist.*

Remark 1. *For $u=0$ and by using the fact that*

$$\sum_{k=0}^{\infty}\frac{(q^{-n},ax;q)_k}{(q;q)_k}q^k={}_2\Phi_1\left[\begin{matrix}q^{-n},ax;\\ 0;\end{matrix}\;q;q\right]=(ax)^n,$$

the assertions (44) *and* (45) *reduce to* (43).

In our proof of Theorem 2, we shall need Theorem 3 and Corollary 2 below.

Theorem 3. *Each of the following assertions holds true:*

$$\mathbb{T}(r,f,g,v,w,uD_x)\left\{\frac{p_n\left(x,\frac{y}{a}\right)\,(cx;q)_\infty}{(ax,bx;q)_\infty}\right\}$$
$$=\frac{(y;q)_n}{a^n}\frac{(cx;q)_\infty}{(ax,bx;q)_\infty}\sum_{k=0}^{\infty}\frac{(q^{-n},ax;q)_k\,q^k}{(y,q;q)_k}$$
$$\cdot\sum_{j,i\geqq 0}\frac{(r,f,g;q)_{j+i}\,u^{j+i}}{(q;q)_i(v,w;q)_{j+i}}\frac{\left(\frac{c}{b},axq^k;q\right)_j}{(cx,q;q)_j}(aq^k)^i\,b^j\tag{46}$$

and

$$\mathbb{E}(r,f,g,v,w,u\theta_x)\left\{\frac{p_n\left(x,\frac{y}{a}\right)(bx,cx;q)_\infty}{(ax;q)_\infty}\right\}$$
$$=\frac{(y;q)_n}{a^n}\frac{(bx,cx;q)_\infty}{(ax;q)_\infty}\sum_{i,j\geqq 0}\frac{(-1)^{j+i}\,q^{\binom{i}{2}}\,(r,f,g;q)_{j+i}}{(q;q)_i\,(v,w;q)_{j+i}}$$
$$\cdot\frac{\left(\frac{bq^{1-k}}{a},\frac{q}{cx};q\right)_j}{\left(\frac{q^{1-k}}{ax},q;q\right)_j}(uc)^{j+i}, \tag{47}$$

provided that $\max\{|ax|,|bx|,|cx|\}<1$.

Corollary 2. *Each of the following assertions holds true:*

$$\mathbb{T}(r,f,g,v,w,uD_x)\left\{\frac{x^n(cx;q)_\infty}{(ax,bx;q)_\infty}\right\}$$
$$=\frac{1}{a^n}\frac{(cx;q)_\infty}{(ax,bx;q)_\infty}\sum_{k=0}^{\infty}\frac{(q^{-n},ax;q)_k\,q^k}{(q;q)_k}$$
$$\cdot\sum_{j,i\geqq 0}\frac{(r,f,g;q)_{j+i}\,u^{j+i}}{(q;q)_i\,(v,w;q)_{j+i}}\frac{\left(\frac{c}{b},axq^k;\right)_j}{(cx,q;q)_j}(aq^k)^i\,b^j \tag{48}$$

and

$$\mathbb{E}(r,f,g,v,w,u\theta_x)\left\{\frac{x^n(cx,bx;q)_\infty}{(ax;q)_\infty}\right\}$$
$$=\frac{1}{a^n}\frac{(bx,cx;q)_\infty}{(ax;q)_\infty}\sum_{k=0}^{\infty}\frac{(q^{-n},ax;q)_k\,q^k}{(q;q)_k}\sum_{i,j\geqq 0}\frac{(-1)^{j+i}q^{\binom{i}{2}}\,(r,f,g;q)_{j+i}}{(q;q)_i\,(v,w;q)_{j+i}}$$
$$\cdot\frac{\left(\frac{bq^{1-k}}{a},\frac{q}{cx};q\right)_j}{\left(\frac{q^{1-k}}{ax},q;q\right)_j}(cu)^{j+i}, \tag{49}$$

provided that $\max\{|ax|,|bx|,|cx|,|cu|\}<1$.

Remark 2. *For $y=0$, the assertions* (46) *and* (47) *reduce to* (48) *and* (49), *respectively.*

Proof of Theorem 3. Upon first setting $x\to ax$ in (55) and then multiplying both sides of the resulting equation by $\frac{(cx;q)_\infty}{(bx;q)_\infty}$, we get

$$\sum_{n=0}^{\infty}\frac{(q^{-n};q)_k\,q^k}{(y,q;q)_k}\frac{(cx;q)_\infty}{(axq^k,bx;q)_\infty}=\frac{(ax)^n\left(\frac{y}{ax};q\right)_n(cx;q)_\infty}{(y;q)_n(ax,bx;q)_\infty}. \tag{50}$$

Now, by applying the operator $\mathbb{T}(r,f,g,v,w,uD_x)$ to both sides of (50), it is easy to see that

$$\begin{aligned}\sum_{k=0}^{\infty}\frac{(q^{-n};q)_k\,q^k}{(y,q;q)_k}\,&\mathbb{T}(r,f,g,v,w,uD_x)\left\{\frac{(cx;q)_\infty}{(axq^k,bx;q)_\infty}\right\}\\&=\frac{a^n}{(y;q)_n}\,\mathbb{T}(r,f,g,v,w,uD_x)\left\{\frac{x^n\left(\frac{y}{ax};q\right)_n(cx;q)_\infty}{(ax,bx;q)_\infty}\right\}\\&=\frac{a^n}{(y;q)_n}\,\mathbb{T}(r,f,g,v,w,uD_x)\left\{\frac{p_n\left(x,\frac{y}{a}\right)(cx;q)_\infty}{(ax,bx;q)_\infty}\right\}.\end{aligned}\tag{51}$$

The proof of the first assertion (46) of Theorem 3 is completed by using the relation (33) in the left-hand side of (51).

The proof of the second assertion (47) of Theorem 3 is much akin to that of the first assertion (46). The details involved are, therefore, being omitted here. □

Proof of Theorem 2. Multiplying both sides of (43) by $\frac{(cx;q)_\infty}{(bx;q)_\infty}$, we find that

$$\sum_{n=0}^{\infty}\frac{(a;q)_n}{(q;q)_n}\frac{x^n\,(cx;q)_\infty}{(ax,bx;q)_\infty}=\frac{(cx;q)_\infty}{(bx,x;q)_\infty}.\tag{52}$$

Equation (44) can be written equivalently as follows:

$$\begin{aligned}\sum_{n=0}^{\infty}\frac{(a;q)_n}{(q;q)_n}\cdot\frac{a^{-n}(cx;q)_\infty}{(bx,ax;q)_\infty}&\sum_{k=0}^{\infty}\frac{(q^{-n},ax;q)_k\,q^k}{(q;q)_k}\\&\cdot\sum_{j,i\geqq 0}\frac{(r,f,g;q)_{j+i}\,u^{j+i}}{(q;q)_i(v,w;q)_{j+i}}\frac{\left(\frac{c}{b},axq^k;q\right)_j}{(cx,q;q)_j}\,(aq^k)^i\,b^j\\=\frac{(cx;q)_\infty}{(bx,x;q)_\infty}&\sum_{j,i=0}\frac{(r,f,g;q)_{j+i}\,u^{j+i}}{(q;q)_i(v,w;q)_{j+i}}\cdot\frac{\left(\frac{c}{b},x;q\right)_j}{(cx,q;q)_j}\,b^j.\end{aligned}\tag{53}$$

If we use $F(r,f,g,v,w,a,u)$ to denote the right-hand side of (53), it is easy to verify that $F(r,f,g,v,w,a,u)$ satisfies (23). By applying (24), we thus find that

$$\begin{aligned}F(r,f,g,v,w,x,u)&=\mathbb{T}(r,f,g,v,w,uD_x)\big\{F(r,f,g,v,w,x,0)\big\}\\&=\mathbb{T}(r,f,g,v,w,uD_x)\left\{\frac{(cx;q)_\infty}{(bx,x;q)_\infty}\right\}\\&=\mathbb{T}(r,f,g,v,w,uD_x)\left\{\sum_{n=0}^{\infty}\frac{(a;q)_n}{(q;q)_n}\frac{x^n\,(cx;q)_\infty}{(ax,bx;q)_\infty}\right\}\\&=\sum_{n=0}^{\infty}\frac{(a;q)_n}{(q;q)_n}\,\mathbb{T}(r,f,g,v,w,uD_x)\left\{\frac{x^n\,(cx;q)_\infty}{(ax,bx;q)_\infty}\right\}.\end{aligned}\tag{54}$$

The proof of the first assertion (44) of Theorem 2 can now be completed by making use of the relation (48).

The proof of the second assertion (45) of Theorem 2 is much akin to that of the first assertion (44). The details involved are, therefore, being omitted here. □

3. Two Generalizations of the q-Chu-Vandermonde Summation Formula

The q-Chu-Vandermonde summation formula is recalled here as follows (see, for example, [1,31]):

$${}_2\Phi_1\left[\begin{matrix}q^{-n},x;\\ y;\end{matrix}\;q;q\right]=\frac{\left(\frac{y}{x};q\right)_n}{(y;q)_n}\,x^n\qquad(n\in\mathbb{N}_0:=\mathbb{N}\cup\{0\}).\tag{55}$$

In this section, we give two generalizations of the q-Chu-Vandermonde summation Formula (55) by applying q-difference equations.

Theorem 4. *The following assertion holds true for $y \neq 0$:*

$$\sum_{k=0}^{n} \frac{(q^{-n}, x; q)_k\, q^k}{(q, y; q)_k}\, {}_3\Phi_2\left[\begin{matrix} r, f, g; \\ & q; uq^k \\ v, w; \end{matrix}\right] = \frac{x^n\left(\frac{y}{x}; q\right)_n}{(y;q)_n} \sum_{k,j \geqq 0} \frac{(r,f,g;q)_{k+j}}{(q;q)_j\,(v,w;q)_{k+j}} \frac{\left(\frac{q^{1-n}}{y}, \frac{qx}{y}; q\right)_k}{\left(\frac{xq^{1-n}}{y}, q; q\right)_k} u^{k+j}\left(\frac{q}{y}\right)^j. \tag{56}$$

We next derive another generalization of the q-Chu-Vandermonde summation Formula (55) as follows.

Theorem 5. *For $m \in \mathbb{N}_0$ and $y \neq 0$, it is asserted that*

$${}_2\Phi_1\left[\begin{matrix} q^{-n}, x; \\ & q; q^{1+m} \\ y; \end{matrix}\right] = \frac{x^n\left(\frac{y}{x}; q\right)_n}{(y;q)_n} \sum_{j=0}^{m} \begin{bmatrix} m \\ j \end{bmatrix}_q \frac{\left(\frac{q^{1-n}}{y}, \frac{qx}{y}; q\right)_{m-j}}{\left(\frac{xq^{1-n}}{y}; q\right)_{m-j}} \left(\frac{q}{y}\right)^j. \tag{57}$$

Remark 3. *For $u = 0$ or $m = 0$, the assertion (56) or (57) reduces to the q-Chu-Vandermonde summation Formula (55). Furthermore, if we first set $i + j = m$ and then extract the coefficients of $\frac{(r,f,g;q)_m}{(v,w;q)_m} u^m$ from the two members of the assertion (56) of Theorem 4, we obtain the transformation formula (57), which leads us to the q-Chu-Vandermonde summation Formula (55) when $m = 0$. Also, upon putting $n = 0$, the assertion (57) reduces to the following identity:*

$$\sum_{j=0}^{m} \begin{bmatrix} m \\ j \end{bmatrix}_q \left(\frac{q}{y}; q\right)_{m-j} \left(\frac{q}{y}\right)^j = 1 \qquad (y \neq 0). \tag{58}$$

Proof of Theorem 4. We first write (55) in the following form:

$$\sum_{k=0}^{n} \frac{(q^{-n};q)_k\, q^k}{(y,q;q)_k} \frac{1}{(xq^k;q)_\infty} = \frac{(-1)^n\, y^n\, q^{\binom{n}{2}}}{(y;q)_n} \frac{\left(\frac{xq^{1-n}}{y}; q\right)_\infty}{\left(x, \frac{qx}{y}; q\right)_\infty}. \tag{59}$$

Equation (56) can be written equivalently as follows:

$$\sum_{k=0}^{\infty} \frac{(q^{-n};q)_k\, q^k}{(q,c;q)_k} \cdot \frac{1}{(xq^k;q)_\infty}\, {}_3\Phi_2\left[\begin{matrix} r, f, g; \\ & q; uq^k \\ v, w; \end{matrix}\right] = \frac{(-1)^n\, y^n\, q^{\binom{n}{2}}}{(y;q)_n} \frac{\left(\frac{xq^{1-n}}{y}; q\right)_\infty}{\left(x, \frac{qx}{y}; q\right)_\infty} \cdot \sum_{i,j \geqq 0} \frac{(r,f,g;q)_{j+i}\, u^{j+i}}{(q;q)_i\,(v,w;q)_{j+i}} \frac{\left(\frac{q^{1-n}}{y}, \frac{qx}{y}; q\right)_j}{\left(\frac{xq^{1-n}}{y}, q; q\right)_j} \left(\frac{q}{y}\right)^i. \tag{60}$$

If we use $G(r,f,g,v,w,x,u)$ to denote the right-hand side of (60), it is easy to observe that $G(r,f,g,v,w,x,u)$ satisfies (23). By using (24), we obtain

$$\begin{aligned}G(r,f,g,v,w,x,u) &= \mathbb{T}(r,f,g,v,w,uD_x)\Big\{G(r,f,g,v,w,x,0)\Big\}\\ &= \mathbb{T}(r,f,g,v,w,uD_x)\left\{\frac{(-1)^n\, y^n\, q^{\binom{n}{2}}}{(y;q)_n}\,\frac{\left(\frac{xq^{1-n}}{y};q\right)_\infty}{\left(x,\frac{qx}{y};q\right)_\infty}\right\}\\ &= \mathbb{T}(r,f,g,v,w,uD_x)\left\{\sum_{k=0}^{\infty}\frac{(q^{-n};q)_k\,q^k}{(y,q;q)_k}\,\frac{1}{(xq^k;q)_\infty}\right\}\\ &= \sum_{k=0}^{n}\frac{(q^{-n};q)_k\,q^k}{(y,q;q)_k}\,\mathbb{T}(r,f,g,v,w,uD_x)\left\{\frac{1}{(xq^k;q)_\infty}\right\}. \end{aligned} \tag{61}$$

Finally, by using the fact that

$$\mathbb{T}(r,f,g,v,w,uD_x)\left\{\frac{1}{(xq^k;q)_\infty}\right\} = \frac{1}{(xq^k;q)_\infty}\,{}_3\Phi_2\left[\begin{matrix} r,f,g; \\ & q;uq^k \\ v,w; \end{matrix}\right], \tag{62}$$

and after some simplification involving $\frac{1}{(x;q)_\infty}$, we get the left-hand side of (56). $\square$

4. New Generalizations of the Andrews-Askey Integral

The following famous formula is known as the Andrews-Askey integral (see, for details, [32]). It was derived from Ramanujan's celebrated ${}_1\Psi_1$-summation formula.

Proposition 2. *(see ([32], Equation (2.1))). For* $\max\{|ac|,|ad|,|bc|,|bd|\} < 1$, *it is asserted that*

$$\int_c^d \frac{\left(\frac{qt}{c},\frac{qt}{d};q\right)_\infty}{(at,bt;q)_\infty}\,\mathrm{d}_q t = \frac{d(1-q)\left(q,\frac{dq}{c},\frac{c}{d},abcd;q\right)_\infty}{(ac,ad,bc,bd;q)_\infty}. \tag{63}$$

The Andrews-Askey integral (63) is indeed an important formula in the theory of q-series (see [16]).
Recently, Cao [33] gave the following two generalizations of the Andrews-Askey integral (63) by the method based upon q-difference equations.

Proposition 3. *(see ([33], Theorems 14 and 15)) For* $N \in \mathbb{N}$ *and* $r = q^{-N}$, *suppose that*

$$\max\left\{|ac|,|ad|,|bc|,|bd|,\left|\frac{qwr}{v}\right|,\left|\frac{q}{v}\right|\right\} < 1.$$

Then

$$\begin{aligned}&\int_c^d \frac{\left(\frac{qt}{c},\frac{qt}{d};q\right)_\infty}{(at,bt;q)_\infty}\,{}_4\Phi_2\left[\begin{matrix} r,w,\frac{c}{t},abcd; \\ & q;\frac{qt}{vbcd} \\ ac,\frac{qwr}{v}; \end{matrix}\right]\mathrm{d}_q t\\ &\quad= \frac{d(1-q)\left(q,\frac{dq}{c},\frac{c}{d},abcd,\frac{qw}{v},\frac{qr}{v};q\right)_\infty}{\left(ac,ad,bc,bd,\frac{qwr}{v},\frac{q}{v};q\right)_\infty}\,{}_2\Phi_1\left[\begin{matrix} w,r; \\ & q;\frac{q}{bc} \\ v; \end{matrix}\right]. \end{aligned} \tag{64}$$

Furthermore, for $N \in \mathbb{N}$ *and* $r = q^{-N}$, *suppose that*

$$\max\left\{|ac|,|ad|,|bc|,|bd|,\left|\frac{v}{w}\right|,\left|\frac{v}{r}\right|\right\} < 1.$$

Then

$$\int_c^d \frac{\left(\frac{qt}{c},\frac{qt}{d};q\right)_\infty}{(at,bt;q)_\infty}\,{}_4\Phi_2\left[\begin{matrix} r,w,\frac{c}{t},\frac{q}{ad}; \\ \frac{q}{at},\frac{qrw}{v}; \end{matrix}\; q;q\right] d_qt = \frac{d(1-q)\left(q,\frac{dq}{c},\frac{c}{d},abcd,\frac{v}{wr},v;q\right)_\infty}{\left(ac,ad,bc,bd,\frac{v}{w},\frac{v}{r};q\right)_\infty}\,{}_2\Phi_1\left[\begin{matrix} w,r; \\ v; \end{matrix}\; q;\frac{vbc}{wr}\right]. \tag{65}$$

In this section, we give the following two generalizations of the Andrews-Askey integral (63) by using the method of q-difference equations.

Theorem 6. *For $M \in \mathbb{N}$ and $r = q^{-M}$, suppose that*

$$\max\left\{|ac|,|ad|,|bc|,|bd|,\left|\frac{q}{bc}\right|\right\} < 1.$$

Then

$$\int_c^d \frac{\left(\frac{qt}{c},\frac{qt}{d};q\right)_\infty}{(at,bt;q)_\infty}\sum_{k=0}^{\infty}\frac{\left(r,f,g,\frac{c}{t},abcd;q\right)_k\left(\frac{qt}{bcd}\right)^k}{(v,w,ac,q;q)_k}\,{}_3\Phi_2\left[\begin{matrix} rq^k,fq^k,gq^k; \\ vq^k,wq^k; \end{matrix}\; q;q\right] \mathrm{d}_qt = \frac{d(1-q)\left(q,\frac{dq}{c},\frac{c}{d},abcd;q\right)_\infty}{(ac,ad,bc,bd;q)_\infty}\,{}_3\Phi_2\left[\begin{matrix} r,f,g; \\ v,w; \end{matrix}\; q;\frac{q}{bc}\right]. \tag{66}$$

Theorem 7. *For $M \in \mathbb{N}$ and $r = q^{-M}$, suppose that* $\max\{|ac|,|ad|,|bc|,|bd|\} < 1$. *Then*

$$\int_c^d \frac{\left(\frac{qt}{c},\frac{qt}{d};q\right)_\infty}{(at,bt;q)_\infty}\sum_{k=0}^{\infty}\frac{\left(r,f,g,\frac{c}{t},\frac{q}{ad};q\right)_k\left(\frac{vw}{rfg}\right)^k}{\left(v,w,\frac{q}{at},q;q\right)_k}\,{}_3\Phi_3\left[\begin{matrix} rq^k,fq^k,gq^k; \\ vq^k,wq^k,0; \end{matrix}\; q;-\frac{vw}{rfg}\right] \mathrm{d}_qt = \frac{d(1-q)\left(q,\frac{dq}{c},\frac{c}{d},abcd;q\right)_\infty}{(ac,ad,bc,bd;q)_\infty}\,{}_3\Phi_3\left[\begin{matrix} r,f,g; \\ v,w,0; \end{matrix}\; q;\frac{vwbc}{rfg}\right]. \tag{67}$$

Remark 4. *For $r = 1$, both (66) and (67) reduce to (63). Moreover, for $r = q^{-N}$, $g = w = 0$ and $u = \frac{q}{bcd}$, the assertion (66) of Theorem 6 reduces to (64). For $r = q^{-N}$, $g = w = 0$ and $u = \frac{v}{rfbcd}$, the assertion (66) of Theorem 7 reduces to (67).*

Proof of Theorems 6 and 7. Equation (66) can be written equivalently as follows:

$$\int_c^d \frac{\left(\frac{qt}{c},\frac{qt}{d};q\right)_\infty}{(bt;q)_\infty}\cdot\frac{(ac;q)_\infty}{(at,abcd;q)_\infty}\sum_{k=0}^{\infty}\frac{\left(r,f,g,\frac{c}{t},abcd;q\right)_k\left(\frac{qt}{bcd}\right)^k}{(v,w,ac,q;q)_k} \cdot{}_3\Phi_2\left[\begin{matrix} rq^k,fq^k,gq^k; \\ vq^k,wq^k; \end{matrix}\; q;q\right] \mathrm{d}_qt = \frac{d(1-q)\left(q,\frac{dq}{c},\frac{c}{d};q\right)_\infty}{(bc,bd;q)_\infty}\cdot\frac{1}{(ad;q)_\infty}\,{}_3\Phi_2\left[\begin{matrix} r,f,g; \\ v,w; \end{matrix}\; q;\frac{q}{bc}\right]. \tag{68}$$

If we use $H(r,f,g,v,w,a,u)$ to denote the right-hand side of (68), it is easy to see that $H(r,f,g,v,w,a,u)$ satisfies (23) with $u=\frac{q}{bcd}$. By making use of (24), we thus find that

$$\begin{aligned}
H(r,f,g,v,w,a,u) &= \mathbb{T}(r,f,g,v,w,uD_a)\Big\{H(r,f,g,v,w,a,0)\Big\} \\
&= \mathbb{T}\left(r,f,g,v,w,\frac{q}{bcd}\,D_a\right)\Big\{H(1,f,g,v,w,a,u)\Big\} \\
&= \mathbb{T}\left(r,f,g,v,w,\frac{q}{bcd}\,D_a\right)\left\{\frac{d(1-q)\left(q,\frac{dq}{c},\frac{c}{d};q\right)_\infty}{(bc,bd;q)_\infty}\cdot\frac{1}{(ad;q)_\infty}\right\} \\
&= \mathbb{T}\left(r,f,g,v,w,\frac{q}{bcd}\,D_a\right)\left\{\int_c^d\frac{\left(\frac{qt}{c},\frac{qt}{d};q\right)_\infty}{(bt;q)_\infty}\cdot\frac{(ac;q)_\infty}{(at,abcd;q)_\infty}\,\mathrm{d}_qt\right\} \\
&= \int_c^d\frac{\left(\frac{qt}{c},\frac{qt}{d};q\right)_\infty}{(bt;q)_\infty}\cdot\mathbb{T}\left(r,f,g,v,w,\frac{q}{bcd}\,D_a\right)\left\{\frac{(ac;q)_\infty}{(at,abcd;q)_\infty}\right\}\mathrm{d}_qt.
\end{aligned}$$

Now, by applying the fact that

$$\begin{aligned}
&\mathbb{T}\left(r,f,g,v,w,\frac{q}{bcd}\,D_a\right)\left\{\frac{(ac;q)_\infty}{(at,abcd;q)_\infty}\right\} \\
&\quad=\frac{(ac;q)_\infty}{(at,abcd;q)_\infty}\sum_{k=0}^{\infty}\frac{\left(r,f,g,\frac{c}{t},abcd;q\right)_k\left(\frac{qt}{bcd}\right)^k}{(v,w,ac,q;q)_k}\,{}_3\Phi_2\left[\begin{matrix}rq^k,fq^k,gq^k;\\ vq^k,wq^k;\end{matrix}\;q;q\right],
\end{aligned}$$

we get the left-hand side of (66).

Similarly, Equation (67) can be written equivalently as follows:

$$\begin{aligned}
&\int_c^d\frac{\left(\frac{qt}{c},\frac{qt}{d};q\right)_\infty}{(bt;q)_\infty}\cdot\frac{(ac;q)_\infty}{(at,abcd;q)_\infty}\sum_{k=0}^{\infty}\frac{\left(r,f,g,\frac{c}{t},\frac{q}{ad};q\right)_k\left(\frac{vw}{rfg}\right)^k}{\left(v,w,\frac{q}{at},q;q\right)_k} \\
&\qquad\cdot{}_3\Phi_3\left[\begin{matrix}rq^k,fq^k,gq^k;\\ vq^k,wq^k,0;\end{matrix}\;q;-\frac{vw}{rfg}\right]\mathrm{d}_qt \\
&\quad=\frac{d(1-q)\left(q,\frac{dq}{c},\frac{c}{d};q\right)_\infty}{(bc,bd;q)_\infty}\cdot\frac{1}{(ad;q)_\infty}\,{}_3\Phi_3\left[\begin{matrix}r,f,g;\\ v,w,0;\end{matrix}\;q;\frac{vwbc}{rfg}\right].
\end{aligned}\tag{69}$$

The proof of the assertion (67) of Theorem 7 is much akin to that of the assertion (66) of Theorem 6 by using Proposition 1 (II). The analogous details involved are, therefore, being omitted here.

The proofs of Theorems 6 and 7 are thus completed. □

5. Concluding Remarks and Observations

In our present investigation, we have introduced a set of two q-operators $\mathbb{T}(a,b,c,d,e,yD_x)$ and $\mathbb{E}(a,b,c,d,e,y\theta_x)$ with to applying them to derive two potentially useful generalizations of the q-binomial theorem, two extensions of the q-Chu-Vandermonde summation formula and two new generalizations of the Andrews-Askey integral by means of the q-difference equations. We have also briefly described relevant connections of various special cases and consequences of our main results with several known results.

It is believed that the q-series and q-integral identities, which we have presented in this paper, as well as the various related recent works cited here, will provide encouragement and motivation for further researches on the topics that are dealt with and investigated in this paper.

Just as we mentioned in Section 1 above, basic (or q-) series and basic (or q-) polynomials, especially the basic (or q-) hypergeometric functions and basic (or q-) hypergeometric polynomials, are applicable particularly in several diverse areas [see also ([3], pp. 350–351)]. In particular, the celebrated

Chu-Vandermonde summation theorem and its known q-extensions, which we have considered in this paper, have already been demonstrated to be useful (see, for details, [1,14,31]). The new q-Chu-Vandermone summation theorems, which we have presented in this paper, are believed to be useful as well.

In conclusion, we find it to be worthwhile to remark that some potential further applications of the methodology and findings, which we have presented here by means of the q-analysis and the q-calculus, can be found in the study of the zeta and q-zeta functions as well as their related functions of Analytic Number Theory (see, for example, [34,35]; see also [14]) and also in the study of analytic and univalent functions of Geometric Function Theory *via* number-theoretic objects (see, for example, [36]).

Author Contributions: Conceptualization, S.A., H.M.S. and J.C.; Formal analysis, H.M.S. and S.A.; Investigation, S.A. and J.C.; Methodology, S.A., H.M.S. and J.C.; Validation, S.A. and H.M.S.; Visualization, S.A.; Writing—Review and Editing, H.M.S. and S.A. All authors have read and agreed to the published version of the manuscript.

References

1. Gasper, G.; Rahman, M. *Basic Hypergeometric Series (with a Foreword by Richard Askey), Encyclopedia of Mathematics and Its Applications*, 2nd ed.; Cambridge University Press: Cambridge, UK; London, UK; New York, NY, USA, 2004; Volume 96.
2. Slater, L.J. *Generalized Hypergeometric Functions*; Cambridge University Press: Cambridge, UK; London, UK; New York, NY, USA, 1966.
3. Srivastava, H.M.; Karlsson, P.W. *Multiple Gaussian Hypergeometric Series*; Ellis Horwood Limited: Chichester, UK, 1985.
4. Koekock, R.; Swarttouw, R.F. *The Askey-Scheme of Hypergeometric Orthogonal Polynomials and its q-Analogue*; Report No. 98-17; Delft University of Technology: Delft, The Netherlands, 1998.
5. Srivastava, H.M. Certain q-polynomial expansions for functions of several variables. I. *IMA J. Appl. Math.* **1983**, *30*, 315–323. [CrossRef]
6. Srivastava, H.M. Certain q-polynomial expansions for functions of several variables. II. *IMA J. Appl. Math.* **1984**, *33*, 205–209. [CrossRef]
7. Srivastava, H.M. Operators of basic (or q-) calculus and fractional q-calculus and their applications in geometric function theory of complex analysis. *Iran. J. Sci. Technol. Trans. A Sci.* **2020**, *44*, 327–344. [CrossRef]
8. Morozov, V.N. Penetration of nonstationary ionospheric electric fields into lower atmospheric layers in the global electric circuit model. *Geomagnet. Aeronom.* **2018**, *58*, 113–118. [CrossRef]
9. Suzuki, A.T. Correspondence between the one-loop three-point vertex and the Y and Δ electric resistor networks. *Canad. J. Phys.* **2014**, *92*, 131–135. [CrossRef]
10. Di Barba, P.; Fattorusso, L.; Versaci, M. Electrostatic field in terms of geometric curvature in membrane MEMS devices. *Commun. Appl. Industr. Math.* **2017**, *8*, 165–184. [CrossRef]
11. Włodarczyk, M.; Zawadzki, A. The application of hypergeometric functions to computing fractional order derivatives of sinusoidal functions. *Bull. Polish Acad. Sci. Tech. Sci.* **2016**, *64*, 243–248. [CrossRef]
12. Chen, W.Y.C.; Fu, A.M.; Zhang, B. The homogeneous q-difference operator. *Adv. Appl. Math.* **2003**, *31*, 659–668. [CrossRef]
13. Cao, J.; Srivastava, H.M. Some q-generating functions of the Carlitz and Srivastava-Agarwal types associated with the generalized Hahn polynomials and the generalized Rogers-Szegö polynomials. *Appl. Math. Comput.* **2013**, *219*, 8398–8406. [CrossRef]
14. Srivastava, H.M.; Choi, J. *Zeta and q-Zeta Functions and Associated Series and Integrals*; Elsevier Science Publishers: Amsterdam, The Netherlands; London, UK; New York, NY, USA, 2012.
15. Srivastava, H.M.; Chaudhary, M.P.; Wakene, F.K. A family of theta-function identities based upon q-binomial theorem and Heine's transformations. *Montes Taurus J. Pure Appl. Math.* **2020**, *8*, 918.
16. Chen, W.Y.C.; Liu, Z.-G. Parameter augmenting for basic hypergeometric series. II. *J. Combin. Theory Ser. A* **1997**, *80*, 175–195. [CrossRef]
17. Saad, H.L.; Sukhi, A.A. Another homogeneous q-difference operator. *Appl. Math. Comput.* **2010**, *215*, 4332–4339. [CrossRef]
18. Srivastava, H.M.; Abdlhusein, M.A. New forms of the Cauchy operator and some of their applications. *Russ. J. Math. Phys.* **2016**, *23*, 124–134. [CrossRef]

19. Roman, S. The theory of the umbral calculus. I. *J. Math. Anal. Appl.* **1982**, *87*, 58–115. [CrossRef]
20. Chen, W.Y.C.; Liu, Z.-G. Parameter augmentation for basic hypergeometric series. I. In *Mathematical Essays in Honor of Gian-Carlo Rota*; Sagan, B.E., Stanley, R.P., Eds.; Birkäuser: Basel, Switzerland; New York, NY, USA, 1998; pp. 111–129.
21. Chen, V.Y.B.; Gu, N.S.S. The Cauchy operator for basic hypergeometric series. *Adv. Appl. Math.* **2008**, *41*, 177–196. [CrossRef]
22. Fang, J.P. Some applications of q-differential operator. *J. Korean Math. Soc.* **2010**, *47*, 223–233. [CrossRef]
23. Zhang, Z.Z.; Wang, J.Z. Finite q-exponential operators with two parameters and their applications. *Acta Math. Sinica* **2010**, *53*, 1007–1018. (In Chinese)
24. Li, N.N.; Tan, W. Two generalized q-exponential operators and their applications. *Adv. Differ. Equ.* **2016**, *2016*, 53. [CrossRef]
25. Cao, J.; Xu, B.; Arjika, S. A note on generalized q-difference equations for general Al-Salam-Carlitz polynomials. *J. Differ. Equ. Appl.* **2020**, submitted.
26. Arjika, S. q-difference equations for homogeneous q-difference operators and their applications. *J. Differ. Equ. Appl.* **2020**, *26*, 987–999. [CrossRef]
27. Srivastava, H.M.; Arjika, S. Generating functions for some families of the generalized Al-Salam-Carlitz q-polynomials. *Adv. Differ. Equ.* **2020**, *2020*, 498. [CrossRef]
28. Srivastava, H.M.; Arjika, S.; Kelil, A.S. Some homogeneous q-difference operators and the associated generalized Hahn polynomials. *Appl. Set Valued Anal. Optim.* **2019**, *1*, 187–201.
29. Liu, Z.-G. Two q-difference equations and q-operator identities. *J. Differ. Equ. Appl.* **2010**, *16*, 1293–1307. [CrossRef]
30. Liu, Z.-G. An extension of the non-terminating ${}_6\phi_5$-summation and the Askey-Wilson polynomials. *J. Differ. Equ. Appl.* **2011**, *17*, 1401–1411. [CrossRef]
31. Andrews, G.E. *q-Series: Their Development and Applications in Analysis, Number Theory, Combinatorics, Physics and Computer Algebra*; CBMS Regional Conference Lecture Series; American Mathematical Society: Ann Arbor, MI, USA, 1986; Volume 66.
32. Andrews, G.E.; Askey, R. Another q-extension of the beta function. *Proc. Am. Math. Soc.* **1981**, *81*, 97–100.
33. Cao, J. A note on generalized q-difference equations for q-beta and Andrews-Askey integral. *J. Math. Anal. Appl.* **2014**, *412*, 841–851. [CrossRef]
34. Srivastava, H.M. The Zeta and related functions: Recent developments. *J. Adv. Engrg. Comput.* **2019**, *3*, 329–354. [CrossRef]
35. Srivastava, H.M. Some general families of the Hurwitz-Lerch Zeta functions and their applications: Recent developments and directions for further researches. *Proc. Inst. Math. Mech. Nat. Acad. Sci. Azerbaijan* **2019**, *45*, 234–269. [CrossRef]
36. Shafiq, M.; Srivastava, H.M.; Khan, N.; Ahmad, Q.Z.; Darus, M.; Kiran, S. An upper bound of the third Hankel determinant for a subclass of q-starlike functions associated with k-Fibonacci numbers. *Symmetry* **2020**, *12*, 1043. [CrossRef]

3

On almost b-Metric Spaces and Related Fixed Point Results

Nabil Mlaiki [1], Katarina Kukić [2], Milanka Gardašević-Filipović [3] and Hassen Aydi [4,5,*]

1 Department of Mathematics and General Sciences, Prince Sultan University, P. O. Box 66833, Riyadh 11586, Saudi Arabia; nmlaiki@psu.edu.sa
2 Faculty for Traffic and Transport Engineering, University of Belgrade, 11000 Belgrade, Serbia; k.mijailovic@sf.bg.ac.rs
3 School of Computing, Union University, 11000 Belgrade, Serbia; mgardasevic@raf.edu.rs
4 Institut Supérieur d'Informatique et des Techniques de Communication, Université de Sousse, H. Sousse 4000, Tunisia
5 China Medical University Hospital, China Medical University, Taichung 40402, Taiwan
* Correspondence: hassen.aydi@isima.rnu.tn

Abstract: In this manuscript, we introduce almost b-metric spaces and prove modifications of fixed point theorems for Reich and Hardy–Rogers type contractions. We present an approach generalizing some fixed point theorems to the case of almost b-metric spaces by reducing almost b-metrics to the corresponding b-metrics. Later, we show that this approach can not work for all kinds of contractions. To confirm this, we present a proof in which the contraction condition is such that it cannot be reduced to corresponding b-metrics.

Keywords: fixed point; Reich contraction; Hardy–Rogers contraction; almost b-metric space

1. Introduction

In [1] Filipović and Kukić considered some classical contraction principles of Kannan [2], Reich [3] and Hardy–Rogers [4] in b-metric spaces and rectangular b-metric spaces without the assumption of continuity of the corresponding metric. The fact that a b-metric d need not be continuous must remind us to use caution in the proofs.

As possibly more general forms of the theorems proven in [1], here we further try to, as many authors before, generalize metric spaces. Plenty of generalizations in previous two decades were done. Starting from 1989, b-metric spaces were introduced in [5]. After, partial b-metric spaces [6], metric-like spaces [7] and b-dislocated metric spaces [8] have been given. For related contraction principles in the setting of above spaces, the readers can see [9–19].

As an attempt to continue in that spirit, we initiate the concept of almost b-metric spaces. The motivation of this initiation comes from [20] where Mitrović, George and Hussain introduced almost rectangular b-metric spaces.

2. Preliminaries

Bakhtin in [5] and Czerwik in [21] introduced b-metric spaces as a generalization of standard metric spaces.

Definition 1 (Ref. [5,21])**.** *Let X be a nonempty set and $s \geq 1$. The function $d_b : X \times X \to [0, +\infty)$ is a b-metric if and only if, for all $\chi, \zeta, \sigma \in X$, we have*

(bM1) $d_b(\chi, \zeta) = 0$ *if and only if* $\chi = \zeta$,
(bM2) $d_b(\chi, \zeta) = d_b(\zeta, \chi)$,
(bM3) $d_b(\chi, \sigma) \leq s(d_b(\chi, \zeta) + d_b(\zeta, \sigma))$.

(X, d_b, s) is said a b-metric space and $s \geq 1$ is its coefficient.

In particular, if $s = 1$ then (X, d) is a standard metric space.

Recall that a sequence $\{\chi_n\}$ in X, b-converges to $\chi \in X$ if and only if $d_b(\chi_n, \chi) \to 0$ as $n \to \infty$. $\{\chi_n\}$ is b-Cauchy if and only if $d_b(\chi_n, \chi_m) \to 0$ as $n, m \to \infty$. If each b-Cauchy sequence is b-convergent in X, then (X, d_b, s) is said to be b-complete.

If in previous definition, we assume that only (bM1) and (bM3) hold, then we denote d_b as d_q and we call (X, d_q, s) a quasi-b-metric space.

In next few lines, we make a brief overview of some well known types of contractions. Let (X, d) be a metric space and $T : X \to X$ be such that

- $d(T\chi, T\zeta) \leq \lambda d(\chi, \zeta)$, $\lambda \in [0, 1)$, a Banach type of contraction;
- $d(T\chi, T\zeta) \leq \lambda\left(d(\chi, T\chi) + d(\zeta, T\zeta)\right)$, $\lambda \in [0, \frac{1}{2})$, a Kannan type of contraction;
- $d(T\chi, T\zeta) \leq \lambda\left(d(\chi, T\zeta) + d(\zeta, T\chi)\right)$, $\lambda \in [0, \frac{1}{2})$, a Chatterjea type of contraction;
- $d(T\chi, T\zeta) \leq \alpha d(\chi, \zeta) + \beta d(\chi, T\chi) + \gamma d(\zeta, T\zeta)$ where $\alpha, \beta, \gamma \geq 0$ with $\alpha + \beta + \gamma < 1$, a Reich type of contraction;
- $d(T\chi, T\zeta) \leq \alpha d(\chi, \zeta) + \beta d(\chi, T\chi) + \gamma d(\zeta, T\zeta) + \delta d(\chi, T\zeta) + \mu d(\zeta, T\chi)$ where $\alpha, \beta, \gamma, \delta, \mu \geq 0$ with $\alpha + \beta + \gamma + \delta + \mu < 1$, a Hardy–Rogers type of contraction.

In [1] Filipović and Kukić proved new theorems with additional conditions that are necessary to prove the theorems without assumption of continuity of b-metric. Here, we cite only formulations of those theorems and for the proofs, we refer on [1].

Theorem 1. *Ref. [1] let T be a self-mapping on a complete b-metric space $(X, d_b, s \geq 1)$ such that*

$$d_b(T\chi, T\zeta) \leq \lambda d_b(\chi, \zeta) + \mu d_b(\chi, T\chi) + \delta d_b(\zeta, T\zeta),$$

for all $\chi, \zeta \in X$, where $\lambda, \mu, \delta \geq 0$ with $\lambda + \mu + \delta < 1$ and

$$\delta < \frac{1}{s}.$$

Then there is a unique fixed point of T.

Theorem 2. *Ref. [1] let $(X, d_b, s \geq 1)$ be a complete b-metric space and $T : X \to X$ be a mapping satisfying*

$$d_b(T\chi, T\zeta) \leq a_1 d_b(\chi, \zeta) + a_2 d_b(\chi, T\chi) + a_3 d_b(\zeta, T\zeta) + a_4 d_b(\chi, T\zeta) + a_5 d_b(\zeta, T\chi),$$

for all $\chi, \zeta \in X$, where a_1, a_2, a_3, a_4, $a_5 \geq 0$ are such that $a_1 + a_2 + a_3 + s(a_4 + a_5) < 1$ and $a_1 > 1 - \frac{2}{s}$. Then T has a unique fixed point.

In the sequel of this paper, we introduce almost-b-metric spaces and present the related previous theorems in this setting. At the end, we also give some results for different type of contractions, where the proofs cannot be reduced to the corresponding b-metrics.

3. Main Results

In this section, let us firstly introduce the concept of almost-b-metric spaces, as a class of quasi-b-metric spaces with the additional requirement that diminishes a lack of symmetry. We set a demand that existence of the left limit of sequence implies the existence of the right limit (bM2l) or that existence of the right limit of sequence implies the existence of the left limit of the same sequence (bM2r). After that, we introduce a couple of examples of almost-b-metrics and also an example of a quasi-b-metric, which is not an almost-b-metric. Finally, we prove Theorems 1 and 2 with the assumption (bM2left) instead of (bM2).

Definition 2. *Let X be a nonempty set and $s \geq 1$. Let $d_{ab} : X \times X \to [0, +\infty)$ be a function such that for all $\chi, \zeta, \sigma, \chi_n \in X$,*

(bM1) $d_{ab}(\chi, \zeta) = 0$ *iff* $\chi = \zeta$,
(bM2l) $d_{ab}(\chi_n, \chi) \to 0,\ n \to \infty$ *implies* $d_{ab}(\chi, \chi_n) \to 0,\ n \to \infty$,
(bM2r) $d_{ab}(\chi, \chi_n) \to 0,\ n \to \infty$ *implies* $d_{ab}(\chi_n, \chi) \to 0,\ n \to \infty$,
(bM3) $d_{ab}(\chi, \zeta) \leq s(d_{ab}(\chi, \sigma) + d_{ab}(\sigma, \zeta))$.

Then (X, d_{ab}, s) is called an

1. *l-almost-b-metric space if (bM1), (bM2l) and (bM3) hold;*
2. *r-almost-b-metric space if (bM1), (bM2r) and (bM3) hold;*
3. *almost-b-metric space if (bM1), (bM2l), (bM2r) and (bM3) hold.*

In the next two examples, we present two quasi-b-metrics, which are also almost-b-metrics.

Example 1. *Let $X = \{0, 1, 2\}$. Choose $\alpha \geq 2$. Consider the b-metric $d_{ab} : X \times X \to [0, +\infty)$ defined by*

$$d_{ab}(0,0) = d_{ab}(1,1) = d_{ab}(2,2) = 0,$$

$$d_{ab}(1,0) = 1, \quad d_{ab}(0,1) = \frac{3}{2},$$

$$d_{ab}(2,1) = 1, \quad d_{ab}(1,2) = \frac{3}{2},$$

$$d_{ab}(2,0) = \alpha, \quad d_{ab}(0,2) = \alpha + 1.$$

Note that d_{ab} satisfies (bM1), (bM3), (bM2l) and (bM2r) (but not (bM2)). For $\alpha > 2$, the ordinary triangle inequality is not verified. Indeed,

$$d_{ab}(0,2) = \alpha + 1 > 3 = \frac{3}{2} + \frac{3}{2} = d_{ab}(0,1) + d_{ab}(1,2).$$

However, the following is satisfied for all $x, y, z \in X$,

$$d_{ab}(x,y) \leq \frac{\alpha+2}{2}(d_{ab}(x,z) + d_{ab}(z,y)).$$

Example 2. *Let $X = [0, +\infty)$ and define $d_{ab} : X \times X \to [0, +\infty)$ as*

$$d_{ab}(x,y) = \begin{cases} (x-y)^3, & x \geq y \\ 4(y-x)^3, & x < y \end{cases}$$

Then $(X, d_{ab}, 4)$ is an almost b-metric space. (bM1), (bM2l) and (bM2r) are obvious. It remains to prove that for all $x, y, z, \in X$,

$$d_{ab}(x,y) \leq 4(d_{ab}(x,z) + d_{ab}(z,y)).$$

Case 1. *$x \geq y$ and $d_{ab}(x,y) = (x-y)^3$. Starting from the inequality $(\alpha+\beta)^3 \leq 4(\alpha^3 + \beta^3)$, we separate the cases:*

$y \leq z \leq x$:

$$\begin{aligned} d_{ab}(x,y) &= (x-y)^3 = (x-z+z-y)^3 \\ &\leq 4((x-z)^3 + (y-z)^3) = 4(d_{ab}(x,z) + d_{ab}(z,y)), \end{aligned}$$

$z \le y \le x$:

$$\begin{aligned} d_{ab}(x,y) &= (x-y)^3 \le 4((x-z)^3 + (y-z)^3) \\ &\le 4((x-z)^3 + 4(y-z)^3) = 4(d_{ab}(x,z) + d_{ab}(z,y)), \end{aligned}$$

$y \le x \le z$:

$$\begin{aligned} d_{ab}(x,y) &= (x-y)^3 \le 4((x-z)^3 + (z-y)^3) \\ &\le 4(4(z-x)^3 + (z-y)^3) = 4(d_{ab}(x,z) + d_{ab}(z,y)). \end{aligned}$$

Case 2. *$x < y$ and $d_{ab}(x,y) = 4(y-x)^3$. Again, we separate the cases:*

$x \le z \le y$:

$$\begin{aligned} d_{ab}(x,y) &= 4(y-x)^3 = 4(y-z+z-x)^3 \\ &\le 4(4(y-z)^3 + 4(z-x)^3) = 4(d_{ab}(x,z) + d_{ab}(z,y)), \end{aligned}$$

$z \le x \le y$:

$$\begin{aligned} d_{ab}(x,y) &= 4(y-x)^3 \le 4 \cdot 4((y-z)^3 + (z-x)^3) \\ &= 4(4(y-z)^3 + 4(z-x)^3) \\ &\le 4(4(y-z)^3 + (x-z)^3) = 4(d_{ab}(x,z) + d_{ab}(z,y)), \end{aligned}$$

$x \le y \le z$:

$$\begin{aligned} d_{ab}(x,y) &= 4(y-x)^3 \le 4 \cdot 4((y-z)^3 + (z-x)^3) \\ &= 4 \cdot (4(y-z)^3 + 4(z-x)^3) \\ &\le 4((z-y)^3 + 4(z-x)^3) = 4(d_{ab}(x,z) + d_{ab}(z,y)). \end{aligned}$$

In the two previous examples, we constructed an almost-b-metric, which is also a quasi-b metric. The next example shows that there is a quasi-b-metric d_q, that it is not an almost-b-metric.

Example 3. *Let $X = \mathbb{R}$ and define $d_q : X \times X \to [0, \infty)$ as*

$$d_q(x,y) = \begin{cases} (x-y)^3, & x \ge y \\ 1, & x < y \end{cases}$$

As in the previous example, (bM3) and (bM1) are obvious. Notice that

$$d_q(\frac{1}{n}, 0) \to 0,\ n \to \infty \quad \text{but} \quad d_q(0, \frac{1}{n}) = 1,$$

so (bM2l) does not hold and it is the same for (bM2r). We conclude that $(X, d_q, 4)$ is a quasi-b-metric space, but it is not an almost-b-metric space.

There are many examples of b-metrics that are not continuous. Here, we modify one of such examples in sense that we do not demand symmetry.

Example 4. *Let $A = \mathbb{N} \cup \{\infty\}$ and define $d_q : A \times A \to [0, +\infty)$:*

$$d_q(x,y) = \begin{cases} 0, & x = y \\ \dfrac{1}{x} - \dfrac{1}{y}, & \text{if } x < y \text{ and one of } x \text{ and } y \text{ is odd and the other} \\ & \text{is odd or } \infty \\ \dfrac{1}{2}\left(\dfrac{1}{y} - \dfrac{1}{x}\right), & \text{if } y < x \text{ and one of } x \text{ and } y \text{ is odd and the other} \\ & \text{is odd or } \infty \\ 3, & \text{if one of } x \text{ and } y \text{ is even and the other is even or } \infty \\ 2, & \text{otherwise.} \end{cases}$$

Then $(A, d_q, \frac{3}{2})$ *is a quasi-b-metric space (it is also an almost-b-metric space). Note that* d_q *is not continuous. Indeed,* $d_q(2n+1, \infty) \to 0$*, when* $n \to \infty$*. But,* $d_q(2n+1, 2) = 2$*, while* $d_q(\infty, 2) = 3$*.*

Here, we introduce some basic concepts for almost-*b*-metric spaces. The following notions are quite standard and also valid in quasi-*b*- metric spaces.

Definition 3. *Let* (X, d_{ab}, s) *be an almost-b-metric space. A sequence* $\{\chi_n\}$ *in X is said to be*

left-Cauchy *if and only if for each* $\varepsilon > 0$ *there is an* $n_0 \in \mathbb{N}$ *such that* $d_{ab}(\chi_n, \chi_m) < \varepsilon$ *for all* $n \geq m > n_0$*, which can be written as* $\lim_{n \geq m \to \infty} d_{ab}(\chi_n, \chi_m) = 0$*,*

right-Cauchy *if and only if for each* $\varepsilon > 0$ *there is* $n_0 \in \mathbb{N}$ *so that* $d_{ab}(\chi_n, \chi_m) < \varepsilon$ *for all* $m \geq n > n_0$*, which can be written as* $\lim_{m \geq n \to \infty} d(\chi_n, \chi_m) = 0$*,*

Cauchy *if and only if for each* $\varepsilon > 0$*, there is* $n_0 \in \mathbb{N}$ *so that* $d_{ab}(\chi_n, \chi_m) < \varepsilon$ *for all* $n, m > n_0$*.*

In a quasi-*b*-metric space, a sequence is Cauchy if and only if it is left-Cauchy and right-Cauchy. The same is satisfied in almost-*b*-metric spaces. An almost-*b*-metric space (X, d_{ab}, s) is left-complete if and only if each left-Cauchy sequence $\{\chi_n\}$ in X satisfies $\lim_{n\to\infty} d_{ab}(\chi_n, \chi) = 0$, right-complete if and only if each right-Cauchy sequence $\{\chi_n\}$ in X satisfies $\lim_{n\to\infty} d_{ab}(\chi, \chi_n) = 0$ and is complete if and only if each Cauchy sequence in X is convergent.

In the next lemma, we will associate a *b*-metric to a given quasi-*b*-metric or an almost-*b*-metric. For some kind of contractions, by virtue of this correlation, the proofs from *b*-metric spaces can easily be translated into quasi-*b*-metric spaces and almost-*b*-metric spaces as their subclass.

Lemma 1. *If* (X, d_q, s) *is a quasi-b-metric space with* $s \geq 1$*, then* (X, l, s) *is a b-metric space, where*

$$l(\chi, \zeta) = \frac{d_q(\chi, \zeta) + d_q(\zeta, \chi)}{2}.$$

Proof. $l(x, y)$ is a *b*-metric.

(bM1) Suppose that $l(x, y) = 0$. Then $\frac{d_q(x,y)+d_q(y,x)}{2} = 0$ and since $d_q(x, y) \geq 0$, we obtain that $d_q(x, y) = d_q(y, x) = 0$ and that is, $x = y$, so we conclude that $l(x, y)$ satisfies (bM1).

(bM2) $l(x, y)$ is symmetric by definition:

$$l(x, y) = \frac{d_q(x, y) + d_q(y, x)}{2} = \frac{d_q(y, x) + d_q(x, y)}{2} = l(y, x).$$

(bM3) For all $x, y, z, \in X$, the following is satisfied:

$$d_q(x, z) \leq s(d_q(x, y) + d_q(y, z)).$$

Simply, by adding the following inequality to the previous

$$d_q(z, x) \leq s(d_q(z, y) + d_q(y, x))$$

and dividing the resulted sum by two, we obtain

$$l(x, z) \leq s(l(x, y) + l(y, z)).$$

□

Remark 1. *If* (X, d_{ab}, s) *is a complete almost-b-metric space, then from (bM2l) and (bM2r), we conclude that* (X, l, s) *is a complete b-metric space.*

The following theorems are modifications of Theorems 1 and 2 for quasi-b metric spaces and almost-b-metric spaces. Since almost-b-metric spaces are contained in quasi-b-metric spaces, we denote a metric by d_q.

Theorem 3. *Let (X, d_q, s) be a b-complete quasi-b-metric space with coefficient $s > 1$ and $T : X \to X$ be a mapping such that*

$$d_q(Tx, Ty) \leq \lambda d_q(x, y) + \mu d_q(x, Tx) + \delta d_q(Ty, y), \tag{1}$$

for all $x, y, z \in X$, where $\lambda, \mu, \delta \geq 0$ and

$$\lambda + 2 \cdot max\{\mu, \delta\} < 1 \quad and \quad max\{\mu, \delta\} < \frac{1}{s}. \tag{2}$$

Then T has a unique fixed point.

Proof. From Lemma 1, we conclude that (X, l, s) is a complete b-metric space. Further, from (1), the b-metric $l(x, y)$ satisfies:

$$\begin{aligned} l(Tx, Ty) &= \frac{d_q(Tx, Ty) + d_Q(Ty, Tx)}{2} \\ &\leq \frac{1}{2}\left(\lambda d_q(x, y) + \mu d_q(x, Tx) + \delta d_q(Ty, y)\right) \\ &+ \frac{1}{2}\left(\lambda d_q(y, x) + \mu d_q(y, Ty) + \delta d_q(Tx, x)\right) \\ &= \lambda l(x, y) + \frac{1}{2}(\mu d_q(x, Tx) + \delta d_q(Tx, x)) + \frac{1}{2}(\mu d_q(y, Ty) + \delta d_q(Ty, y)) \\ &\leq \lambda l(x, y) + \frac{1}{2} \cdot max\{\mu, \delta\}(d_q(x, Tx) + d_q(Tx, x)) \\ &+ \frac{1}{2} \cdot max\{\mu, \delta\}(d_q(y, Ty) + d_q(Ty, y)) \\ &= \lambda l(x, y) + max\{\mu, \delta\} l(x, Tx) + max\{\mu, \delta\} l(y, Ty). \end{aligned}$$

Now, from Theorem 1, we conclude that T has a unique fixed point. □

In the next result, we propose a Hardy–Rogers type contraction for quasi-b metric spaces and almost-b-metric spaces.

Theorem 4. *Let (X, d_q, s) be a complete quasi-b-metric space with coefficient $s > 1$ and $T : X \to X$ be a mapping satisfying*

$$d_q(Tx, Ty) \leq a_1 d_q(x, y) + a_2 d_q(x, Tx) + a_3 d_q(Ty, y) + a_4 d_q(x, Ty) + a_5 d_q(Tx, y), \tag{3}$$

for all $x, y \in X$, where a_1, a_2, a_3, a_4, $a_5 \geq 0$ with $a_1 + 2 \cdot max\{a_2, a_3\} + 2s \cdot max\{a_4, a_5\} < 1$ and $a_1 > max\{0, 1 - \frac{2}{s}\}$. Then T has a unique fixed point.

Proof. From Lemma 1, we conclude that (X, l, s) is a complete b-metric space. Starting from (3), we obtain for any $x, y \in X$,

$$\begin{aligned} 2l(Tx, Ty) &= d_q(Tx, Ty) + d_q(Ty, Tx) \\ &\leq a_1 d_q(x, y) + a_1 d_q(y, x) + a_2 d_q(x, Tx) + a_2 d_q(y, Ty) + a_3 d_q(Ty, y) \\ &+ a_3 d_q(Tx, x) + a_4 d_q(x, Ty) + a_4 d_q(y, Tx) + a_5 d_q(Tx, y) + a_5 d_q(Ty, x). \end{aligned}$$

Further, we get

$$\begin{aligned} l(Tx,Ty) &\le a_1 l(x,y) + \frac{1}{2}\left(a_2 d_q(x,Tx) + a_3 d_q(Tx,x)\right) + \frac{1}{2}\left(a_2 d_q(y,Ty) + a_3 d_q(Ty,y)\right) \\ &+ \frac{1}{2}\left(a_4 d_q(x,Ty) + a_5 d_q(Ty,x)\right) + \frac{1}{2}\left(a_4 d_q(y,Tx) + a_5 d_q(Tx,y)\right) \\ &\le a_1 l(x,y) + max\{a_2,a_3\} l(x,Tx) + max\{a_2,a_3\} l(y,Ty) \\ &+ max\{a_4,a_5\} l(x,Ty) + max\{a_4,a_5\} l(y,Tx). \end{aligned}$$

From Theorem 2 and conditions from Theorem 4, we conclude that self-mapping T on the complete b-metric space (X,l,s) has an unique fixed point, say x^*. Finally, according to Theorem 2, the result follows. □

It is not difficult to see that Theorems 3 and 4 are also satisfied for $s = 1$. To be specific, then $(X,d,1)$ is a quasi-metric space, (X,l) is a metric space, while condition (2) reduces to the well known condition $\lambda + \mu + \delta < 1$ for Reich type contractions, and similar for Hardy–Rogers type contractions.

The following results slightly differ from previous in a sense that we use properties (bM2l) and (bM2r). Before we state our result, we prove an auxiliary lemma that we use it in the proof. Since the lemma is satisfied in the quasi-b-metric spaces, it is also valid in almost-b-metric spaces, so again we denote it by d_q (having in mind that it is also valid for d_{ab}).

Lemma 2. *Let $\{\chi_n\}$ be a sequence in a quasi-b-metric space $(X,d_q,s\ge 1)$ such that*

$$d_q(\chi_n,\chi_{n+1}) \le \lambda \cdot d_q(\chi_{n-1},\chi_n), \tag{4}$$

for some $\lambda \in [0,\frac{1}{s})$ and each $n \in N$. Then $\{\chi_n\}$ is a right-Cauchy sequence.

Proof. From (4), we get

$$d_q(\chi_n,\chi_{n+1}) \le \lambda^n d_q(\chi_0,\chi_1). \tag{5}$$

Let $n,m \in \mathbb{N}$ with $n < m$. Then

$$\begin{aligned} & d_q(\chi_n,\chi_m) \\ \le\ & s\left(d_q(\chi_n,\chi_{n+1}) + d_q(\chi_{n+1},\chi_m)\right) \\ =\ & s d_q(\chi_n,\chi_{n+1}) + s d_q(\chi_{n+1},\chi_m) \\ \le\ & s d_q(\chi_n,\chi_{n+1}) + s^2 d_q(\chi_{n+1},\chi_{n+2}) + s^2 d_q(\chi_{n+2},\chi_m) \\ \le\ & s d_q(\chi_n,\chi_{n+1}) + s^2 d_q(\chi_{n+1},\chi_{n+2}) + s^3 d_q(\chi_{n+2},\chi_{n+3}) + \ldots \\ +\ & s^{m-n-1} d_q(\chi_{m-2},\chi_{m-1}) + s^{m-n-1} d_q(\chi_{m-1},\chi_m) \\ \le\ & \left[s\lambda^n + s^2\lambda^{n+1} + s^3\lambda^{n+2} + \ldots + s^{m-n-1}\lambda^{m-2}\right] d_q(\chi_0,\chi_1) \\ +\ & s^{m-n-1}\lambda^{m-1} d_q(\chi_0,\chi_1) \\ =\ & s\lambda^n\left(1 + (s\lambda) + (s\lambda)^2 + \ldots + (s\lambda)^{m-n-2}\right) d_q(\chi_0,\chi_1) + \frac{(s\lambda)^{m-1}}{s^n} d_q(\chi_0,\chi_1) \\ \le\ & \left(\frac{s\lambda^n}{1-s\lambda} + \frac{(s\lambda)^{m-1}}{s^n}\right) d_q(\chi_0,\chi_1) \to 0\ (m > n \to \infty). \end{aligned}$$

Since $s\lambda < 1$, we have

$$d_q(\chi_n,\chi_m) \to 0, m > n, n \to \infty \text{ or equivalently } \lim_{m>n\to\infty} d_q(\chi_n,\chi_m) = 0,$$

that is, $\{\chi_n\}$ is right-Cauchy. □

The following result is analogue to Lemma 2 for left- Cauchy sequences.

Lemma 3. *Let $\{\chi_n\}$ be a sequence in a quasi-b-metric space $(X, d_q, s \geq 1)$ such that*

$$d_q(\chi_{n+1}, \chi_n) \leq \lambda \cdot d_q(\chi_n, \chi_{n-1}) \tag{6}$$

for some $\lambda \in [0, \frac{1}{s})$ and each $n \in N$. Then $\{\chi_n\}$ is a left-Cauchy sequence.

Proof. The proof follows the same steps as in Lemma 2, where, starting from (6), the condition (5) is replaced by

$$d_q(\chi_{n+1}, \chi_n) \leq \lambda^n d_q(\chi_1, \chi_0). \tag{7}$$

Let $n, m \in \mathbb{N}$ with $n > m$. Then

$$\begin{aligned}
& d_q\left(\chi_n, \chi_m\right) \\
\leq\ & s\left(d_q\left(\chi_n, \chi_{m+1}\right)+d_q\left(\chi_{m+1}, \chi_m\right)\right) \\
=\ & s d_q\left(\chi_{m+1}, \chi_m\right)+s d_q\left(\chi_n, \chi_{m+1}\right) \\
\leq\ & s d_q\left(\chi_{m+1}, \chi_m\right)+s^2 d_q\left(\chi_n, \chi_{m+2}\right)+s^2 d_q\left(\chi_{m+2}, \chi_{m+1}\right) \\
\leq\ & s d_q\left(\chi_{m+1}, \chi_m\right)+s^2 d\left(\chi_{m+2}, \chi_{m+1}\right)+\ldots \\
+\ & s^{n-m-1}\left(d_q(\chi_n, \chi_{n-1})+d_q(\chi_{n-1}, \chi_{n-2})\right) \\
\leq\ & \left[s \lambda^m+s^2 \lambda^{m+1}+s^3 \lambda^{m+2}+\ldots+s^{n-m-1} \lambda^{n-2}\right] d_q\left(\chi_1, \chi_0\right) \\
+\ & s^{n-m-1} \lambda^{n-1} d_q\left(\chi_1, \chi_0\right) \\
=\ & s \lambda^m\left(1+(s \lambda)+(s \lambda)^2+\ldots+(s \lambda)^{n-m-2}\right) d_q\left(\chi_1, \chi_0\right)+\frac{(s \lambda)^{n-1}}{s^m} d_q\left(\chi_1, \chi_0\right) \\
\leq\ & \left(\frac{s \lambda^m}{1-s \lambda}+\frac{(s \lambda)^{n-1}}{s^m}\right) d_q\left(\chi_1, \chi_0\right) \rightarrow 0 \ (n>m \rightarrow \infty).
\end{aligned}$$

Since $s\lambda < 1$, we conclude that

$$d_q(\chi_n, \chi_m) \to 0, n > m, m \to \infty \text{ or equivalently } \lim_{n>m\to\infty} d_q(\chi_n, \chi_m) = 0,$$

that is, $\{\chi_n\}$ is left-Cauchy. □

Remark 2. *It is not hard to see that Lemma 2 and Lemma 3 hold if $\lambda \in [\frac{1}{s}, 1)$. For details, see Lemma 5 in [22].*

In the proof of the next theorem, we use the assumption (bM2r), hence we state it an almost-*b*-metric, and so denote the metric by d_{ab}.

Theorem 5. *Let (X, d_{ab}, s) be a right-complete r-almost b-metric space with coefficient $s > 1$ and $T : X \to X$ be a mapping satisfying*

$$d_{ab}(Tx, Ty) \leq k \cdot max\{d_{ab}(x, y), d_{ab}(x, Tx), d_{ab}(y, Ty)\}, \tag{8}$$

for all $x, y \in X$, where k is such that $0 \leq k < \frac{1}{s}$. Then T has a unique fixed point.

Proof. At the beginning of the proof, let us consider uniqueness of a possible fixed point. To prove that the fixed point is unique, if it exists, suppose that T has two distinct fixed points $x^*, y^* \in X$. Then we get

$$\begin{aligned} d_{ab}(x^*, y^*) &= d_{ab}(Tx^*, Ty^*) \\ &\leq k \cdot max\{d_{ab}(x^*, y^*), d_{ab}(x^*, Tx^*), d_{ab}(y^*, Ty^*)\} \\ &= kd_{ab}(x^*, y^*) < d_{ab}(x^*, y^*), \end{aligned}$$

which is a contradiction.

For an arbitrary $\chi_0 \in X$, consider the sequence $\chi_n = T\chi_{n-1} = T^n\chi_0$, $n \in \mathbb{N}$. If $\chi_n = \chi_{n+1}$ for some n, then χ_n is the unique fixed point of T. We suppose that $d_{ab}(\chi_n, \chi_{n+1}) > 0$ for all $n \in \mathbb{N}$.

We start from (8) for $d_{ab}(\chi_n, \chi_{n+1})$. Then for any $n \in \mathbb{N}$, we get

$$\begin{aligned} d_{ab}(\chi_n, \chi_{n+1}) &= d_{ab}(T\chi_{n-1}, T\chi_n) \\ &\leq k \cdot max\{d_{ab}(\chi_{n-1}, \chi_n), d_{ab}(\chi_{n-1}, T\chi_{n-1}), d_{ab}(\chi_n, T\chi_n)\} \\ &= k \cdot max\{d_{ab}(\chi_{n-1}, \chi_n), d_{ab}(\chi_{n-1}, \chi_n), d_{ab}(\chi_n, \chi_{n+1})\} \\ &= k \cdot max\{d_{ab}(\chi_{n-1}, \chi_n), d_{ab}(\chi_n, \chi_{n+1})\}. \end{aligned} \tag{9}$$

If $d_{ab}(\chi_{m-1}, \chi_m) \leq d_{ab}(\chi_m, \chi_{m+1})$ for some $m \in \mathbb{N}$, then from (9) we get

$$d_{ab}(\chi_m, \chi_{m+1}) \leq k \cdot d_{ab}(\chi_m, \chi_{m+1}) < d_{ab}(\chi_m, \chi_{m+1})$$

which is a contradiction. So, we have

$$d_{ab}(\chi_n, \chi_{n+1}) \leq k \cdot d_{ab}(\chi_{n-1}, \chi_n) \quad \text{for all} \quad n \in \mathbb{N}. \tag{10}$$

From (10) and Lemma 2 we can easily conclude that for some $n_0 \in \mathbb{N}$,

$$d_{ab}(\chi_n, \chi_m) < \varepsilon$$

for all $m \geq n > n_0$, so $\{\chi_n\}$ is a right-Cauchy sequence.

Since $(X, d_{ab}, s > 1)$ is a right-complete r-almost-b-metric space, we get that the sequence $\{\chi_n\}$ right converges to $x^* \in X$, i.e., $d_{ab}(x, \chi_n) \to 0$ as $n \to \infty$. (bM2r) implies that $d_{ab}(\chi_n, x^*) \to 0$ as $n \to \infty$.

The end of the proof is analogue to the standard case. From (bM3) and (8), we obtain

$$\begin{aligned} \frac{1}{s} d_{ab}(x^*, Tx^*) &\leq d_{ab}(x^*, \chi_{n+1}) + d_{ab}(\chi_{n+1}, Tx^*) \\ &= d_{ab}(x^*, \chi_{n+1}) + d_{ab}(T\chi_n, Tx^*) \\ &\leq d_{ab}(x^*, \chi_{n+1}) + k \cdot max\{d_{ab}(\chi_n, x^*), d_{ab}(\chi_n, T\chi_n), d_{ab}(x^*, Tx^*)\} \\ &\to k \cdot d_{ab}(x^*, Tx^*), \ n \to \infty. \end{aligned}$$

Finally, $x^* = Tx^*$. In the last inequality, we used property (bM2r) to obtain that $d_{ab}(\chi_n, x^*) \to 0$ as $n \to \infty$ and also that $d_{ab}(\chi_n, T\chi_n) = d_{ab}(\chi_n, \chi_{n+1}) \to 0$ as $n \to \infty$ since $\{\chi_n\}$ is a right-Cauchy sequence. □

From the previous theorem, we can draw several corollaries that are analogous to Banach, Kannan and Reich type contraction principles, respectively.

Corollary 1. *Let (X, d_{ab}, s) be a right-complete r-almost b-metric space with coefficient $s > 1$ and $T : X \to X$ be such that*

Banach contraction:

$$d_{ab}(Tx, Ty) \leq k \cdot d_{ab}(x, y)$$

for all $x, y \in X$ *where* $0 \le k < \frac{1}{s}$.

Kannan contraction:

$$d_{ab}(Tx, Ty) \le k_1 d_{ab}(x, fx) + k_2 d_{ab}(y, fy)$$

for all $x, y \in X$ *where* $k_1, k_2 \ge 0$ *such that* $k_1 + k_2 < \frac{1}{s}$.

Reich contraction:

$$d_{ab}(Tx, Ty) \le k_1 d_{ab}(x, y) + k_2 d_{ab}(x, fx) + k_3 d_{ab}(y, fy),$$

for all $x, y \in X$ *where* $k_1, k, k_3 \ge 0$ *such that* $k_1 + k_2 + k_3 < \frac{1}{s}$.

Then T *has a unique fixed point.*

The next result is analogue to Theorem 5 for left-complete l-almost b-metric spaces.

Theorem 6. *Let* (X, d_{ab}, s) *be a left-complete l-almost b-metric space with* $s > 1$ *and* $T : X \to X$ *be such that*

$$d_{ab}(Tx, Ty) \le k \cdot max\{d_{ab}(x, y), d_{ab}(Tx, x), d_{ab}(Ty, y)\}, \tag{11}$$

for all $x, y \in X$ *where* $0 \le k < \frac{1}{s}$. *Then* T *has a unique fixed point.*

Proof. The uniqueness of a possible fixed point is obtained the same way as in proof of Theorem 5.

For arbitrary $\chi_0 \in X$, consider the sequence $\chi_n = T\chi_{n-1} = T^n\chi_0$, $n \in \mathbb{N}$. If $\chi_n = \chi_{n+1}$ for some n, then χ_n is a unique fixed point of T. Hence, we suppose that $d_{ab}(\chi_{n+1}, \chi_n) > 0$ for all $n \in \mathbb{N}$.

We start from (11) for $d_{ab}(\chi_{n+1}, \chi_n)$. Then for any $n \in \mathbb{N}$, using same considerations as in previous proof, we get

$$\begin{aligned} d_{ab}(\chi_{n+1}, \chi_n) &= d_{ab}(T\chi_n, T\chi_{n-1}) \\ &\le k \cdot max\{d_{ab}(\chi_n, \chi_{n-1}), d_{ab}(T\chi_n, \chi_n), d_{ab}(T\chi_{n-1}, \chi_{n-1})\} \\ &\le k \cdot d_{ab}(\chi_n, \chi_{n-1}). \end{aligned} \tag{12}$$

From (12) and Lemma 3, we can easily conclude that for some $n_0 \in \mathbb{N}$,

$$d_{ab}(\chi_n, \chi_m) < \varepsilon$$

for all $n \ge m > n_0$, so $\{\chi_n\}$ is a left-Cauchy sequence.

Since $(X, d_{ab}, s > 1)$ is a left-complete l-almost-b-metric space, we get that the sequence $\{\chi_n\}$ left converges to $x^* \in X$, i.e., $d_{ab}(\chi_n, x^*) \to 0$, $n \to \infty$. (bM2l) implies that $d_{ab}(x^*, \chi_n) \to 0$ as $n \to \infty$.

Finally, from (bM3) and (11), we obtain

$$\begin{aligned} \frac{1}{s} d_{ab}(Tx^*, x^*) &\le d_{ab}(Tx^*, \chi_{n+1}) + d_{ab}(\chi_{n+1}, x^*) \\ &= d_{ab}(Tx^*, T\chi_n) + d_{ab}(\chi_{n+1}, x^*) \\ &\le k \cdot max\{d_{ab}(x^*, \chi_n), d_{ab}(Tx^*, x^*), d_{ab}(T\chi_n, \chi_n)\} + d_{ab}(\chi_{n+1}, x^*) \\ &\to k \cdot d_{ab}(Tx^*, x^*), \ n \to \infty, \end{aligned}$$

and so $x^* = Tx^*$. In the last inequality, we used property (bM2l) that implies $d_{ab}(x^*, \chi_n) \to 0$, $n \to \infty$ and also that $d_{ab}(T\chi_n, \chi_n) = d_{ab}(\chi_{n+1}, \chi_n) \to 0, n \to \infty$ since $\{\chi_n\}$ is a left-Cauchy sequence. □

The previous considerations should convince the readers that many generalizations of contraction principles may be obtained in almost-b-spaces, which are introduced here, and present a proper subclass of quasi-b-metric spaces. As another benefit of this paper, we point out the principle applied in Theorems 3 and 4 that elegantly proves some contractions in quasi-b-metric spaces.

Finally, we state some open questions in the context of almost-b-metric spaces (respectively quasi-b-metric spaces). If $s = 1$, we have appropriate unresolved questions in the context of quasi-metric spaces. We present formulations for the case of a right-complete r-almost b-metric space, noting that similar issues remain open in left-complete l-almost b-metric spaces.

Problem 1. *(Generalized Ćirić type contraction of first order) Let $(X, d_{ab}, s \geq 1)$ be a right-complete r-almost b-metric space and $T : X \to X$ be a mapping satisfying*

$$d_{ab}(Tx, Ty) \leq k \max \left\{ d_{ab}(x, y), \frac{d_{ab}(x, Tx) + d_{ab}(y, Ty)}{2s}, \frac{d_{ab}(x, Ty) + d_{ab}(y, Tx)}{2s} \right\},$$

for all $x, y \in X$ where $0 \leq k < \frac{1}{s}$. Then T has a unique fixed point.

Problem 2. *(Generalized Ćirić type contraction of second order) Let $(X, d_{ab}, s \geq 1)$ be a right-complete r-almost b-metric space and $T : X \to X$ be a mapping satisfying*

$$d_{ab}(Tx, Ty) \leq k \max \left\{ d_{ab}(x, y), d_{ab}(x, Tx), d_{ab}(y, Ty), \frac{d_{ab}(x, Ty) + d_{ab}(y, Tx)}{2s} \right\},$$

for all $x, y \in X$ where $0 \leq k < \frac{1}{s}$. Then T has a unique fixed point.

Problem 3. *(Quasicontraction of Ćirić type) Let $(X, d_{ab}, s \geq 1)$ be a right-complete r-almost b-metric space and $T : X \to X$ be such that*

$$d_{ab}(Tx, Ty) \leq k \max \{ d_{ab}(x, y), d_{ab}(x, Tx), d_{ab}(y, Ty), d_{ab}(x, Ty), d_{ab}(y, Tx) \},$$

for all $x, y \in X$ where $0 \leq k < \frac{1}{s}$. Then T has a unique fixed point.

Author Contributions: All authors contributed equally and significantly in writing this article. All authors read and approved the final manuscript.

Acknowledgments: The first author would like to thank Prince Sultan University for funding this work through research group Nonlinear Analysis Methods in Applied Mathematics (NAMAM) group number RG-DES-2017-01-17. The research of the second author (K.K.) was partially supported by the Serbian Ministry of Science and Technological Development, Project TR36002.

References

1. Filipović, M.G.; Kukić, K. Some results about contraction principles in bMS and RbMS without assumption of b-metric continuity. *Fixed Point Theory* **2019**, submitted.
2. Kannan, R. Some results on fixed points. *Bull. Calcutta Math. Soc.* **1968**, *60*, 71–76.
3. Reich, S. Some remarks concerning contraction mappings. *Can. Math. Bull.* **1971**, *14*, 121–124. [CrossRef]
4. Hardy, G.E.; Rogers, T.D. A generalization of a fixed point theorem of Reich. *Can. Math. Bull.* **1973**, *16*, 201–206. [CrossRef]
5. Bakhtin, I.A. The contraction mapping principle in quasimetric spaces. *Funct. Anal. Ulianowsk Gos. Ped. Inst.* **1989**, *30*, 26–37.
6. Shukla, S. Partial b-metric spaces and fixed point theorems. *Mediterr. J. Math.* **2013**, *11*, 703–711. [CrossRef]
7. Amini-Harandi, A. Metric-like spaces, partial metric spaces and fixed points. *Fixed Point Theory Appl.* **2012**, *2012*, 204. [CrossRef]
8. Hussain, N.; Roshan, J.R.; Parvaneh, V.; Abbas, M. Common fixed point results for weak contractive mappings in ordered b-dislocated metric spaces with applications. *J. Inequal. Appl.* **2013**, *2013*, 486. [CrossRef]

9. Aydi, H.; Bota, M.F.; Karapinar, E.; Moradi, S. A common fixed point for weak ϕ-contractions on b-metric spaces. *Fixed Point Theory* **2012**, *13*, 337–346.
10. Aydi, H.; Felhi, A.; Sahmim, S. Common fixed points via implicit contractions on b-metric-like spaces. *J. Nonlinear Sci. Appl.* **2017**, *10*, 1524–1537. [CrossRef]
11. Aydi, H.; Felhi, A.; Sahmim, S. On common fixed points for (α, ψ)-contractions and generalized cyclic contractions in b-metric-like spaces and consequences. *J. Nonlinear Sci. Appl.* **2016**, *9*, 2492–2510. [CrossRef]
12. Aydi, H.; Karapinar, E.; Bota, M.F.; Mitrović, S. A fixed point theorem for set-valued quasi-contractions in b-metric spaces. *Fixed Point Theory Appl.* **2012**, *2012*, 88. [CrossRef]
13. Faraji, H.; Savić, D.; Radenović, S. Fixed point theorems for Geraghty contraction type mappings in b-metric spaces and applications. *Axioms* **2019**, *8*, 34. [CrossRef]
14. Karapınar, E.; Czerwik, S.; Aydi, H. (α, ψ)-Meir-Keeler contraction mappings in generalized b-metric spaces. *J. Funct. Spaces* **2018**, *2018*, 3264620. [CrossRef]
15. Hussain, A.; Kanwal, T.; Adeel, M.; Radenović, S. Best proximity point results in b-metric spaces and application to nonlinear fractional differential equation. *Mathematics* **2018**, *6*, 221. [CrossRef]
16. Kirk, W.; Shahzad, N. *Fixed Point Theory in Distance Spaces*; Springer International Publishing: Cham, Switzerland, 2014; xii+173p.
17. Patle, P.R.; Vujaković, L.; Radenović, S.; Patel, D.K. Topology induced by θ-metric and multivalued mappings. *Symmetry* **2019**, *7*, 144.
18. Vujaković, J.; Aydi, H.; Radenović, S.; Mukheimer, A. Some remarks and new results in ordered partial b-metric spaces. *Mathematics* **2019**, *7*, 334. [CrossRef]
19. Vujaković, J.; Kishore, G.N.V.; Rao, K.P.R.; Radenoviić, S.; Sadik, S.K. Existence and unique coupled solution in S_b-metric spaces by rational contraction with application. *Mathematics* **2019**, *7*, 313. [CrossRef]
20. Mitrović, Z.D.; George, R.; Hussain, N. Some remarks on contraction mappings in rectangular b-metric spaces. *BSPM* **2018**, in press. [CrossRef]
21. Czerwik, S. Contraction mappings in b-metric spaces. *Acta. Math. Inf. Univ. Ostrav* **1993**, *1*, 5–11.
22. Miculescu, R.; Mihail, A. New fixed points theorems for set-valued contractions in b-metric spaces. *J. Fixed Point Theory Appl.* **2017**, *19*, 2153–2163. [CrossRef]

4

Some Summation Theorems for Generalized Hypergeometric Functions

Mohammad Masjed-Jamei [1,2,*] and Wolfram Koepf [1]

1 Department of Mathematics, University of Kassel, Heinrich-Plett-Str. 40, D-34132 Kassel, Germany; koepf@mathematik.uni-kassel.de
2 Department of Mathematics, K. N. Toosi University of Technology, P.O. Box 16315-1618, 11369 Tehran, Iran
* Correspondence: mmjamei@kntu.ac.ir or mmjamei@yahoo.com

Abstract: Essentially, whenever a generalized hypergeometric series can be summed in terms of gamma functions, the result will be important as only a few such summation theorems are available in the literature. In this paper, we apply two identities of generalized hypergeometric series in order to extend some classical summation theorems of hypergeometric functions such as Gauss, Kummer, Dixon, Watson, Whipple, Pfaff–Saalschütz and Dougall formulas and also obtain some new summation theorems in the sequel.

Keywords: generalized hypergeometric series; Gauss and confluent hypergeometric functions; classical summation theorems of hypergeometric functions

---

1. Introduction

Let $\mathbb{R}$ and $\mathbb{C}$ denote the sets of real and complex numbers and z be a complex variable. For real or complex parameters a and b, the generalized binomial coefficient:

$$\binom{a}{b} = \frac{\Gamma(a+1)}{\Gamma(b+1)\Gamma(a-b+1)} = \binom{a}{a-b} \qquad (a,b \in \mathbb{C}),$$

in which:

$$\Gamma(z) = \int_0^\infty x^{z-1}e^{-x}dx,$$

denotes the well-known gamma function for $Re(z) > 0$, can be reduced to the particular case:

$$\binom{a}{n} = \frac{(-1)^n(-a)_n}{n!},$$

where $(a)_b$ denotes the Pochhammer symbol [1] given by:

$$(a)_b = \frac{\Gamma(a+b)}{\Gamma(a)} = \begin{cases} 1 & (b=0,\ a \in \mathbb{C}\backslash\{0\}), \\ a(a+1)...(a+n-1) & (b \in \mathbb{N},\ a \in \mathbb{C}). \end{cases} \tag{1}$$

Based on Pochhammer's symbol (1), the generalized hypergeometric functions [2]:

$${}_pF_q\left(\begin{matrix} a_1, & \dots & ,a_p \\ b_1, & \dots & ,b_q \end{matrix} \middle| z \right) = \sum_{k=0}^{\infty} \frac{(a_1)_k...(a_p)_k}{(b_1)_k...(b_q)_k} \frac{z^k}{k!}, \tag{2}$$

are indeed a Taylor series expansion for a function, say f, as $\sum_{k=0}^{\infty} c_k^* z^k$ with $c_k^* = f^{(k)}(0)/k!$, for which the ratio of successive terms can be written as:

$$\frac{c_{k+1}^*}{c_k^*} = \frac{(k+a_1)(k+a_2)...(k+a_p)}{(k+b_1)(k+b_2)...(k+b_q)(k+1)}.$$

According to the ratio test [3,4], the series (2) is convergent for any $p \leq q+1$. In fact, it converges in $|z| < 1$ for $p = q+1$, converges everywhere for $p < q+1$ and converges nowhere ($z \neq 0$) for $p > q+1$. Moreover, for $p = q+1$, it absolutely converges for $|z| = 1$ if the condition:

$$A^* = \mathrm{Re}\left(\sum_{j=1}^{q} b_j - \sum_{j=1}^{q+1} a_j\right) > 0,$$

holds and is conditionally convergent for $|z| = 1$ and $z \neq 1$ if $-1 < A^* \leq 0$ and is finally divergent for $|z| = 1$ and $z \neq 1$ if $A^* \leq -1$.

There are two important cases of the series (2) arising in many physics problems [5,6]. The first case (convergent in $|z| \leq 1$) is the Gauss hypergeometric function:

$$y = {}_2F_1\left(\begin{matrix} a,\ b \\ c \end{matrix}\,\middle|\, z\right) = \sum_{k=0}^{\infty} \frac{(a)_k (b)_k}{(c)_k} \frac{z^k}{k!},$$

with the integral representation:

$${}_2F_1\left(\begin{matrix} a,\ b \\ c \end{matrix}\,\middle|\, z\right) = \frac{\Gamma(c)}{\Gamma(b)\Gamma(c-b)} \int_0^1 t^{b-1}(1-t)^{c-b-1}(1-tz)^{-a} dt,$$

$$(\mathrm{Re}\, c > \mathrm{Re}\, b > 0;\ |\arg(1-z)| < \pi). \quad (3)$$

Replacing $z = 1$ in (3) directly leads to the well-known Gauss identity [1]:

$${}_2F_1\left(\begin{matrix} a,\ b \\ c \end{matrix}\,\middle|\, 1\right) = \frac{\Gamma(c)\Gamma(c-a-b)}{\Gamma(c-a)\Gamma(c-b)} \qquad \mathrm{Re}(c-a-b) > 0. \quad (4)$$

The second case, which converges everywhere, is the Kummer confluent hypergeometric function:

$$y = {}_1F_1\left(\begin{matrix} b \\ c \end{matrix}\,\middle|\, z\right) = \sum_{k=0}^{\infty} \frac{(b)_k}{(c)_k} \frac{z^k}{k!},$$

with the integral representation:

$${}_1F_1\left(\begin{matrix} b \\ c \end{matrix}\,\middle|\, z\right) = \frac{\Gamma(c)}{\Gamma(b)\Gamma(c-b)} \int_0^1 t^{b-1}(1-t)^{c-b-1} e^{zt} dt,$$

$$(\mathrm{Re}\, c > \mathrm{Re}\, b > 0;\ |\arg(1-z)| < \pi).$$

Essentially, whenever a generalized hypergeometric series can be summed in terms of gamma functions, the result will be important as only a few such summation theorems are available in the literature; see, e.g., [7–13]. In this sense, the classical summation theorems such as Kummer and Gauss for ${}_2F_1$, Dixon, Watson, Whipple and Pfaff–Saalschütz for ${}_3F_2$, Whipple for ${}_4F_3$, Dougall for ${}_5F_4$ and Dougall for ${}_7F_6$ are well known [1,14]. In this paper, we apply two identities of generalized hypergeometric functions in order to obtain some new summation theorems and extend the above-mentioned classical theorems. For this purpose, we should first recall the classical theorems as follows.

* Kummer's theorem ([1], p. 108):

$$ {}_2F_1\left(\begin{matrix} a,\ b \\ 1+a-b \end{matrix}\ \middle|\ -1\right) = \frac{\Gamma(1+a-b)\Gamma(1+(a/2))}{\Gamma(1-b+(a/2))\Gamma(1+a)}. \tag{5} $$

* Second Gauss theorem ([1], p. 108):

$$ {}_2F_1\left(\begin{matrix} a,\ b \\ (a+b+1)/2 \end{matrix}\ \middle|\ \frac{1}{2}\right) = \frac{\sqrt{\pi}\,\Gamma((a+b+1)/2)}{\Gamma((a+1)/2)\Gamma((b+1)/2)}. \tag{6} $$

* Bailey's theorem ([1], p. 108):

$$ {}_2F_1\left(\begin{matrix} a,\ 1-a \\ b \end{matrix}\ \middle|\ \frac{1}{2}\right) = \frac{\Gamma(b/2)\Gamma((b+1)/2)}{\Gamma((a+b)/2)\Gamma((b-a+1)/2)}. \tag{7} $$

* Dixon's theorem ([1], p. 108):

$$ {}_3F_2\left(\begin{matrix} a,\ b,\ c \\ 1+a-b,\ 1+a-c \end{matrix}\ \middle|\ 1\right) = \frac{\Gamma(1+a/2)\Gamma(1+a-b)\Gamma(1+a-c)\Gamma(1-b-c+a/2)}{\Gamma(1+a)\Gamma(1-b+a/2)\Gamma(1-c+a/2)\Gamma(1+a-b-c)}. \tag{8} $$

* Watson's theorem ([1], p. 108):

$$ {}_3F_2\left(\begin{matrix} a,\ b,\ c \\ (a+b+1)/2,\ 2c \end{matrix}\ \middle|\ 1\right) = \frac{\sqrt{\pi}\,\Gamma(1+c/2)\Gamma((a+b+1)/2)\Gamma(c-(a+b-1)/2)}{\Gamma((a+1)/2)\Gamma((b+1)/2)\Gamma(c-(a-1)/2)\Gamma(c-(b-1)/2)}. \tag{9} $$

* Whipple's theorem ([1], p. 108):

$$ {}_3F_2\left(\begin{matrix} a,\ 1-a,\ c \\ c,\ 2b-c+1 \end{matrix}\ \middle|\ 1\right) = \frac{\pi\, 2^{1-2b}\Gamma(c)\Gamma(2b-c+1)}{\Gamma((a+c)/2)\Gamma(b+(a-c+1)/2)\Gamma((1-a+c)/2)\Gamma(b+1-(a+c)/2)}. \tag{10} $$

* Pfaff–Saalschütz theorem ([1], p. 108):

$$ {}_3F_2\left(\begin{matrix} a,\ b,\ -n \\ c,\ 1+a+b-c-n \end{matrix}\ \middle|\ 1\right) = \frac{(c-a)_n(c-b)_n}{(c)_n(c-a-b)_n}. \tag{11} $$

* Second Whipple theorem ([1], p. 108):

$$ {}_4F_3\left(\begin{matrix} a,\ 1+a/2,\ b,\ c \\ a/2,\ a-b+1,\ a-c+1 \end{matrix}\ \middle|\ -1\right) = \frac{\Gamma(a-b+1)\Gamma(a-c+1)}{\Gamma(a+1)\Gamma(a-b-c+1)}. \tag{12} $$

* Dougall's theorem ([1], p. 108):

$$ {}_5F_4\left(\begin{matrix} a,\ 1+a/2,\ c,\ d,\ e \\ a/2,\ a-c+1,\ a-d+1,\ a-e+1 \end{matrix}\ \middle|\ 1\right) $$

$$= \frac{\Gamma(a-c+1)\Gamma(a-d+1)\Gamma(a-e+1)\Gamma(a-c-d-e+1)}{\Gamma(a+1)\Gamma(a-d-e+1)\Gamma(a-c-e+1)\Gamma(a-c-d+1)}. \quad (13)$$

* Second Dougall theorem ([1], p. 108):

$${}_7F_6\left(\begin{matrix} a,\ 1+a/2,\ b,\ c,\ d,\ 1+2a-b-c+n,\ -n \\ a/2,\ a-b+1,\ a-c+1,\ a-d+1,\ b+c+d-a-n,\ a+1+n \end{matrix}\,\middle|\, 1\right)$$

$$= \frac{(a+1)_n(a-b-c+1)_n(a-b-d+1)_n(a-c-d+1)_n}{(a+1-b)_n(a+1-c)_n(a+1-d)_n(a+1-b-c-d)_n}. \quad (14)$$

In order to derive the first identity and only for simplicity, we will use the following symbol for representing finite sums of hypergeometric series:

$${}_p\overset{(m)}{F}_q\left(\begin{matrix} a_1, & \dots & ,a_p \\ b_1, & \dots & ,b_q \end{matrix}\,\middle|\, z\right) = \sum_{k=0}^{m} \frac{(a_1)_k \dots (a_p)_k}{(b_1)_k \dots (b_q)_k} \frac{z^k}{k!}.$$

For instance, we have:

$${}_p\overset{(-1)}{F}_q(z) = 0,\ {}_p\overset{(0)}{F}_q(z) = 1 \text{ and } {}_p\overset{(1)}{F}_q(z) = 1 + \frac{a_1 \dots a_p}{b_1 \dots b_q} z.$$

2. First Hypergeometric Identity

Let m, n be two natural numbers so that $n \le m$. By referring to Relation (1), since:

$$\frac{(n)_k}{(m)_k} = \frac{\Gamma(k+n)\Gamma(m)}{\Gamma(k+m)\Gamma(n)} = \frac{\Gamma(m)}{\Gamma(n)} \frac{1}{(k+n)(k+n+1)\dots(k+m-1)}, \quad (15)$$

substituting (15) in a special case of (2) yields:

$$\begin{aligned} {}_pF_q\left(\begin{matrix} a_1, & \dots & ,a_{p-1}, & n \\ b_1, & \dots & ,b_{q-1}, & m \end{matrix}\,\middle|\, z\right) &= \frac{\Gamma(m)}{\Gamma(n)} \sum_{k=0}^{\infty} \frac{(a_1)_k \dots (a_{p-1})_k}{(b_1)_k \dots (b_{q-1})_k} \frac{(k+1)(k+2)\dots(k+n-1)\, z^k}{(k+m-1)!} \\ &= \frac{\Gamma(m)}{\Gamma(n)} \sum_{j=m-1}^{\infty} \frac{(a_1)_{j-m+1} \dots (a_{p-1})_{j-m+1}}{(b_1)_{j-m+1} \dots (b_{q-1})_{j-m+1}} \frac{(j+2-m)(j+3-m)\dots(j-m+n)\, z^{j-m+1}}{j!}. \quad (16) \end{aligned}$$

Relation (16) shows that we encounter a complicated computational problem that cannot be easily evaluated. However, some particular cases such as $n = 1$ and $n = 2$ can be directly computed. We leave other cases as open problems.

The case $n = 1$ leads to a known result in the literature [14], because:

$$
{}_pF_q\left(\begin{array}{cc} a_1, \ldots, a_{p-1}, & 1 \\ b_1, \ldots, b_{q-1}, & m \end{array} \middle| z\right) = \Gamma(m) \sum_{j=m-1}^{\infty} \frac{(a_1)_{j-m+1} \ldots (a_{p-1})_{j-m+1}}{(b_1)_{j-m+1} \ldots (b_{q-1})_{j-m+1}} \frac{z^{j-m+1}}{j!}
$$

$$
= \Gamma(m) \left(\sum_{j=0}^{\infty} \frac{(a_1)_{j-m+1} \ldots (a_{p-1})_{j-m+1}}{(b_1)_{j-m+1} \ldots (b_{q-1})_{j-m+1}} \frac{z^{j-m+1}}{j!} - \sum_{j=0}^{m-2} \frac{(a_1)_{j-m+1} \ldots (a_{p-1})_{j-m+1}}{(b_1)_{j-m+1} \ldots (b_{q-1})_{j-m+1}} \frac{z^{j-m+1}}{j!} \right), \quad (17)
$$

and since:

$$
(a)_{j-m+1} = \frac{\Gamma(a-m+1)}{\Gamma(a)} (a-m+1)_j,
$$

Relation (17) is simplified as:

$$
{}_pF_q\left(\begin{array}{cc} a_1, \ldots, a_{p-1}, & 1 \\ b_1, \ldots, b_{q-1}, & m \end{array} \middle| z\right)
$$

$$
= \frac{\Gamma(b_1) \ldots \Gamma(b_{q-1})}{\Gamma(a_1) \ldots \Gamma(a_{p-1})} \frac{\Gamma(a_1-m+1) \ldots \Gamma(a_{p-1}-m+1)}{\Gamma(b_1-m+1) \ldots \Gamma(b_{q-1}-m+1)} \frac{(m-1)!}{z^{m-1}}
$$

$$
\times \left({}_{p-1}F_{q-1}\left(\begin{array}{c} a_1-m+1, \ldots, a_{p-1}-m+1 \\ b_1-m+1, \ldots, b_{q-1}-m+1 \end{array} \middle| z\right) - {}_{p-1}F_{q-1}^{(m-2)}\left(\begin{array}{c} a_1-m+1, \ldots, a_{p-1}-m+1 \\ b_1-m+1, \ldots, b_{q-1}-m+1 \end{array} \middle| z\right) \right). \quad (18)
$$

However, the interesting point is that using Relation (18), we can obtain various special cases that extend all classical summation theorems as follows.

Special Case 1. When $p = 3, q = 2$ and $x = 1$, Relation (18) is simplified as:

$$
{}_3F_2\left(\begin{array}{c} a, \ b, \ 1 \\ c, \ m \end{array} \middle| 1\right) = \frac{\Gamma(m)\Gamma(c)\Gamma(a-m+1)\Gamma(b-m+1)}{\Gamma(a)\Gamma(b)\Gamma(c-m+1)}
$$

$$
\times \left(\frac{\Gamma(c-m+1)\Gamma(c-a-b+m-1)}{\Gamma(c-a)\Gamma(c-b)} - {}_2F_1^{(m-2)}\left(\begin{array}{c} a-m+1, \ b-m+1 \\ c-m+1 \end{array} \middle| 1\right) \right). \quad (19)
$$

For $m = 1$, Relation (19) exactly gives Formula (4), while for $m = 2, 3$, we have:

$$
{}_3F_2\left(\begin{array}{c} a, \ b, \ 1 \\ c, \ 2 \end{array} \middle| 1\right) = \frac{c-1}{(a-1)(b-1)} \left(\frac{\Gamma(c-1)\Gamma(c-a-b+1)}{\Gamma(c-a)\Gamma(c-b)} - 1 \right),
$$

and:

$$
{}_3F_2\left(\begin{array}{c} a, \ b, \ 1 \\ c, \ 3 \end{array} \middle| 1\right) = \frac{2(c-2)_2}{(a-2)_2(b-2)_2}
$$

$$
\times \left(\frac{\Gamma(c-2)\Gamma(c-a-b+2)}{\Gamma(c-a)\Gamma(c-b)} - \frac{ab+c-2a-2b+2}{c-2} \right).
$$

These two formulas are given in [14].

Special Case 2. When $p = 3, q = 2$ and $x = -1$, by noting the Kummer theorem (5), Relation (18) is simplified as:

$$
{}_3F_2\left(\begin{array}{c} a,\ b,\ 1 \\ a-b+m,\ m \end{array}\middle| -1\right) = (-1)^{m-1}\frac{\Gamma(m)\Gamma(a-b+m)\Gamma(a-m+1)\Gamma(b-m+1)}{\Gamma(a)\Gamma(b)\Gamma(a-b+1)}
$$
$$
\times\left(\frac{\Gamma(a-b+1)\Gamma(1+(a-m+1)/2)}{\Gamma(2+a-m)\Gamma(m-b+(a-m+1)/2)} - {}_2F_1^{(m-2)}\left(\begin{array}{c} a-m+1,\ b-m+1 \\ a-b+1 \end{array}\middle| -1\right)\right). \quad (20)
$$

For $m = 1$, Relation (20) exactly gives the Kummer formula, while for $m = 2, 3$, we have:

$$
{}_3F_2\left(\begin{array}{c} a,\ b,\ 1 \\ a-b+2,\ 2 \end{array}\middle| -1\right) = \frac{a-b+1}{(a-1)(b-1)}\left(1 - \frac{\Gamma(a-b+1)\Gamma(1+(a-1)/2)}{\Gamma(a)\Gamma(-b+2+(a-1)/2)}\right),
$$

and:

$$
{}_3F_2\left(\begin{array}{c} a,\ b,\ 1 \\ a-b+3,\ 3 \end{array}\middle| -1\right) = \frac{2(a-b+1)_2}{(a-2)_2(b-2)_2}\left(\frac{\Gamma(a-b+1)\Gamma(a/2)}{\Gamma(a-1)\Gamma(-b+2+a/2)} - \frac{3a+b-ab-3}{a-b+1}\right).
$$

Special Case 3. When $p = 3, q = 2$ and $x = 1/2$, by noting the second kind of Gauss Formula (6), Relation (18) is simplified as:

$$
{}_3F_2\left(\begin{array}{c} a,\ b,\ 1 \\ (a+b+1)/2,\ m \end{array}\middle| \frac{1}{2}\right) = (2)^{m-1}
$$
$$
\times\frac{\Gamma(m)\Gamma((a+b+1)/2)\Gamma(a-m+1)\Gamma(b-m+1)}{\Gamma(a)\Gamma(b)\Gamma(-m+1+(a+b+1)/2)}
$$
$$
\times\left(\frac{\sqrt{\pi}\,\Gamma(-m+1+(a+b+1)/2)}{\Gamma(1+(a-m)/2)\,\Gamma(1+(b-m)/2)} - {}_2F_1^{(m-2)}\left(\begin{array}{c} a-m+1,\ b-m+1 \\ -m+1+(a+b+1)/2 \end{array}\middle| \frac{1}{2}\right)\right). \quad (21)
$$

For $m = 1$, Relation (21) exactly gives the second kind of Gauss formula, while for $m = 2, 3$, we have:

$$
{}_3F_2\left(\begin{array}{c} a,\ b,\ 1 \\ (a+b+1)/2,\ 2 \end{array}\middle| \frac{1}{2}\right) = \frac{a+b-1}{(a-1)(b-1)}\left(\frac{\sqrt{\pi}\,\Gamma(-1+(a+b+1)/2)}{\Gamma(a/2)\,\Gamma(b/2)} - 1\right),
$$

and:

$$
{}_3F_2\left(\begin{array}{c} a,\ b,\ 1 \\ (a+b+1)/2,\ 3 \end{array}\middle| \frac{1}{2}\right) = \frac{2(a+b-1)(a+b-3)}{(a-2)_2(b-2)_2}
$$
$$
\times\left(\frac{\sqrt{\pi}\,\Gamma((a+b-3)/2)}{\Gamma((a-1)/2)\,\Gamma((b-1)/2)} - \frac{ab-a-b+1}{a+b-3}\right).
$$

Special Case 4. When $p = 3, q = 2$ and $x = 1/2$, by noting the Bailey theorem (7), Relation (18) is simplified as:

$$ {}_3F_2\left(\begin{matrix} a,\ 2m-a-1,\ 1 \\ b,\ m \end{matrix}\ \middle|\ \frac{1}{2}\right) = (2)^{m-1}\frac{\Gamma(m)\Gamma(b)\Gamma(a-m+1)\Gamma(m-a)}{\Gamma(a)\Gamma(2m-a-1)\Gamma(b-m+1)} $$

$$ \times\left(\frac{\Gamma((b-m+1)/2)\Gamma((b-m+2)/2)}{\Gamma(-m+1+(a+b)/2)\,\Gamma((b-a+1)/2)} - {}_2F_1^{(m-2)}\left(\begin{matrix} a-m+1,\ m-a \\ b-m+1 \end{matrix}\ \middle|\ \frac{1}{2}\right)\right). \tag{22} $$

For $m=1$, Relation (22) exactly gives the Bailey formula, while for $m=2,3$, we have:

$$ {}_3F_2\left(\begin{matrix} a,\ 3-a,\ 1 \\ b,\ 2 \end{matrix}\ \middle|\ \frac{1}{2}\right) = \frac{2(1-b)}{(1-a)_2}\left(\frac{\Gamma((b-1)/2)\Gamma(b/2)}{\Gamma(-1+(a+b)/2)\,\Gamma((b-a+1)/2)} - 1\right), $$

and:

$$ {}_3F_2\left(\begin{matrix} a,\ 5-a,\ 1 \\ b,\ 3 \end{matrix}\ \middle|\ \frac{1}{2}\right) = \frac{8(b-2)_2}{(a-4)_4} $$

$$ \times\left(\frac{\Gamma((b-1)/2)\Gamma((b-2)/2)}{\Gamma(-2+(a+b)/2)\,\Gamma((b-a+1)/2)} - \frac{5a-a^2+2b-10}{2(b-2)}\right). $$

Special Case 5. When $p=4, q=3$ and $x=1$, by noting the Dixon theorem (8), Relation (18) is simplified as:

$$ {}_4F_3\left(\begin{matrix} a,\ b,\ c,\ 1 \\ a-b+m,\ a-c+m,\ m \end{matrix}\ \middle|\ 1\right) $$

$$ = \frac{\Gamma(m)\Gamma(a-b+m)\Gamma(a-c+m)\Gamma(a+1-m)\Gamma(b+1-m)\Gamma(c+1-m)}{\Gamma(a)\Gamma(b)\Gamma(c)\Gamma(a-b+1)\Gamma(a-c+1)} $$

$$ \times\left(\begin{matrix} \frac{\Gamma((a+3-m)/2)\Gamma(a-b+1)\Gamma(a-c+1)\Gamma(-b-c+(a+3m-1)/2)}{\Gamma(a+2-m)\Gamma(-b+(a+m+1)/2)\Gamma(-c+(a+m+1)/2)\Gamma(a-b-c+m)} \\ - {}_3F_2^{(m-2)}\left(\begin{matrix} a-m+1,\ b-m+1,\ c-m+1 \\ a-b+1,\ a-c+1 \end{matrix}\ \middle|\ 1\right) \end{matrix}\right). \tag{23} $$

For $m=1$, Relation (23) exactly gives the Dixon formula, while for $m=2,3$, we have:

$$ {}_4F_3\left(\begin{matrix} a,\ b,\ c,\ 1 \\ a-b+2,\ a-c+2,\ 2 \end{matrix}\ \middle|\ 1\right) = \frac{(a-b+1)(a-c+1)}{(a-1)(b-1)(c-1)} $$

$$ \times\left(\frac{\Gamma((a+1)/2)\Gamma(a-b+1)\Gamma(a-c+1)\Gamma(-b-c+(a+5)/2)}{\Gamma(a)\Gamma(-b+(a+3)/2)\Gamma(-c+(a+3)/2)\Gamma(a-b-c+2)} - 1\right), $$

and:

$$ {}_4F_3\left(\begin{array}{c} a,\ b,\ c,\ 1 \\ a-b+3,\ a-c+3,\ 3 \end{array}\middle|\ 1\right) = \frac{2(a-b+1)_2(a-c+1)_2}{(a-2)_2(b-2)_2(c-2)_2} $$

$$ \times\left(\frac{\Gamma(a/2)\Gamma(a-b+1)\Gamma(a-c+1)\Gamma((a/2)-b-c+4)}{\Gamma(a-1)\Gamma((a/2)-b+2)\Gamma((a/2)-c+2)\Gamma(a-b-c+3)} - \frac{(a-2)(b-2)(c-2)}{(a-b+1)(a-c+1)} - 1\right). $$

Special Case 6. When $p = 4, q = 3$ and $x = 1$, by noting the Watson theorem (9), Relation (18) is simplified as:

$$ {}_4F_3\left(\begin{array}{c} a,\ b,\ c,\ 1 \\ (a+b+1)/2,\ 2c+1-m,\ m \end{array}\middle|\ 1\right) $$

$$ = \frac{\Gamma(m)\Gamma((a+b+1)/2)\Gamma(2c+1-m)\Gamma(a+1-m)\Gamma(b+1-m)\Gamma(c+1-m)}{\Gamma(a)\Gamma(b)\Gamma(c)\Gamma(-m+(a+b+3)/2)\Gamma(2c-2m+2)} $$

$$ \times\left(\begin{array}{c} \frac{\sqrt{\pi}\,\Gamma(c-m+(3/2))\Gamma(-m+(a+b+3)/2)\Gamma(c-(a+b-1)/2)}{\Gamma(1+(a-m)/2)\Gamma(1+(b-m)/2)\Gamma(c+1-(a+m)/2)\Gamma(c+1-(b+m)/2)} \\ -\ {}_3^{(m-2)}F_2\left(\begin{array}{c} a-m+1,\ b-m+1,\ c-m+1 \\ -m+1+(a+b+1)/2,\ 2c-2m+2 \end{array}\middle|\ 1\right) \end{array}\right). \qquad (24) $$

For $m = 1$, Relation (24) exactly gives the Watson formula, while for $m = 2, 3$, we have:

$$ {}_4F_3\left(\begin{array}{c} a,\ b,\ c,\ 1 \\ (a+b+1)/2,\ 2c-1,\ 2 \end{array}\middle|\ 1\right) = \frac{a+b-1}{(a-1)(b-1)} $$

$$ \times\left(\frac{\sqrt{\pi}\,\Gamma(c-(1/2))\Gamma((a+b-1)/2)\Gamma(c-(a+b-1)/2)}{\Gamma(a/2)\Gamma(b/2)\Gamma(c-(a/2))\Gamma(c-(b/2))} - 1\right), $$

and:

$$ {}_4F_3\left(\begin{array}{c} a,\ b,\ c,\ 1 \\ (a+b+1)/2,\ 2c-2,\ 3 \end{array}\middle|\ 1\right) = \frac{(2c-3)(a+b-1)(a+b-3)}{(a-2)_2(b-2)_2(c-1)} $$

$$ \times\left(\frac{\sqrt{\pi}\,\Gamma(c-(3/2))\Gamma((a+b-3)/2)\Gamma(c-(a+b-1)/2)}{\Gamma((a-1)/2)\Gamma((b-1)/2)\Gamma(c-(a+1)/2)\Gamma(c-(b+1)/2)} - \frac{(a-2)(b-2)}{a+b-3} - 1\right). $$

Special Case 7. When $p = 4, q = 3$ and $x = 1$, by noting the Whipple theorem (10), Relation (18) is simplified as:

$$
{}_4F_3\left(\begin{matrix} a,\ 2m-1-a,\ b,\ 1 \\ c,\ 2b-c+1,\ m \end{matrix}\ \middle|\ 1\right)
= \frac{\Gamma(m)\Gamma(c)\Gamma(2b-c+1)\Gamma(a+1-m)\Gamma(m-a)\Gamma(b+1-m)}{\Gamma(a)\Gamma(2m-1-a)\Gamma(b)\Gamma(c+1-m)\Gamma(2b-c-m+2)}
\times\left(\begin{matrix}\dfrac{\pi\, 2^{2m-2b-1}\Gamma(c-m+1)\Gamma(2b-c+2-m)}{\Gamma(-m+1+(a+c)/2)\Gamma(-m+1+b+(a-c+1)/2)\Gamma((1-a+c)/2)\Gamma(b+1-(a+c)/2)} \\ -\ {}^{(m-2)}_{\ \ 3}F_2\left(\begin{matrix} a-m+1,\ m-a,\ b-m+1 \\ c-m+1,\ 2b-c-m+2 \end{matrix}\ \middle|\ 1\right)\end{matrix}\right). \quad (25)
$$

For $m=1$, Relation (25) exactly gives the Whipple formula, while for $m=2,3$, we have:

$$
{}_4F_3\left(\begin{matrix} a,\ 3-a,\ b,\ 1 \\ c,\ 2b-c+1,\ 2 \end{matrix}\ \middle|\ 1\right) = \frac{(c-1)(c-2b)}{(a-2)_2(b-1)}\times
\left(\frac{\pi\, 2^{3-2b}\Gamma(c-1)\Gamma(2b-c)}{\Gamma(-1+(a+c)/2)\Gamma(b+(a-c-1)/2)\Gamma((1-a+c)/2)\Gamma(b+1-(a+c)/2)}-1\right),
$$

and:

$$
{}_4F_3\left(\begin{matrix} a,\ 5-a,\ b,\ 1 \\ c,\ 2b-c+1,\ 3 \end{matrix}\ \middle|\ 1\right) = \frac{2(c-2)_2(2b-c-1)_2}{(a-4)_4(b-2)_2}
\times\left(\frac{\pi\, 2^{5-2b}\Gamma(c-2)\Gamma(2b-c-1)}{\Gamma(-2+(a+c)/2)\Gamma(b+(a-c-3)/2)\Gamma((1-a+c)/2)\Gamma(b+1-(a+c)/2)}
-\frac{(a-2)(3-a)(b-2)}{(c-2)(2b-c-1)}-1\right).
$$

Special Case 8. When $p=4, q=3$ and $x=1$, by noting the Pfaff–Saalschütz theorem (11), Relation (18) is simplified as:

$$
{}_4F_3\left(\begin{matrix} a,\ b,\ -n+m-1,\ 1 \\ c,\ 1+a+b-c-n,\ m \end{matrix}\ \middle|\ 1\right) = \frac{(m-1)!(1-c)_{m-1}(c-a-b+n)_{m-1}}{(1-a)_{m-1}(1-b)_{m-1}(n+2-m)_{m-1}}
\times\left(\frac{(c-a)_n(c-b)_n}{(c+1-m)_n(c-a-b+m-1)_n} - {}^{(m-2)}_{\ \ 3}F_2\left(\begin{matrix} a-m+1,\ b-m+1,\ -n \\ c-m+1,\ 2+a+b-c-m-n \end{matrix}\ \middle|\ 1\right)\right). \quad (26)
$$

For $m=1$, Relation (26) exactly gives the Pfaff–Saalschütz formula, while for $m=2,3$, we have:

$$
{}_4F_3\left(\begin{matrix} a,\ b,\ -n+1,\ 1 \\ c,\ 1+a+b-c-n,\ 2 \end{matrix}\ \middle|\ 1\right) = \frac{(1-c)(c-a-b+n)}{n\,(1-a)(1-b)}
\times\left(\frac{(c-a)_n(c-b)_n}{(c-1)_n(c-a-b+1)_n}-1\right),
$$

and:

$$ {}_4F_3\left(\begin{array}{c} a,\ b,\ -n+2,\ 1 \\ c,\ 1+a+b-c-n,\ 3 \end{array}\middle|\, 1\right) = \frac{2(1-c)_2(c-a-b+n)_2}{n(1-a)_2(1-b)_2(n-1)_2} \times\left(\frac{(c-a)_n(c-b)_n}{(c-2)_n(c-a-b+2)_n} + \frac{n(a-2)(b-2)}{(c-2)(a+b-c-n-1)} - 1\right). $$

Special Case 9. When $p = 5, q = 4$ and $x = -1$, by noting the second theorem of Whipple (12), Relation (18) is simplified as:

$$ {}_5F_4\left(\begin{array}{c} a,\ (a+m+1)/2,\ b,\ c,\ 1 \\ (a+m-1)/2, a-b+m,\ a-c+m,\ m \end{array}\middle|\, -1\right) = (-1)^{m-1}\times $$

$$ \frac{\Gamma(m)\Gamma((a+m-1)/2)\Gamma(a-b+m)\Gamma(a-c+m)\Gamma(a-m+1)\Gamma((a-m+3)/2)\Gamma(b-m+1)\Gamma(c-m+1)}{\Gamma(a)\Gamma((a+m+1)/2)\Gamma(b)\Gamma(c)\Gamma((a-m+1)/2)\Gamma(a-b+1)\Gamma(a-c+1)} $$

$$ \times\left(\frac{\Gamma(1+a-b)\Gamma(1+a-c)}{\Gamma(2-m+a)\Gamma(m+a-b-c)} - {}_4^{(m-2)}F_3\left(\begin{array}{c} a-m+1,\ (a-m+3)/2,\ b-m+1,\ c-m+1 \\ (a-m+1)/2,\ a-b+1,\ a-c+1 \end{array}\middle|\, -1\right)\right). \quad (27) $$

For $m = 1$, Relation (27) exactly gives the Whipple formula, while for $m = 2, 3$, we have:

$$ {}_5F_4\left(\begin{array}{c} a,\ (a+3)/2,\ b,\ c,\ 1 \\ (a+1)/2, a-b+2,\ a-c+2,\ 2 \end{array}\middle|\, -1\right) $$

$$ = \frac{4(a-b+1)(a-c+1)}{(a^2-1)(a-1)(b-1)(c-1)}\left(1 - \frac{\Gamma(1+a-b)\Gamma(1+a-c)}{\Gamma(a)\Gamma(2+a-b-c)}\right), $$

and:

$$ {}_5F_4\left(\begin{array}{c} a,\ (a+4)/2,\ b,\ c,\ 1 \\ (a+2)/2, a-b+3,\ a-c+3,\ 3 \end{array}\middle|\, -1\right) = \frac{2(a-b+1)_2(a-c+1)_2}{(a+2)(a-1)(b-2)_2(c-2)_2} $$

$$ \times\left(\frac{\Gamma(1+a-b)\Gamma(1+a-c)}{\Gamma(a-1)\Gamma(3+a-b-c)} + \frac{a(b-2)(c-2)}{(a-b+1)(a-c+1)} - 1\right). $$

Special Case 10. When $p = 6, q = 5$ and $x = 1$, by noting the Dougall theorem (13), Relation (18) is simplified as:

$$ {}_6F_5\left(\begin{array}{c} a,\ (a+m+1)/2,\ c,\ d,\ e,\ 1 \\ (a+m-1)/2, a-c+m,\ a-d+m,\ a-e+m,\ m \end{array}\middle|\, 1\right) $$

$$ = \Gamma(m)\Gamma((a+m-1)/2)\Gamma(a-c+m) $$

$$ \times\frac{\Gamma(a-d+m)\Gamma(a-e+m)\Gamma(a-m+1)\Gamma((a-m+3)/2)\Gamma(c-m+1)\Gamma(d-m+1)\Gamma(e-m+1)}{\Gamma(a)\Gamma((a+m+1)/2)\Gamma(c)\Gamma(d)\Gamma(e)\Gamma((a-m+1)/2)\Gamma(a-c+1)\Gamma(a-d+1)\Gamma(a-e+1)} $$

$$ \times\left(\begin{array}{c} \frac{\Gamma(a-c+1)\Gamma(a-d+1)\Gamma(a-e+1)\Gamma(a-c-d-e+2m-1)}{\Gamma(a+2-m)\Gamma(a-d-e+m)\Gamma(a-c-e+m)\Gamma(a-c-d+m)} \\ -\ {}_5^{(m-2)}F_4\left(\begin{array}{c} a-m+1,\ (a-m+3)/2,\ c-m+1, d-m+1,\ e-m+1 \\ (a-m+1)/2,\ a-c+1,\ a-d+1,\ a-e+1 \end{array}\middle|\, 1\right) \end{array}\right). \quad (28) $$

For $m = 1$, Relation (28) exactly gives the Dougall formula, while for $m = 2, 3$, we have:

$$
{}_6F_5\left(\left.\begin{matrix} a,\ (a+3)/2,\ c,\ d,\ e,\ 1 \\ (a+1)/2, a-c+2,\ a-d+2,\ a-e+2,\ 2 \end{matrix}\right| 1\right)
= \frac{(a-c+1)(a-d+1)(a-e+1)}{(c-1)(d-1)(e-1)(a+1)}
\times\left(\frac{\Gamma(a-c+1)\Gamma(a-d+1)\Gamma(a-e+1)\Gamma(a-c-d-e+3)}{\Gamma(a)\Gamma(a-d-e+2)\Gamma(a-c-e+2)\Gamma(a-c-d+2)} - 1\right),
$$

and:

$$
{}_6F_5\left(\left.\begin{matrix} a,\ (a+4)/2,\ c,\ d,\ e,\ 1 \\ (a+2)/2, a-c+3,\ a-d+3,\ a-e+3,\ 3 \end{matrix}\right| 1\right)
= \frac{2\,(a-c+1)_2(a-d+1)_2(a-e+1)_2}{(a-1)(a+2)(c-2)_2(d-2)_2(e-2)_2}
\times\left(\frac{\Gamma(a-c+1)\Gamma(a-d+1)\Gamma(a-e+1)\Gamma(a-c-d-e+5)}{\Gamma(a-1)\Gamma(a-d-e+3)\Gamma(a-c-e+3)\Gamma(a-c-d+3)}
- \frac{a(c-2)(d-2)(e-2)}{(a-c+1)(a-d+1)(a-e+1)} - 1\right).
$$

Special Case 11. When $p = 8, q = 7$ and $x = 1$, by noting the second theorem of Dougall (14), Relation (18) is simplified as:

$$
{}_8F_7\left(\left.\begin{matrix} a,\ (a+m+1)/2,\ b,c,\ d,\ 2a-b-c-d+2m-1+n,\ m-n-1,\ 1 \\ (a+m-1)/2,\ a-b+m,\ a-c+m,\ a-d+m,\ b+c+d-a+1-m-n, a+n+1,\ m \end{matrix}\right| 1\right)
= (-1)^{m-1}(m-1)!\times
$$

$$
\frac{((3-a-m)/2)_{m-1}(1-a+b-m)_{m-1}(1-a+c-m)_{m-1}(1-a+d-m)_{m-1}(m+n+a-b-c-d)_{m-1}(-a-n)_{m-1}}{((1-a-m)/2)_{m-1}(1-a)_{m-1}(1-b)_{m-1}(1-c)_{m-1}(1-d)_{m-1}(b+c+d-2a+2-2m-n)_{m-1}(n+2-m)_{m-1}}
$$

$$
\times\left(\begin{matrix} \frac{(a-m+2)_n(a-b-c+m)_n(a-b-d+m)_n(a-c-d+m)_n}{(a-b+1)_n(a-c+1)_n(a-d+1)_n(a-b-c-d+2m-1)_n} \\ - {}_7F_6^{(m-2)}\left(\left.\begin{matrix} a-m+1,\ (a-m+3)/2,\ b-m+1, c-m+1,\ d-m+1,\ 2a-b-c-d+m+n, -n \\ (a-m+1)/2,\ a-b+1,\ a-c+1,\ a-d+1,\ b+c+d-a+2-2m-n, a-m+n+2 \end{matrix}\right| 1\right) \end{matrix}\right). \tag{29}
$$

For $m = 1$, Relation (29) exactly gives the Dougall formula, while for $m = 2, 3$, we have:

$$
{}_8F_7\left(\begin{array}{c} a,\ (a+3)/2,\ b,c,\ d,\ 2a-b-c-d+3+n,\ -n+1,\ 1 \\ (a+1)/2,\ a-b+2,\ a-c+2,\ a-d+2,\ b+c+d-a-1-n, a+n+1,\ 2 \end{array}\middle|\ 1\right)
$$

$$
= \frac{(-a+b-1)(-a+c-1)(-a+d-1)(n+2+a-b-c-d)(a+n)}{n\,(1+a)(1-b)(1-c)(1-d)(b+c+d-2a-2-n)}
$$

$$
\times\left(1-\frac{(a)_n(a-b-c+2)_n(a-b-d+2)_n(a-c-d+2)_n}{(a-b+1)_n(a-c+1)_n(a-d+1)_n(a-b-c-d+3)_n}\right),
$$

and:

$$
{}_8F_7\left(\begin{array}{c} a,\ (a+4)/2,\ b,c,\ d,\ 2a-b-c-d+5+n,\ -n+2,\ 1 \\ (a+2)/2,\ a-b+3,\ a-c+3,\ a-d+3,\ b+c+d-a-2-n, a+n+1,\ 3 \end{array}\middle|\ 1\right)
$$

$$
= \frac{(a-2)(-a+b-2)_2(-a+c-2)_2(-a+d-2)_2(3+n+a-b-c-d)_2(-a-n)_2}{(a+2)(1-a)_2(1-b)_2(1-c)_2(1-d)_2(b+c+d-2a-4-n)_2(n-1)_2}
$$

$$
\times\left(\begin{array}{l} \frac{(a-1)_n(a-b-c+3)_n(a-b-d+3)_n(a-c-d+3)_n}{(a-b+1)_n(a-c+1)_n(a-d+1)_n(a-b-c-d+5)_n} \\ \\ +\frac{na(b-2)(c-2)(d-2)(2a-b-c-d+3+n)}{(a-b+1)(a-c+1)(a-d+1)(b+c+d-a-n-4)(n+a-1)}-1 \end{array}\right).
$$

Remark 1. *There are two further special cases, which however do not belong to classical summation theorems. When $p=q=1$, Relation (18) is simplified as:*

$$
{}_1F_1\left(\begin{array}{c} 1 \\ m \end{array}\middle|\ z\right) = \frac{(m-1)!}{z^{m-1}}\left(e^z-\sum_{j=0}^{m-2}\frac{z^j}{j!}\right),
$$

and when $p=q+1=2$, it yields:

$$
{}_2F_1\left(\begin{array}{c} a,\ 1 \\ m \end{array}\middle|\ z\right) = \frac{(m-1)!}{z^{m-1}}\frac{\Gamma(a-m+1)}{\Gamma(a)}\left((1-z)^{m-a-1}-\sum_{j=0}^{m-2}(a-m+1)_j\frac{z^j}{j!}\right).
$$

Similarly, for the case $n=2$, Relation (16) changes to:

$$
{}_pF_q\left(\begin{array}{ccc} a_1, \dots a_{p-1}, & 2 \\ b_1, \dots b_{q-1}, & m \end{array}\middle|\ z\right) = \frac{\Gamma(m)}{z^{m-1}}\sum_{j=m-1}^{\infty}\frac{(a_1)_{j-m+1}\cdots(a_{p-1})_{j-m+1}}{(b_1)_{j-m+1}\cdots(b_{q-1})_{j-m+1}}\frac{(j+2-m)\,z^j}{j!}
$$

$$
= \frac{\Gamma(m)}{z^{m-1}}\left(z\sum_{r=m-2}^{\infty}\frac{(a_1)_{j-m+2}\cdots(a_{p-1})_{j-m+2}}{(b_1)_{j-m+2}\cdots(b_{q-1})_{j-m+2}}\frac{z^r}{r!}+(2-m)\sum_{j=m-1}^{\infty}\frac{(a_1)_{j-m+1}\cdots(a_{p-1})_{j-m+1}}{(b_1)_{j-m+1}\cdots(b_{q-1})_{j-m+1}}\frac{z^j}{j!}\right)
$$

$$
= \frac{(m-1)!}{z^{m-2}}\frac{\Gamma(b_1)\dots\Gamma(b_{q-1})}{\Gamma(a_1)\dots\Gamma(a_{p-1})}\frac{\Gamma(a_1-m+2)\dots\Gamma(a_{p-1}-m+2)}{\Gamma(b_1-m+2)\dots\Gamma(b_{q-1}-m+2)}
$$

$$\times\left({}_{p-1}F_{q-1}\left(\begin{array}{c} a_1-m+2,\ \ldots,\ a_{p-1}-m+2 \\ b_1-m+2,\ \ldots,\ b_{q-1}-m+2 \end{array}\middle| z\right)-{}_{p-1}^{(m-3)}F_{q-1}\left(\begin{array}{c} a_1-m+2,\ \ldots,\ a_{p-1}-m+2 \\ b_1-m+2,\ \ldots,\ b_{q-1}-m+2 \end{array}\middle| z\right)\right)$$

$$+\ (2-m)\frac{(m-1)!}{z^{m-1}}\frac{\Gamma(b_1)\ldots\Gamma(b_{q-1})}{\Gamma(a_1)\ldots\Gamma(a_{p-1})}\frac{\Gamma(a_1-m+1)\ldots\Gamma(a_{p-1}-m+1)}{\Gamma(b_1-m+1)\ldots\Gamma(b_{q-1}-m+1)}$$

$$\times\left({}_{p-1}F_{q-1}\left(\begin{array}{c} a_1-m+1,\ \ldots,\ a_{p-1}-m+1 \\ b_1-m+1,\ \ldots,\ b_{q-1}-m+1 \end{array}\middle| z\right)-{}_{p-1}^{(m-2)}F_{q-1}\left(\begin{array}{c} a_1-m+1,\ \ldots,\ a_{p-1}-m+1 \\ b_1-m+1,\ \ldots,\ b_{q-1}-m+1 \end{array}\middle| z\right)\right).$$

Therefore:

$$ {}_pF_q\left(\begin{array}{cc} a_1,\ \ldots\ a_{p-1}, & 2 \\ b_1,\ \ldots\ b_{q-1}, & m \end{array}\middle| z\right)=\frac{(m-1)!}{z^{m-1}}\frac{\Gamma(b_1)\ldots\Gamma(b_{q-1})}{\Gamma(a_1)\ldots\Gamma(a_{p-1})}\frac{\Gamma(a_1-m+1)\ldots\Gamma(a_{p-1}-m+1)}{\Gamma(b_1-m+1)\ldots\Gamma(b_{q-1}-m+1)}$$
$$\times\left(\begin{array}{l} \frac{(a_1-m+1)\ldots(a_{p-1}-m+1)}{(b_1-m+1)\ldots(b_{q-1}-m+1)}z \\ \times\left({}_{p-1}F_{q-1}\left(\begin{array}{c} a_1-m+2,\ \ldots,\ a_{p-1}-m+2 \\ b_1-m+2,\ \ldots,\ b_{q-1}-m+2 \end{array}\middle| z\right)-{}_{p-1}^{(m-3)}F_{q-1}\left(\begin{array}{c} a_1-m+2,\ \ldots,\ a_{p-1}-m+2 \\ b_1-m+2,\ \ldots,\ b_{q-1}-m+2 \end{array}\middle| z\right)\right) \\ -(m-2)\left({}_{p-1}F_{q-1}\left(\begin{array}{c} a_1-m+1,\ \ldots,\ a_{p-1}-m+1 \\ b_1-m+1,\ \ldots,\ b_{q-1}-m+1 \end{array}\middle| z\right)-{}_{p-1}^{(m-2)}F_{q-1}\left(\begin{array}{c} a_1-m+1,\ \ldots,\ a_{p-1}-m+1 \\ b_1-m+1,\ \ldots,\ b_{q-1}-m+1 \end{array}\middle| z\right)\right) \end{array}\right). \quad (30)$$

For instance, if $m=3$, Relation (30) reads as:

$$ {}_pF_q\left(\begin{array}{cc} a_1,\ \ldots,a_{p-1}, & 2 \\ b_1,\ \ldots,b_{q-1}, & 3 \end{array}\middle| z\right)=\frac{2}{z^2}\frac{\Gamma(b_1)\ldots\Gamma(b_{q-1})}{\Gamma(a_1)\ldots\Gamma(a_{p-1})}\frac{\Gamma(a_1-2)\ldots\Gamma(a_{p-1}-2)}{\Gamma(b_1-2)\ldots\Gamma(b_{q-1}-2)}$$

$$\times\left(\frac{(a_1-2)\ldots(a_{p-1}-2)}{(b_1-2)\ldots(b_{q-1}-2)}z\,{}_{p-1}F_{q-1}\left(\begin{array}{c} a_1-1,\ \ldots,\ a_{p-1}-1 \\ b_1-1,\ \ldots,\ b_{q-1}-1 \end{array}\middle| z\right)\right.$$

$$\left.-{}_{p-1}F_{q-1}\left(\begin{array}{c} a_1-2,\ \ldots,\ a_{p-1}-2 \\ b_1-2,\ \ldots,\ b_{q-1}-2 \end{array}\middle| z\right)+1\right).$$

Hence, for $p=q+1=3$ and $z=1$, we have:

$$ {}_3F_2\left(\begin{array}{c} a,\ b,\ 2 \\ c,\ 3 \end{array}\middle| 1\right)=\frac{2}{(a-2)_2(b-2)_2}$$
$$\times\left((c-2)_2+\frac{\Gamma(c)\Gamma(c-a-b+1)}{\Gamma(c-a)\Gamma(c-b)}(ab-a-b-c+3)\right).$$

3. Second Hypergeometric Identity

By noting Relation (1), first it is not difficult to verify that:

$$(a+m)_k=\frac{(a)_{k+m}}{(a)_m}. \quad (31)$$

Now, if the identity (31) is applied in a special case of (2), we obtain:

$$
{}_pF_q\left(\begin{matrix} a_1+m, \ldots a_{p-1}+m, & 1 \\ b_1+m, \quad \ldots & ,b_q+m \end{matrix}\,\middle|\, z\right) = \sum_{k=0}^{\infty} \frac{(a_1+m)_k \ldots (a_{p-1}+m)_k}{(b_1+m)_k \ldots (b_q+m)_k} z^k
$$

$$
= \frac{(b_1)_m \ldots (b_q)_m}{(a_1)_m \ldots (a_{p-1})_m} \sum_{k=0}^{\infty} \frac{(a_1)_{k+m} \ldots (a_{p-1})_{k+m}}{(b_1)_{k+m} \ldots (b_{q-1})_{k+m}} z^k
$$

$$
= \frac{(b_1)_m \ldots (b_q)_m}{(a_1)_m \ldots (a_{p-1})_m} \sum_{j=m}^{\infty} \frac{(a_1)_j \ldots (a_{p-1})_j}{(b_1)_j \ldots (b_{q-1})_j} z^{j-m}
$$

$$
= \frac{(b_1)_m \ldots (b_q)_m}{(a_1)_m \ldots (a_{p-1})_m} z^{-m} \left(\sum_{j=0}^{\infty} \frac{(a_1)_j \ldots (a_{p-1})_j}{(b_1)_j \ldots (b_{q-1})_j} z^j - \sum_{j=0}^{m-1} \frac{(a_1)_j \ldots (a_{p-1})_j}{(b_1)_j \ldots (b_{q-1})_j} z^j \right),
$$

leading to the second identity:

$$
{}_pF_q\left(\begin{matrix} a_1+m, \ldots a_{p-1}+m, & 1 \\ b_1+m, \quad \ldots & ,b_q+m \end{matrix}\,\middle|\, z\right) = \frac{(b_1)_m \ldots (b_q)_m}{(a_1)_m \ldots (a_{p-1})_m} z^{-m}
$$

$$
\times \left({}_pF_q\left(\begin{matrix} a_1, \ldots a_{p-1}, & 1 \\ b_1, \quad \ldots & ,b_q \end{matrix}\,\middle|\, z\right) - \overset{(m-1)}{{}_pF_q}\left(\begin{matrix} a_1, \ldots a_{p-1}, & 1 \\ b_1, \quad \ldots & ,b_q \end{matrix}\,\middle|\, z\right)\right), \tag{32}
$$

which is equivalent to:

$$
{}_pF_q\left(\begin{matrix} a_1, \ldots a_{p-1}, & 1 \\ b_1, \quad \ldots & ,b_q \end{matrix}\,\middle|\, z\right) = \frac{(b_1-m)_m \ldots (b_q-m)_m}{(a_1-m)_m \ldots (a_{p-1}-m)_m} z^{-m}
$$

$$
\times \left({}_pF_q\left(\begin{matrix} a_1-m, \ldots a_{p-1}-m, & 1 \\ b_1-m, \quad \ldots & ,b_q-m \end{matrix}\,\middle|\, z\right)\right.
$$

$$
\left. - \overset{(m-1)}{{}_pF_q}\left(\begin{matrix} a_1-m, \ldots a_{p-1}-m, & 1 \\ b_1-m, \quad \ldots & ,b_q-m \end{matrix}\,\middle|\, z\right)\right). \tag{33}
$$

Once again, the interesting point is that by using Relation (32) or (33), various special cases can be considered as follows.

Special Case 12. When $p = 2, q = 1$ and $x = -1$, by noting the Kummer formula and Relation (32), we get:

$$
{}_2F_1\left(\begin{matrix} b+m, \ 1 \\ 2-b+m \end{matrix}\,\middle|\, -1\right)
$$

$$
= (-1)^m \frac{(2-b)_m}{(b)_m} \left(\frac{\sqrt{\pi}}{2} \frac{\Gamma(2-b)}{\Gamma(-b+(3/2))} - \overset{(m-1)}{{}_2F_1}\left(\begin{matrix} b, \ 1 \\ 2-b \end{matrix}\,\middle|\, -1\right)\right). \tag{34}
$$

For instance, if $m = 1, 2$, Relation (34) is simplified as:

$$ {}_2F_1\left(\begin{array}{c} b+1,\ 1 \\ 3-b \end{array} \middle| -1 \right) = -\frac{\sqrt{\pi}}{2} \frac{\Gamma(3-b)}{\Gamma((3/2)-b)} + \frac{2}{b} - 1, $$

and:

$$ {}_2F_1\left(\begin{array}{c} b+2,\ 1 \\ 4-b \end{array} \middle| -1 \right) = \frac{\sqrt{\pi}}{2b(b+1)} \frac{\Gamma(4-b)}{\Gamma((3/2)-b)} - \frac{2(b-1)(b-3)}{b(b+1)}. $$

Special Case 13. When $p = 2, q = 1$ and $x = 1/2$, by noting the second kind of Gauss formula and Relation (32), we get:

$$ {}_2F_1\left(\begin{array}{c} a+m,\ 1 \\ (a/2)+m+1 \end{array} \middle| \frac{1}{2} \right) = \frac{(1+(a/2))_m}{(a)_m} 2^m \left(\frac{\sqrt{\pi}\,\Gamma(1+(a/2))}{\Gamma((a+1)/2)} - {}_2F_1^{(m-1)}\left(\begin{array}{c} a,\ 1 \\ (a/2)+1 \end{array} \middle| \frac{1}{2} \right) \right). \tag{35} $$

For instance, if $m = 1, 2$, Relation (35) is simplified as:

$$ {}_2F_1\left(\begin{array}{c} a+1,\ 1 \\ (a/2)+2 \end{array} \middle| \frac{1}{2} \right) = \frac{a+2}{a} \left(\sqrt{\pi}\frac{\Gamma(1+(a/2))}{\Gamma((a+1)/2)} - 1 \right), $$

and:

$$ {}_2F_1\left(\begin{array}{c} a+2,\ 1 \\ (a/2)+3 \end{array} \middle| \frac{1}{2} \right) = \sqrt{\pi}\frac{(a+2)(a+4)}{a(a+1)} \frac{\Gamma(1+(a/2))}{\Gamma((a+1)/2)} - 2(1+\frac{4}{a}). $$

Special Case 14. When $p = 3, q = 2$ and $x = 1$, by noting the Dixon formula and Relation (32), we get:

$$ {}_3F_2\left(\begin{array}{c} b+m,\ c+m,\ 1 \\ 2-b+m,\ 2-c+m \end{array} \middle| 1 \right) = \frac{(2-b)_m(2-c)_m}{(b)_m(c)_m} $$

$$ \times \left(\frac{\sqrt{\pi}}{2} \frac{\Gamma(2-b)\Gamma(2-c)\Gamma(-b-c+(3/2))}{\Gamma(-b+(3/2))\Gamma(-c+(3/2))\Gamma(2-b-c)} - {}_3F_2^{(m-1)}\left(\begin{array}{c} b,\ c,\ 1 \\ 2-b,\ 2-c \end{array} \middle| 1 \right) \right). \tag{36} $$

For instance, if $m = 1, 2$, Relation (36) is simplified as:

$$ {}_3F_2\left(\begin{array}{c} b+1,\ c+1,\ 1 \\ 3-b,\ 3-c \end{array} \middle| 1 \right) = \frac{\sqrt{\pi}}{2bc} \frac{\Gamma(3-b)\Gamma(3-c)\Gamma((3/2)-b-c)}{\Gamma((3/2)-b)\Gamma((3/2)-c)\Gamma(2-b-c)} - \frac{(b-2)(c-2)}{bc}, $$

and:

$$ {}_3F_2\left(\begin{array}{c} b+2,\ c+2,\ 1 \\ 4-b,\ 4-c \end{array} \middle| 1 \right) = \frac{\sqrt{\pi}}{2(b)_2(c)_2} \frac{\Gamma(4-b)\Gamma(4-c)\Gamma((3/2)-b-c)}{\Gamma((3/2)-b)\Gamma((3/2)-c)\Gamma(2-b-c)} - \frac{2(bc-b-c-2)(b-3)(c-3)}{(b)_2(c)_2}. $$

Special Case 15. When $p = 3, q = 2$ and $x = 1$, by noting the Watson formula and Relation (32), we get:

$$ {}_3F_2\left(\begin{array}{c} b+m,\ c+m,\ 1 \\ (b/2)+m+1,\ 2c+m \end{array} \middle| 1\right) = \frac{(1+(b/2))_m (2c)_m}{(b)_m (c)_m} $$

$$ \times \left(\frac{\sqrt{\pi}\,\Gamma(1+(c/2))\Gamma(1+(b/2))\Gamma(c-(b/2))}{\Gamma((b+1)/2)\Gamma(c)\Gamma(c-(b-1)/2)} - {}_3F_2^{(m-1)}\left(\begin{array}{c} b,\ c,\ 1 \\ (b/2)+1,\ 2c \end{array} \middle| 1\right)\right). \quad (37) $$

For instance, if $m = 1, 2$, Relation (37) is simplified as:

$$ {}_3F_2\left(\begin{array}{c} b+1,\ c+1,\ 1 \\ (b/2)+2,\ 2c+1 \end{array} \middle| 1\right) = (1+\frac{2}{b})(\frac{\sqrt{\pi}\,\Gamma(1+(c/2))\Gamma(1+(b/2))\Gamma(c-(b/2))}{\Gamma((b+1)/2)\Gamma(c)\Gamma(c-(b-1)/2)} - 1), $$

and:

$$ {}_3F_2\left(\begin{array}{c} b+2,\ c+2,\ 1 \\ (b/2)+3,\ 2c+2 \end{array} \middle| 1\right) $$

$$ = \frac{2\sqrt{\pi}c(2c+1)}{b(b+1)} \frac{\Gamma((c/2)+1)\Gamma((b/2)+3)\Gamma(c-(b/2))}{\Gamma((b+1)/2)\Gamma(c+2)\Gamma(c-(b-1)/2)} - \frac{(b+4)(2c+1)}{b(c+1)}. $$

Special Case 16. When $p = 3, q = 2$ and $x = 1$, by noting the Whipple formula and Relation (32), we get:

$$ {}_3F_2\left(\begin{array}{c} a+m,\ 1-a+m,\ 1 \\ c+m,\ 3-c+m \end{array} \middle| 1\right) = \frac{(c)_m(3-c)_m}{(a)_m(1-a)_m} $$

$$ \times \left(\frac{\pi\,\Gamma(c)\Gamma(3-c)}{2\,\Gamma((a+c)/2)\Gamma((a-c+3)/2)\Gamma((1-a+c)/2)\Gamma(2-(a+c)/2)}\right. $$

$$ \left. - {}_3F_2^{(m-1)}\left(\begin{array}{c} a,\ 1-a,\ 1 \\ c,\ 3-c \end{array} \middle| 1\right)\right). \quad (38) $$

For instance, if $m = 1, 2$, Relation (38) is simplified as:

$$ {}_3F_2\left(\begin{array}{c} a+1,\ 2-a,\ 1 \\ c+1,\ 4-c \end{array} \middle| 1\right) $$

$$ = \frac{\pi}{2a(1-a)} \frac{\Gamma(c+1)\Gamma(4-c)}{\Gamma((a+c)/2)\Gamma((a-c+3)/2)\Gamma((1-a+c)/2)\Gamma(2-(a+c)/2)} - \frac{c(c-3)}{a(a-1)}, $$

and:

$$ {}_3F_2\left(\begin{array}{c} a+2,\ 3-a,\ 1 \\ c+2,\ 5-c \end{array} \middle| 1\right) $$

$$ = \frac{\pi}{2(a-2)_4} \frac{\Gamma(c+2)\Gamma(5-c)}{\Gamma((a+c)/2)\Gamma((a-c+3)/2)\Gamma((1-a+c)/2)\Gamma(2-(a+c)/2)} $$

$$-\frac{(c+1)(4-c)(c(3-c)+a(1-a))}{(a-2)_4}.$$

Special Case 17. When $p = 3, q = 2$ and $x = 1$, by noting the Pfaff–Saalschutz formula and Relation (32), we get:

$$ {}_3F_2\left(\begin{array}{c} b+m,\ \ -n+m,\ 1 \\ c+m,\ 2+b-c-n+m \end{array}\middle|\, 1\right)$$

$$=\frac{(c)_m(2+b-c-n)_m}{(b)_m(-n)_m}\left(\frac{(c-1)(c-1-b+n)}{(c-1-b)(c-1+n)}-{}_3F_2^{(m-1)}\left(\begin{array}{c} b,\ -n,\ 1 \\ c,\ 2+b-c-n \end{array}\middle|\, 1\right)\right). \quad (39)$$

For instance, if $m = 1, 2$, Relation (39) is simplified as:

$$ {}_3F_2\left(\begin{array}{c} b+1,\ -n+1,\ 1 \\ c+1,\ 3+b-c-n \end{array}\middle|\, 1\right)=\frac{c(c-2-b+n)}{nb}\left(\frac{(c-1)(c-1-b+n)}{(c-1-b)(c-1+n)}-1\right),$$

and:

$$ {}_3F_2\left(\begin{array}{c} b+2,\ -n+2,\ 1 \\ c+2,\ 4+b-c-n \end{array}\middle|\, 1\right)$$

$$=\frac{(c-1)c(c+1)(c-1-b+n)(c-2-b+n)(c-3-b+n)}{n(n-1)b(b+1)(c-1-b)(c-1+n)}$$

$$-\frac{(c+1)(c-3-b+n)(nb+c(c-2-b+n))}{n(n-1)b(b+1)}.$$

Special Case 18. When $p = 4, q = 3$ and $x = -1$, by noting the Whipple formula and Relation (32), we get:

$$ {}_4F_3\left(\begin{array}{c} m+(3/2),\ b+m,\ c+m,\ 1 \\ m+(1/2),\ 2-b+m,\ 2-c+m \end{array}\middle|\, -1\right)=\frac{(-1)^m}{2m+1}\frac{(2-b)_m(2-c)_m}{(b)_m(c)_m}$$

$$\times\left(\frac{\Gamma(2-b)\Gamma(2-c)}{\Gamma(2-b-c)}-{}_4F_3^{(m-1)}\left(\begin{array}{c} 3/2,\ b,\ c,\ 1 \\ 1/2,\ 2-b,\ 2-c \end{array}\middle|\, -1\right)\right). \quad (40)$$

For instance, if $m = 1, 2$, Relation (40) is simplified as:

$$ {}_4F_3\left(\begin{array}{c} 5/2,\ b+1,\ c+1,\ 1 \\ 3/2,\ 3-b,\ 3-c \end{array}\middle|\, -1\right)=\frac{1}{3bc}((b-2)(c-2)-\frac{\Gamma(3-b)\Gamma(3-c)}{\Gamma(2-b-c)}),$$

and:

$$ {}_4F_3\left(\begin{array}{c} 7/2,\ b+2,\ c+2,\ 1 \\ 5/2,\ 4-b,\ 4-c \end{array}\middle|\, -1\right)=\frac{(2-b)_2(2-c)_2}{5\,(b)_2(c)_2}$$

$$\times\left(\frac{\Gamma(2-b)\Gamma(2-c)}{\Gamma(2-b-c)}+2\frac{bc+b+c-2}{(b-2)(c-2)}\right).$$

Special Case 19. When $p=5, q=4$ and $x=1$, by noting the Dougall formula and Relation (32), we get:

$$ {}_5F_4\left(\begin{matrix} m+(3/2),\ c+m,\ d+m,\ e+m,\ 1 \\ m+(1/2),\ 2-c+m,\ 2-d+m,\ 2-e+m \end{matrix}\ \middle|\ 1\right) $$

$$ =\frac{1}{2m+1}\frac{(2-c)_m(2-d)_m(2-e)_m}{(c)_m(d)_m(e)_m} $$

$$ \times\left(\frac{\Gamma(2-c)\Gamma(2-d)\Gamma(2-e)\Gamma(2-c-d-e)}{\Gamma(2-d-e)\Gamma(2-c-e)\Gamma(2-c-d)}-{}_5F_4^{(m-1)}\left(\begin{matrix} 3/2,\ c,\ d,\ e,\ 1 \\ 1/2,\ 2-c,\ 2-d,\ 2-e \end{matrix}\ \middle|\ 1\right)\right). \tag{41} $$

For instance, if $m=1,2$, Relation (41) is simplified as:

$$ {}_5F_4\left(\begin{matrix} 5/2,\ c+1,\ d+1,\ e+1,\ 1 \\ 3/2,\ 3-c,\ 3-d,\ 3-e \end{matrix}\ \middle|\ 1\right)=\frac{(2-c)(2-d)(2-e)}{3\,cde} $$

$$ \times\left(\frac{\Gamma(2-c)\Gamma(2-d)\Gamma(2-e)\Gamma(2-c-d-e)}{\Gamma(2-d-e)\Gamma(2-c-e)\Gamma(2-c-d)}-1\right), $$

and:

$$ {}_5F_4\left(\begin{matrix} 7/2,\ c+2,\ d+2,\ e+2,\ 1 \\ 5/2,\ 4-c,\ 4-d,\ 4-e \end{matrix}\ \middle|\ 1\right)=\frac{(2-c)_2(2-d)_2(2-e)_2}{5\,(c)_2(d)_2(e)_2} $$

$$ \times\left(\frac{\Gamma(2-c)\Gamma(2-d)\Gamma(2-e)\Gamma(2-c-d-e)}{\Gamma(2-d-e)\Gamma(2-c-e)\Gamma(2-c-d)}-\frac{3\,cde}{(2-c)(2-d)(2-e)}-1\right). $$

Special Case 20. When $p=7, q=6$ and $x=1$, by noting the Dougall formula and Relation (32), we get:

$$ {}_7F_6\left(\begin{matrix} m+(3/2),\ b+m, c+m,\ d+m,\ 3-b-c+n+m,\ m-n,\ 1 \\ m+(1/2), 2-b+m, 2-c+m, 2-d+m,\ b+c+d-n-1+m, n+2+m \end{matrix}\ \middle|\ 1\right) $$

$$ =\frac{1}{2m+1}\frac{(2-b)_m(2-c)_m(2-d)_m(b+c+d-n-1)_m(n+2)_m}{(b)_m(c)_m(d)_m(3-b-c+n)_m(-n)_m} $$

$$ \times\left(\frac{(2)_n(2-b-c)_n(2-b-d)_n(2-c-d)_n}{(2-b)_n(2-c)_n(2-d)_n(2-b-c-d)_n}\right. $$

$$ \left.-{}_7F_6^{(m-1)}\left(\begin{matrix} 3/2,\ b, c,\ d,\ 3-b-c+n,\ -n,\ 1 \\ 1/2, 2-b, 2-c, 2-d,\ b+c+d-n-1, n+2 \end{matrix}\ \middle|\ 1\right)\right). \tag{42} $$

For instance, if $m = 1, 2$, Relation (42) is simplified as:

$$
{}_7F_6\left(\begin{matrix} 5/2,\ b+1, c+1,\ d+1,\ 3-b-c+n+1, 1-n,\ 1 \\ 3/2, 3-b, 3-c, 3-d,\ b+c+d-n, n+3 \end{matrix}\ \middle|\ 1\right)
$$

$$
= \frac{(2-b)(2-c)(2-d)(b+c+d-n-1)(n+2)}{3\,nbcd(3-b-c+n)}
$$

$$
\times\left(1 - \frac{(2)_n(2-b-c)_n(2-b-d)_n(2-c-d)_n}{(2-b)_n(2-c)_n(2-d)_n(2-b-c-d)_n}\right),
$$

and:

$$
{}_7F_6\left(\begin{matrix} 7/2,\ b+2, c+2,\ d+2,\ 5-b-c+n, 2-n,\ 1 \\ 5/2, 4-b, 4-c, 4-d,\ b+c+d-n+1, n+4 \end{matrix}\ \middle|\ 1\right)
$$

$$
= \frac{(2-b)_2(2-c)_2(2-d)_2(b+c+d-n-1)_2(n+2)_2}{5\,(b)_2(c)_2(d)_2(3-b-c+n)_2(-n)_2}
$$

$$
\times\left(\frac{(2)_n(2-b-c)_n(2-b-d)_n(2-c-d)_n}{(2-b)_n(2-c)_n(2-d)_n(2-b-c-d)_n}\right.
$$

$$
\left. + \frac{3\,bcd\,n\,(3-b-c+n)}{(2-b)(2-c)(2-d)(b+c+d-n-1)(n+2)} - 1\right).
$$

4. Conclusions

In this paper, we applied two identities for generalized hypergeometric series in order to extend some classical summation theorems of hypergeometric functions such as Gauss, Kummer, Dixon, Watson, Whipple, Pfaff–Saalschütz and Dougall formulas and then obtained some new summation theorems using the second introduced hypergeometric identity.

Author Contributions: Both authors have the same contribution in all sections.

Acknowledgments: This research was funded by the Alexander von Humboldt Foundation under Grant Number Ref 3.4-IRN-1128637-GF-E.

References

1. Koepf, W. *Hypergeometric Summation: An Algorithmic Approach to Summation and Special Function Identities*, 2nd ed.; Springer: London, UK, 2014.
2. Slater, L.J. *Generalized Hypergeometric Functions*; Cambridge University Press: Cambridge, UK, 1966.
3. Andrews, G.E.; Askey, R.; Roy, R. Special Functions. In *Encyclopedia of Mathematics and Its Applications*; Cambridge University Press: Cambridge, UK, 1999; Volume 71.
4. Arfken, G. *Mathematical Methods for Physicists*; Academic Press: New York, NY, USA, 1985.
5. Mathai, A.M.; Saxena, R.K. *Generalized Hypergeometric Functions with Applications in Statistics and Physical Sciences*; Lecture Notes in Mathematics; Springer: Berlin/Heidelberg, Germany; New York, NY, USA, 1973; Volume 348.
6. Nikiforov, A.F.; Uvarov, V.B. *Special Functions of Mathematical Physics. A Unified Introduction with Applications*; Birkhauser: Basel, Switzerland, 1988.

7. Bailey, W.N. Products of generalized hypergeometric series. *Proc. Lond. Math. Soc.* **1928**, *2*, 242–254. [CrossRef]
8. Karlsson, P.W. Hypergeometric functions with integral parameter differences. *J. Math. Phys.* **1971**, *12*, 270–271. [CrossRef]
9. Miller, A.R. Certain summation and transformation formulas for generalized hypergeometric series. *J. Comput. Appl. Math.* **2009**, *231*, 964–972. [CrossRef]
10. Miller, A.R.; Srivastava, H.M. Karlsson-Minton summation theorems for the generalized hypergeometric series of unit argument. *Integral Transform. Spec. Funct.* **2010**, *21*, 603–612. [CrossRef]
11. Minton, B.M. Generalized hypergeometric function of unit argument. *J. Math. Phys.* **1970**, *11*, 1375–1376. [CrossRef]
12. Rosengren, H. Karlsson-Minton type hypergeometric functions on the root system C_n. *J. Math. Anal. Appl.* **2003**, *281*, 332–345. [CrossRef]
13. Srivastava, H.M. Generalized hypergeometric functions with integral parameter differences. *Indag. Math.* **1973**, *76*, 38–40. [CrossRef]
14. Prudnikov, A.P.; Brychkov, Y.A.; Marichev, O.I. *More Special Functions*; Integrals and Series; Gordon and Breach Science Publishers: Amsterdam, The Netherlands, 1990; Volume 3.

5

A New Identity for Generalized Hypergeometric Functions and Applications

Mohammad Masjed-Jamei [1] and Wolfram Koepf [2,*]

1 Department of Mathematics, K.N.Toosi University of Technology, P.O.Box 16315-1618, 11369 Tehran, Iran; mmjamei@kntu.ac.ir or mmjamei@yahoo.com
2 Institute of Mathematics, University of Kassel, Heinrich-Plett Str. 40, 34132 Kassel, Germany
* Correspondence: koepf@mathematik.uni-kassel.de

Abstract: We establish a new identity for generalized hypergeometric functions and apply it for first- and second-kind Gauss summation formulas to obtain some new summation formulas. The presented identity indeed extends some results of the recent published paper (*Some summation theorems for generalized hypergeometric functions*, Axioms, 7 (2018), Article 38).

Keywords: generalized hypergeometric functions; Gauss and confluent hypergeometric functions; summation theorems of hypergeometric functions

1. Introduction

Let $\mathbb{R}$ and $\mathbb{C}$ denote the sets of real and complex numbers and z be a complex variable. For real or complex parameters a and b, the generalized binomial coefficient

$$\binom{a}{b} = \frac{\Gamma(a+1)}{\Gamma(b+1)\Gamma(a-b+1)} = \binom{a}{a-b} \qquad (a, b \in \mathbb{C}),$$

in which

$$\Gamma(z) = \int_0^\infty x^{z-1} e^{-x} dx,$$

denotes the well-known gamma function for $\mathrm{Re}(z) > 0$, can be reduced to the particular case

$$\binom{a}{n} = \frac{(-1)^n (-a)_n}{n!},$$

where $(a)_b$ denotes the Pochhammer symbol [1] given by

$$(a)_b = \frac{\Gamma(a+b)}{\Gamma(a)} = \begin{cases} 1 & (b = 0,\ a \in \mathbb{C}\backslash\{0\}), \\ a(a+1)...(a+b-1) & (b \in \mathbb{C},\ a \in \mathbb{C}). \end{cases} \tag{1}$$

By referring to the symbol (1), the generalized hypergeometric functions [2]

$${}_pF_q\left(\begin{matrix} a_1, & \dots & , a_p \\ b_1, & \dots & , b_q \end{matrix} \,\middle|\, z \right) = \sum_{k=0}^{\infty} \frac{(a_1)_k \dots (a_p)_k}{(b_1)_k \dots (b_q)_k} \frac{z^k}{k!}, \tag{2}$$

are indeed a Taylor series expansion for a function, say f, as $\sum_{k=0}^{\infty} c_k^* z^k$ with $c_k^* = f^{(k)}(0)/k!$ for which the ratio of successive terms can be written as

$$\frac{c_{k+1}^*}{c_k^*} = \frac{(k+a_1)(k+a_2)...(k+a_p)}{(k+b_1)(k+b_2)...(k+b_q)(k+1)}.$$

According to the ratio test [3,4], the series (2) is convergent for any $p \leq q+1$. In fact, it converges in $|z| < 1$ for $p = q+1$, converges everywhere for $p < q+1$ and converges nowhere ($z \neq 0$) for $p > q+1$. Moreover, for $p = q+1$ it absolutely converges for $|z| = 1$ if the condition

$$A^* = \operatorname{Re}\left(\sum_{j=1}^{q} b_j - \sum_{j=1}^{q+1} a_j\right) > 0,$$

holds and is conditionally convergent for $|z| = 1$ and $z \neq 1$ if $-1 < A^* \leq 0$ and is finally divergent for $|z| = 1$ and $z \neq 1$ if $A^* \leq -1$.

There are two important cases of the series (2) arising in many physics problems [5,6]. The first case (convergent in $|z| \leq 1$) is the Gauss hypergeometric function

$$y = {}_2F_1\left(\begin{matrix} a, \ b \\ c \end{matrix} \,\middle|\, z\right) = \sum_{k=0}^{\infty} \frac{(a)_k (b)_k}{(c)_k} \frac{z^k}{k!},$$

with the integral representation

$${}_2F_1\left(\begin{matrix} a, \ b \\ c \end{matrix} \,\middle|\, z\right) = \frac{\Gamma(c)}{\Gamma(b)\Gamma(c-b)} \int_0^1 t^{b-1}(1-t)^{c-b-1}(1-tz)^{-a} dt,$$

$$(\operatorname{Re} c > \operatorname{Re} b > 0; \ |\arg(1-z)| < \pi), \quad (3)$$

Replacing $z = 1$ in (3) directly leads to the well-known Gauss identity

$${}_2F_1\left(\begin{matrix} a, \ b \\ c \end{matrix} \,\middle|\, 1\right) = \frac{\Gamma(c)\Gamma(c-a-b)}{\Gamma(c-a)\Gamma(c-b)} \qquad \operatorname{Re}(c-a-b) > 0. \quad (4)$$

The second case, which converges everywhere, is the Kummer confluent hypergeometric function

$$y = {}_1F_1\left(\begin{matrix} b \\ c \end{matrix} \,\middle|\, z\right) = \sum_{k=0}^{\infty} \frac{(b)_k}{(c)_k} \frac{z^k}{k!},$$

with the integral representation

$${}_1F_1\left(\begin{matrix} b \\ c \end{matrix} \,\middle|\, z\right) = \frac{\Gamma(c)}{\Gamma(b)\Gamma(c-b)} \int_0^1 t^{b-1}(1-t)^{c-b-1} e^{zt} dt,$$

$$(\operatorname{Re} c > \operatorname{Re} b > 0; \ |\arg(1-z)| < \pi).$$

In this paper, we explicitly obtain the simplified form of the hypergeometric series

$${}_pF_q\left(\begin{matrix} a_1, \ \dots, a_{p-1}, & m+1 \\ b_1, \ \dots, b_{q-1}, & n+1 \end{matrix} \,\middle|\, z\right),$$

when m, n are two natural numbers and $m < n$.

2. A New Identity for Generalized Hypergeometric Functions

Let m, n be two natural numbers so that $m < n$. By noting (1), since

$$\frac{(m+1)_k}{(n+1)_k} = \frac{\Gamma(k+m+1)\Gamma(n+1)}{\Gamma(k+n+1)\Gamma(m+1)} = \frac{n!}{m!}\frac{1}{(k+m+1)(k+m+2)...(k+n)},$$

so, we have

$$\frac{(m+1)_k}{k!(n+1)_k} = \frac{\Gamma(k+m+1)\Gamma(n+1)}{k!\,\Gamma(k+n+1)\Gamma(m+1)} = \frac{n!}{m!}\frac{(k+1)_m}{(k+n)!}. \tag{5}$$

Hence, substituting (5) into a special case of (2) yields

$$\begin{aligned} {}_pF_q\left(\begin{array}{ccc} a_1, & ..., a_{p-1}, & m+1 \\ b_1, & ..., b_{q-1}, & n+1 \end{array} \middle| z\right) &= \frac{n!}{m!}\sum_{k=0}^{\infty}\frac{(a_1)_k...(a_{p-1})_k}{(b_1)_k...(b_{q-1})_k} z^k \frac{(k+1)_m}{(k+n)!} \\ &= \frac{n!}{m!}\sum_{j=n}^{\infty}\frac{(a_1)_{j-n}...(a_{p-1})_{j-n}}{(b_1)_{j-n}...(b_{q-1})_{j-n}} z^{j-n}\frac{(j+1-n)_m}{j!}. \end{aligned} \tag{6}$$

In [7], two particular cases of (6) for $m = 0$ and $m = 1$ were considered and other cases have been left as open problems. In this section, we wish to consider those open problems and solve them for any arbitrary value of m. For this purpose, since

$$(a)_{j-n} = \frac{\Gamma(a-n)}{\Gamma(a)}(a-n)_j = (-1)^n\frac{(a-n)_j}{(1-a)_n},$$

relation (6) is simplified as

$$\begin{aligned} {}_pF_q\left(\begin{array}{ccc} a_1, & ..., a_{p-1}, & m+1 \\ b_1, & ..., b_{q-1}, & n+1 \end{array} \middle| z\right) &= \frac{n!}{m!}\frac{(-1)^{n(p-q)}}{z^n}\frac{(1-b_1)_n...(1-b_{q-1})_n}{(1-a_1)_n...(1-a_{p-1})_n} \\ &\times \sum_{j=n}^{\infty}\frac{(a_1-n)_j...(a_{p-1}-n)_j}{(b_1-n)_j...(b_{q-1}-n)_j}\frac{z^j}{j!}(j+1-n)_m. \end{aligned} \tag{7}$$

It is clear in (7) that

$$\sum_{j=n}^{\infty}\frac{(a_1-n)_j...(a_{p-1}-n)_j}{(b_1-n)_j...(b_{q-1}-n)_j}\frac{z^j}{j!}(j+1-n)_m = \sum_{j=0}^{\infty}(.) - \sum_{j=0}^{n-1}(.) = S_1^* - S_2^*. \tag{8}$$

To evaluate $S_1^* = \sum_{j=0}^{\infty}(.)$, we can directly use Chu-Vandermonde identity, which is a special case of Gauss identity (4), i.e.,

$${}_2F_1\left(\begin{array}{cc} -m, & q-p \\ & q \end{array} \middle| 1\right) = \frac{(p)_m}{(q)_m}. \tag{9}$$

Now if in (9), $p = j-n+1$ and $q = -n+1$, we have

$$(j-n+1)_m = (1-n)_m {}_2F_1\left(\begin{array}{cc} -m, & -j \\ & 1-n \end{array} \middle| 1\right) = (1-n)_m\sum_{k=0}^{m}\frac{(-m)_k(-j)_k}{(1-n)_k k!}. \tag{10}$$

Hence, replacing (10) in S_1^* gives

$$\begin{aligned} S_1^* &= \sum_{j=0}^{\infty} \frac{(a_1-n)_j ... (a_{p-1}-n)_j}{(b_1-n)_j ... (b_{q-1}-n)_j} \frac{z^j}{j!} (1-n)_m \sum_{k=0}^{m} \frac{(-m)_k (-j)_k}{(1-n)_k k!} \\ &= (1-n)_m \sum_{k=0}^{m} \frac{(-m)_k}{(1-n)_k k!} \left(\sum_{j=k}^{\infty} \frac{(a_1-n)_j ... (a_{p-1}-n)_j}{(b_1-n)_j ... (b_{q-1}-n)_j} z^j \frac{(-j)_k}{j!} \right). \end{aligned} \tag{11}$$

It is important to note in the second equality of (11) that $(-j)_k = 0$ for any $j = 0, 1, 2, \ldots, k-1$. Therefore, the lower index is starting from $j = k$ instead of $j = 0$. Now since

$$\frac{(-j)_k}{j!} = \frac{(-1)^k}{(j-k)!},$$

relation (11) is simplified as

$$\begin{aligned} S_1^* &= (1-n)_m \sum_{k=0}^{m} \frac{(-m)_k}{(1-n)_k k!} \left(\sum_{j=k}^{\infty} \frac{(a_1-n)_j ... (a_{p-1}-n)_j}{(b_1-n)_j ... (b_{q-1}-n)_j} z^j \frac{(-1)^k}{(j-k)!} \right) \\ &= (1-n)_m \sum_{k=0}^{m} \frac{(-m)_k (-z)^k}{(1-n)_k k!} \left(\sum_{r=0}^{\infty} \frac{(a_1-n)_{r+k} ... (a_{p-1}-n)_{r+k}}{(b_1-n)_{r+k} ... (b_{q-1}-n)_{r+k}} \frac{z^r}{r!} \right). \end{aligned} \tag{12}$$

On the other hand, the well-known identity

$$(a)_{r+k} = (a)_k (a+k)_r,$$

simplifies (12) as

$$\begin{aligned} S_1^* &= (1-n)_m \sum_{k=0}^{m} \frac{(-m)_k (a_1-n)_k ... (a_{p-1}-n)_k}{(1-n)_k (b_1-n)_k ... (b_{q-1}-n)_k} \frac{(-z)^k}{k!} \\ &\quad \times \left(\sum_{r=0}^{\infty} \frac{(a_1-n+k)_r ... (a_{p-1}-n+k)_r}{(b_1-n+k)_r ... (b_{q-1}-n+k)_r} \frac{z^r}{r!} \right) \\ &= (1-n)_m \sum_{k=0}^{m} \frac{(-m)_k (a_1-n)_k ... (a_{p-1}-n)_k}{(1-n)_k (b_1-n)_k ... (b_{q-1}-n)_k} \frac{(-z)^k}{k!} \\ &\quad \times {}_{p-1}F_{q-1} \left(\begin{matrix} a_1-n+k, & ... & a_{p-1}-n+k \\ b_1-n+k, & ... & b_{q-1}-n+k \end{matrix} \middle| \, z \right). \end{aligned}$$

To compute the finite sum $S_2^* = \sum_{j=0}^{n-1} (.)$ in (8), we can directly use the identity

$$(j-n+1)_m = \frac{(-n+1)_m (-n+1+m)_j}{(-n+1)_j},$$

to get

$$
\begin{aligned}
S_2^* &= \sum_{j=0}^{n-1} \frac{(a_1-n)_j \dots (a_{p-1}-n)_j}{(b_1-n)_j \dots (b_{q-1}-n)_j} \frac{z^j}{j!} (j+1-n)_m \\
&= (1-n)_m \sum_{j=0}^{n-1} \frac{(a_1-n)_j \dots (a_{p-1}-n)_j}{(b_1-n)_j \dots (b_{q-1}-n)_j} \frac{z^j}{j!} \frac{(-n+1+m)_j}{(-n+1)_j} \\
&= (1-n)_m \, {}_pF_q \left(\begin{array}{ccc} a_1-n, \dots a_{p-1}-n, & -(n-1-m) \\ b_1-n, \dots b_{q-1}-n, & -(n-1) \end{array} \middle| \, z \right). \qquad (13)
\end{aligned}
$$

Finally, by noting the identity

$$
\frac{(-n+1)_m}{m!} = (-1)^m \binom{n-1}{m},
$$

the main result of this paper is obtained as follows.

Main Theorem. *If m, n are two natural numbers so that $m < n$, then*

$$
\begin{aligned}
{}_pF_q \left(\begin{array}{cc} a_1, \dots a_{p-1}, & m+1 \\ b_1, \dots b_{q-1}, & n+1 \end{array} \middle| \, z \right) &= n! \binom{n-1}{m} \frac{(-1)^{n(p-q)+m}}{z^n} \frac{(1-b_1)_n \dots (1-b_{q-1})_n}{(1-a_1)_n \dots (1-a_{p-1})_n} \\
\times & \left\{ \begin{array}{l} \sum_{k=0}^{m} \frac{(-m)_k (a_1-n)_k \dots (a_{p-1}-n)_k}{(1-n)_k (b_1-n)_k \dots (b_{q-1}-n)_k} \, {}_{p-1}F_{q-1} \left(\begin{array}{ccc} a_1-n+k, & \dots & a_{p-1}-n+k \\ b_1-n+k, & \dots & b_{q-1}-n+k \end{array} \middle| \, z \right) \frac{(-z)^k}{k!} \\ -\, {}_pF_q \left(\begin{array}{cc} a_1-n, \dots a_{p-1}-n, & -(n-1-m) \\ b_1-n, \dots b_{q-1}-n, & -(n-1) \end{array} \middle| \, z \right) \end{array} \right\}, \qquad (14)
\end{aligned}
$$

where $\{a_k\}_{k=1}^{p-1} \notin \{1, 2, \dots, n\}$ and $\{b_k\}_{k=1}^{q-1} \notin \{n, n-1, \dots, n-m+1\}$.

Note that the case $m > n$ in (14) leads to a particular case of Karlsson-Minton identity, see e.g., [8,9].

3. Some Special Cases of the Main Theorem

Essentially whenever a generalized hypergeometric series can be summed in terms of gamma functions, the result will be important as only a few such summation theorems are available in the literature. In this sense, the classical summation theorems such as Kummer and Gauss for ${}_2F_1$, Dixon, Watson, Whipple and Pfaff-Saalschutz for ${}_3F_2$, Whipple for ${}_4F_3$, Dougall for ${}_5F_4$ and Dougall for ${}_7F_6$ are well known [1,10]. In this section, we consider some special cases of the above main theorem to obtain new hypergeometric summation formulas.

Special case 1. Note that if $m = 0$, the first equality of (13) reads as

$$
S_2^* = \sum_{j=0}^{n-1} \frac{(a_1-n)_j \dots (a_{p-1}-n)_j}{(b_1-n)_j \dots (b_{q-1}-n)_j} \frac{z^j}{j!}.
$$

Hence, the main theorem is simplified as

$$ {}_pF_q\left(\begin{matrix} a_1, \dots a_{p-1}, & 1 \\ b_1, \dots b_{q-1}, & n+1 \end{matrix} \,\middle|\, z\right) = n!\frac{(-1)^{n(p-q)}}{z^n}\frac{(1-b_1)_n\dots(1-b_{q-1})_n}{(1-a_1)_n\dots(1-a_{p-1})_n} $$
$$ \times\left({}_{p-1}F_{q-1}\left(\begin{matrix} a_1-n, \dots, a_{p-1}-n \\ b_1-n, \dots, b_{q-1}-n \end{matrix} \,\middle|\, z\right) - \sum_{j=0}^{n-1}\frac{(a_1-n)_j\dots(a_{p-1}-n)_j}{(b_1-n)_j\dots(b_{q-1}-n)_j}\frac{z^j}{j!}\right), $$

which is a known result in the literature [10] (p. 439).

Special case 2. For $n = m+1$, relation (13) gives $S_2^* = (-1)^m m!$ and the main theorem therefore reads (for $m+1 \to m$) as

$$ {}_pF_q\left(\begin{matrix} a_1, \dots a_{p-1}, & m \\ b_1, \dots b_{q-1}, & m+1 \end{matrix} \,\middle|\, z\right) = (-1)^{m(p-q+1)}\frac{m!}{z^m}\frac{(1-b_1)_m\dots(1-b_{q-1})_m}{(1-a_1)_m\dots(1-a_{p-1})_m}\times $$
$$ \left\{1-\sum_{k=0}^{m-1}\frac{(a_1-m)_k\dots(a_{p-1}-m)_k}{(b_1-m)_k\dots(b_{q-1}-m)_k}\,{}_{p-1}F_{q-1}\left(\begin{matrix} a_1-m+k, & \dots & a_{p-1}-m+k \\ b_1-m+k, & \dots & b_{q-1}-m+k \end{matrix} \,\middle|\, z\right)\frac{(-z)^k}{k!}\right\}. $$

For instance, we have [7]

$$ {}_pF_q\left(\begin{matrix} a_1, \dots a_{p-1}, & 2 \\ b_1, \dots b_{q-1}, & 3 \end{matrix} \,\middle|\, z\right) = \frac{2}{z^2}\frac{(1-b_1)_2\dots(1-b_{q-1})_2}{(1-a_1)_2\dots(1-a_{p-1})_2} $$
$$ \times\left(\frac{(a_1-2)\dots(a_{p-1}-2)}{(b_1-2)\dots(b_{q-1}-2)}z\,{}_{p-1}F_{q-1}\left(\begin{matrix} a_1-1, \dots, a_{p-1}-1 \\ b_1-1, \dots, b_{q-1}-1 \end{matrix} \,\middle|\, z\right)\right. $$
$$ \left. -\,{}_{p-1}F_{q-1}\left(\begin{matrix} a_1-2, \dots, a_{p-1}-2 \\ b_1-2, \dots, b_{q-1}-2 \end{matrix} \,\middle|\, z\right)+1\right). $$

As a very particular case, replacing $p = 3$ and $q = 2$ in the above relation yields

$$ {}_3F_2\left(\begin{matrix} a, \ b, \ 2 \\ c, \ 3 \end{matrix} \,\middle|\, 1\right) $$
$$ = \frac{2}{(a-2)_2(b-2)_2}\left((c-2)_2+\frac{\Gamma(c)\Gamma(c-a-b+1)}{\Gamma(c-a)\Gamma(c-b)}(ab-a-b-c+3)\right). $$

Special case 3. For $p = q = 1$, the main theorem is simplified as

$$ {}_1F_1\left(\begin{matrix} m+1 \\ n+1 \end{matrix} \,\middle|\, z\right) $$
$$ = n!\binom{n-1}{m}\frac{(-1)^m}{z^n}\left(e^z{}_1F_1\left(\begin{matrix} -m \\ -(n-1) \end{matrix} \,\middle|\, -z\right) - {}_1F_1\left(\begin{matrix} -(n-1-m) \\ -(n-1) \end{matrix} \,\middle|\, z\right)\right). $$

For instance, by referring to the special case 1, we have [7,10]

$$ {}_1F_1\left(\begin{matrix} 1 \\ m \end{matrix} \,\middle|\, z\right) = \frac{(m-1)!}{z^{m-1}}\left(e^z - \sum_{j=0}^{m-2}\frac{z^j}{j!}\right). $$

Special case 4. For $p = 2$ and $q = 1$, the main theorem is simplified as

$$
{}_2F_1\left(\begin{array}{c} a,\ m+1 \\ n+1 \end{array}\middle|\, z\right) = n!\binom{n-1}{m}\frac{(-1)^{n+m}}{z^n}\frac{1}{(1-a)_n}
\times\left\{(1-z)^{n-a}{}_2F_1\left(\begin{array}{c} a-n,\ -m \\ -(n-1) \end{array}\middle|\, \frac{z}{z-1}\right) - {}_2F_1\left(\begin{array}{c} a-n,\ -(n-1-m) \\ -(n-1) \end{array}\middle|\, z\right)\right\},
$$

in which we have used the relation ${}_1F_0\left(\begin{array}{c} a \\ - \end{array}\middle|\, z\right) = (1-z)^{-a}$. For instance, by referring to the special case 1, we have [7,10]

$$
{}_2F_1\left(\begin{array}{c} a,\ 1 \\ m \end{array}\middle|\, z\right) = \frac{(m-1)!}{z^{m-1}}\frac{\Gamma(1-a)}{\Gamma(m-a)}\left((1-z)^{m-a-1} - \sum_{j=0}^{m-2}(a-m+1)_j\frac{z^j}{j!}\right).
$$

Special case 5. For $p = 3$ and $q = 2$, the main theorem is simplified as

$$
{}_3F_2\left(\begin{array}{c} a_1,\ a_2,\ m+1 \\ b_1,\ n+1 \end{array}\middle|\, z\right) = n!\binom{n-1}{m}\frac{(-1)^{n+m}}{z^n}\frac{(1-b_1)_n}{(1-a_1)_n(1-a_2)_n}
\times\left\{\begin{array}{l} \sum_{k=0}^{m}\frac{(-m)_k(a_1-n)_k(a_2-n)_k}{(1-n)_k(b_1-n)_k}{}_2F_1\left(\begin{array}{c} a_1-n+k,\ a_2-n+k \\ b_1-n+k \end{array}\middle|\, z\right)\frac{(-z)^k}{k!} \\ -{}_3F_2\left(\begin{array}{c} a_1-n,\ a_2-n,\ -(n-1-m) \\ b_1-n,\ -(n-1) \end{array}\middle|\, z\right) \end{array}\right\}. \quad (15)
$$

As a particular case and by noting the first kind of Gauss formula (4), if $z = 1$ is replaced in (15) then we get

$$
{}_3F_2\left(\begin{array}{c} a_1,\ a_2,\ m+1 \\ b_1,\ n+1 \end{array}\middle|\, 1\right) = (-1)^{n+m}n!\binom{n-1}{m}\frac{(1-b_1)_n}{(1-a_1)_n(1-a_2)_n}
\times\left\{\begin{array}{l} \sum_{k=0}^{m}\frac{(-m)_k(a_1-n)_k(a_2-n)_k}{(1-n)_k(b_1-n)_k}\frac{\Gamma(b_1-n+k)\Gamma(b_1-a_1-a_2+n-k)}{\Gamma(b_1-a_1)\Gamma(b_1-a_2)}\frac{(-1)^k}{k!} \\ -{}_3F_2\left(\begin{array}{c} a_1-n,\ a_2-n,\ -(n-1-m) \\ b_1-n,\ -(n-1) \end{array}\middle|\, 1\right) \end{array}\right\}.
$$

Therefore, we get

$$
{}_3F_2\left(\begin{array}{c} a_1,\ a_2,\ m+1 \\ b_1,\ n+1 \end{array}\middle|\, 1\right) = \binom{n-1}{m}\frac{(-1)^m n!}{(1-a_1)_n(1-a_2)_n}
\times\left\{\begin{array}{l} (b_1-a_1-a_2)_n{}_2F_1\left(\begin{array}{c} a_1,\ a_2 \\ b_1 \end{array}\middle|\, 1\right){}_3F_2\left(\begin{array}{c} a_1-n,\ a_2-n,\ -m \\ 1-n+a_1+a_2-b_1,\ 1-n \end{array}\middle|\, 1\right) \\ -(-1)^n(1-b_1)_n{}_3F_2\left(\begin{array}{c} a_1-n,\ a_2-n,\ -(n-1-m) \\ b_1-n,\ 1-n \end{array}\middle|\, 1\right) \end{array}\right\}. \quad (16)
$$

As a numerical example for the result (16), we have

$$
{}_3F_2\left(\begin{array}{c} 1/5,\ 3/10,\ 2 \\ 4/5,\ 5 \end{array}\middle|\, 1\right) = \frac{72}{(4/5)_4(7/10)_4}\times
\left((1/5)_4\sum_{k=0}^{2}\frac{(-2)_k(-19/5)_k(-37/10)_k}{(-3)_k(-16/5)_k k!}\right.
\left. -(3/10)_4\frac{\Gamma(4/5)\Gamma(3/10)}{\Gamma(3/5)\Gamma(1/2)}\sum_{k=0}^{1}\frac{(-1)_k(-19/5)_k(-37/10)_k}{(-3)_k(-33/10)_k k!}\right).
$$

It is clear that the right-hand side of this equality can be easily computed and therefore the infinite series in the left-hand side has been evaluated.

Similarly, by noting the second kind of Gauss formula [1]

$$
{}_2F_1\left(\begin{array}{c} a,\ b \\ (a+b+1)/2 \end{array}\middle|\, \frac{1}{2}\right) = \frac{\sqrt{\pi}\,\Gamma((a+b+1)/2)}{\Gamma((a+1)/2)\Gamma((b+1)/2)},
$$

relation (15) takes the form

$$
{}_3F_2\left(\begin{array}{c} a_1,\ a_2,\ m+1 \\ b_1,\ n+1 \end{array}\middle|\, \frac{1}{2}\right) = (-1)^{n+m}2^n n!\binom{n-1}{m}\frac{(1-b_1)_n}{(1-a_1)_n(1-a_2)_n}
\times\left\{\begin{array}{l} \sqrt{\pi}\sum_{k=0}^{m}\frac{(-m)_k(a_1-n)_k(a_2-n)_k}{(1-n)_k(b_1-n)_k}\frac{\Gamma(-n+k+b_1)}{\Gamma((a_1-n+k+1)/2)\Gamma((a_2-n+k+1)/2)}\frac{(-1)^k}{2^k k!} \\ \qquad -{}_3F_2\left(\begin{array}{c} a_1-n,\ a_2-n,\ -(n-1-m) \\ b_1-n,\ -(n-1) \end{array}\middle|\, \frac{1}{2}\right) \end{array}\right\},
$$

where $b_1 = (a_1+a_2+1)/2$.

Author Contributions: Both authors have contributed the same amout in all sections.

References

1. Koepf, W. *Hypergeometric Summation: An Algorithmic Approach to Summation and Special Function Identities*, 2nd ed.; Springer Universitext; Springer: London, UK, 2014.
2. Slater, L.J. *Generalized Hypergeometric Functions*; Cambridge University Press: Cambridge, UK, 1966.
3. Andrews, G.E.; Askey, R.; Roy, R. *Special Functions, Encyclopedia of Mathematics and Its Applications*; Cambridge University Press: Cambridge, UK, 1999; Voulme 71.
4. Arfken, G. *MathematicaL Methods for Physicists*; Academic Press Inc.: New York, NY, USA, 1985.
5. Mathai, A.M.; Saxena, R.K. Generalized hypergeometric functions with applications in statistics and physical sciences. In *Lecture Notes in Mathematics*; Springer: Berlin/Heidelberg, Germany; New York, NY, USA, 1973; Volume 348.
6. Nikiforov, A.F.; Uvarov, V.B. *Special Functions of Mathematical Physics*; Birkhäuser: Basel, Switzerland, 1988.
7. Masjed-Jamei, M.; Koepf, W. Some summation theorems for generalized hypergeometric functions. *Axioms* **2018**, *7*, 38. [CrossRef]
8. Karlsson, P.W. Hypergeometric functions with integral parameter differences. *J. Math. Phys.* **1971**, *12*, 270–271. [CrossRef]
9. Minton, B. Generalized hypergeometric function of unit argument. *J. Math. Phys.* **1970**, *11*, 1375–1376. [CrossRef]
10. Prudnikov, A.P.; Brychkov, Y.A.; Marichev, O.I. *Integrals and Series. Vol. 3. More Special Functions*; Gordon and Breach Science Publishers: Amsterdam, The Netherlands, 1990.

6

A New Type of Generalization on W—Asymptotically $\mathcal{J}_\lambda$—Statistical Equivalence with the Number of α

Hafize Gümüş [1,*] and Nihal Demir [2]

1 Department of Math Education, Faculty of Eregli Education, Necmettin Erbakan University, Konya 42060, Turkey
2 İkizli Middle School, Konya 42060, Turkey; nihaldemircise@gmail.com
* Correspondence: hgumus@konya.edu.tr

Abstract: In our paper, by using the concept of $W-$asymptotically $\mathcal{J}-$ statistical equivalence of order α which has been previously defined, we present the definitions of $W-$asymptotically $\mathcal{J}_\lambda-$statistical equivalence of order α, $W-$strongly asymptotically $\mathcal{J}_\lambda-$statistical equivalence of order α, and $W-$strongly Cesáro asymptotically $\mathcal{J}-$statistical equivalence of order α where $0 < \alpha \leq 1$. We also extend these notions with a sequence of positive real numbers, $p = (p_k)$, and we investigate how our results change if p is constant.

Keywords: $\mathcal{J}-$statistical convergence; asymptotic equivalence; set sequences; $\lambda = (\lambda_n)$ sequence

1. Introduction and Background

To make it easier to understand, we prefer to give introduction section in five parts. In first part, we give the main definitions related to statistical convergence: $\lambda-$statistical convergence, $\mathcal{J}$—convergence, $\mathcal{J}$—statistical convergence, and $\mathcal{J}_\lambda-$ statistical convergence. In the second part, we mention asymptotic equivalence and S_λ^L-asymptotic equivalence. In the third part, we explain set sequences, and we give some important definitions for set sequences in the Wijsman sense. In the fourth part, we explain how statistical convergence and $\mathcal{J}-$convergence were expanded using the number of α $(0 < \alpha < 1)$. Finally, in last part, we explain the purpose and innovations of our study.

1.1. $\mathcal{J}_\lambda-$*Statistical Convergence*

Statistical convergence is a concept which was formally introduced by Fast [1] and Steinhaus [2], independently. Later on, Schoenberg reintroduced this concept in his own study [3]. This new type of convergence has been used in different areas by several authors in references [4–8]. Statistical convergence is based on the definition of the natural density of the set $K \subseteq \mathbb{N}$ and we define the natural density of K by $d(K) = \lim_{n\to\infty} \frac{|K_n|}{n}$. In this definition, $K_n = \{k \in K : k \leq n\}$ and $|K_n|$ gives the number of elements in K_n.

Using this information, we say that a sequence (x) of real numbers is statistically convergent to the number L if $d(K(\varepsilon)) = d(\{k \leq n : |x_k - L| \geq \varepsilon\}) = 0$. In this case we write $st - \lim x_k = L$, and usually, S denotes the set of all statistical convergent sequences.

Let $\lambda = (\lambda_n)$ be a positive number sequence which is non-decreasing and tending to ∞. Also, for this sequence $\lambda_{n+1} \leq \lambda_n + 1,\ \lambda_1 = 1$. We denote the set of this kind of sequence by Λ, and we have the interval $I_n = [n - \lambda_n + 1, n]$. Mursaleen [9] defined $\lambda-$statistical convergence such that

$$\lim_{n\to\infty} \frac{1}{\lambda_n} |\{k \in I_n : |x_k - L| \geq \varepsilon\}| = 0$$

for any $\varepsilon > 0$, and he denoted this new method by S_λ. On the other hand, Kostyrko, Šalát and Wilezyński [10] introduced a new type of convergence which is defined in a metric space and is called $\mathcal{J}$—convergence. This type of convergence is based on the definition of an ideal $\mathcal{J}$ in $\mathbb{N}$.

A family of sets, $\mathcal{J} \subseteq 2^{\mathbb{N}}$, is an ideal if the following properties are provided:

(i) $\emptyset \in \mathcal{J}$; (ii) $A, B \in \mathcal{J}$ implies $A \cup B \in \mathcal{J}$; and (iii) for each $A \in \mathcal{J}$ and each $B \subseteq A$ implies $B \in \mathcal{J}$.

We say that $\mathcal{J}$ is non-trivial if $\mathbb{N} \notin \mathcal{J}$ and $\mathcal{J}$ is admissible if $\{n\} \in \mathcal{J}$ for every $n \in \mathbb{N}$.

A family of sets, $\mathcal{F} \subseteq 2^{\mathbb{N}}$, is a filter if the following properties are provided:

(i) $\emptyset \notin \mathcal{F}$; (ii) if $A, B \in \mathcal{F}$ then we have $A \cap B \in \mathcal{F}$; and (iii) for each $A \in \mathcal{F}$ and each $A \subseteq B$, we have $B \in \mathcal{F}$.

If $\mathcal{J}$ is an ideal in $\mathbb{N}$, then we have,

$$F(\mathcal{J}) = \{A \subset \mathbb{N} : \mathbb{N} \backslash A \in \mathcal{J}\}$$

is a filter in $\mathbb{N}$.

Definition 1. *([10]) A sequence of reals $x = (x_k)$ is $\mathcal{J}-$ convergent to $L \in \mathbb{R}$ if and only if the set*

$$A_\varepsilon = \{k \in \mathbb{N} : |x_k - L| \geq \varepsilon\} \in \mathcal{J}$$

for each $\varepsilon > 0$. In this case, we say that L is the $\mathcal{J}-$limit of the sequence (x).

$\mathcal{J}$—convergence generalizes many types of convergence such as usual convergence and statistical convergence. If we choose the ideals $\mathcal{J}_f = \{A \subset \mathbb{N} : A \text{ is finite}\}$ and $\mathcal{J}_d = \{A \subset \mathbb{N} : d(A) = 0\}$, then we obtain usual convergence and statistical convergence, respectively.

Based on the statistical convergence and $\mathcal{J}-$convergence, an important role was located in this area, $\mathcal{J}$—statistical convergence, which was introduced by Das, Savaş and Ghosal [11] as follows:

Definition 2. *([11]) A sequence $x = (x_k)$ is $\mathcal{J}-$statistically convergent to L if*

$$\left\{n \in \mathbb{N} : \frac{1}{n} |\{k \leq n : |x_k - L| \geq \varepsilon\}| \geq \delta\right\} \in \mathcal{J}$$

for every $\varepsilon > 0$ and $\delta > 0$.

1.2. Asymptotic Equivalence

Asymptotic equivalence was first introduced by Pobyvanets [12] and some main definitions and asymptotic reguler matrices were given by Marouf [13]. Bilgin [14] defined $f-$Asymptotically equivalent sequences, and on the other hand, asymptotically statistically equivalent sequences were presented by Patterson [15]. Gümüş and Savaş [16] gave the definition of $\mathcal{J}$—asymptotically $\lambda-$statistically equivalent sequences by using the $\lambda = (\lambda_n)$ sequence, and they were also interested in some inclusion relations between other related spaces.

According to Marouf, if $x = (x_k)$ and $y = (y_k)$ are two non-negative sequences, we say that they are asymptotically equivalent if

$$\lim_k \frac{x_k}{y_k} = 1.$$

This is denoted by $x \sim y$.

Definition 3. *([16]) Let $\mathcal{J}$ be an admissible ideal and $\lambda \in \Lambda$. Two number sequences $x = (x_k)$ and $y = (y_k)$ are $S^L(\mathcal{J})-$asymptotically equivalent of multiple L (or $\mathcal{J}-$ asymptotically $\lambda-$statistically equivalent) if every $\delta, \varepsilon > 0$,*

$$\left\{n \in \mathbb{N} : \frac{1}{\lambda_n} \left| \left\{k \in I_n : \left| \frac{x_k}{y_k} - L \right| \geq \varepsilon \right\} \right| \geq \delta \right\} \in \mathcal{J}.$$

1.3. Set Sequences

In recent years, studies on set sequences has become popular. Firstly, usual convergence has been extended to convergence of sequences of sets. The first definitions of this subject were based on Baronti and Papini's [17] work in 1986. Now, we revisit the definitions of convergence, boundedness, and the Cesáro summability of set sequences. Throughout the paper, (X, ρ) is a metric space, and $A, A_k, B_k \subseteq X$ represents non-empty closed subsets of X for all $k \in \mathbb{N}$.

(X, ρ) is a metric, $x \in X$ is a point in X, and A is any non-empty subset of X. The distance from x to A is defined by

$$\rho(x, A) = \inf_{a \in A} \rho(x, A).$$

Definition 4. *([17]) In any metric space, the set sequence $\{A_k\}$ is Wijsman convergent to A if*

$$\lim_{k \to \infty} \rho(x, A_k) = \rho(x, A)$$

for each $x \in X$. We write for this case $W - \lim_{k \to \infty} A_k = A$.

We would like to give a well known example of this subject.

Example 1. *In the (x, y)-plane, consider the $A_k = \{(x, y) : x^2 + y^2 + 2kx = 0\}$ sequence of circles. We can easily see that for $k \to \infty$, this sequence is Wijsman convergent to the y-axis, i.e., $A = \{(x, y) : x = 0\}$.*

Definition 5. *([17]) In any metric space, the set sequence $\{A_k\}$ is bounded if*

$$\sup_k \rho(x, A_k) < \infty$$

for each $x \in X$. This is shown as $\{A_k\} \in L_\infty$.

Definition 6. *([17]) In any metric space, the set sequence $\{A_k\}$ is Wijsman Cesáro summable to A if*

$$\lim_{n \to \infty} \frac{1}{n} \sum_{k=1}^{n} \rho(x, A_k) = d(x, A)$$

for each $x \in X$, and $\{A_k\}$ is Wijsman strongly Cesáto summable to A if

$$\lim_{n \to \infty} \frac{1}{n} \sum_{k=1}^{n} |\rho(x, A_k) - \rho(x, A)| = 0$$

for each $x \in X$.

Nuray and Rhodes [18] introduced Wijsman statistical convergence for set sequences by combining statistical convergence with this new concept. Similarly, Kisi and Nuray [19] defined Wijsman $\mathcal{J}-$convergence for set sequences with an ideal $\mathcal{J}$.

Definition 7. *([18]) Let (X,ρ) be a metric space. For any non-empty closed subsets, $A, A_k \subseteq X$, we say that the sequence $\{A_k\}$ is Wijsman statistically convergent to A if $\{d(x,A_k)\}$ is statistically convergent to $d(x,A)$, i.e., $\varepsilon > 0$ and $x \in X$,*

$$\lim_{n\to\infty} \frac{1}{n}|\{k \leq n : |\rho(x,A_k) - \rho(x,A)| \geq \varepsilon\}| = 0.$$

In this case, we write $st - \lim_W A_k = A$ or $A_k \to A(WS)$. We denote the set of all Wijsman statistically convergent sequences by WS.

Definition 8. *([19]) Let (X,ρ) be a metric space and $\mathcal{J} \subseteq 2^{\mathbb{N}}$ be a proper ideal in $\mathbb{N}$. For any non-empty closed subsets, $A, A_k \subset X$, we say that the sequence $\{A_k\}$ is Wijsman $\mathcal{J}-$convergent to A, if for each $\varepsilon > 0$, and $x \in X$,, the set is*

$$A(x,\varepsilon) = \{k \in \mathbb{N} : |\rho(x,A_k) - \rho(x,A)| \geq \varepsilon\} \in \mathcal{J}.$$

In this case, we write $\mathcal{I}_W - \lim A_k = A$ or $A_k \to A(\mathcal{I}_W)$, and we denote the set of all Wijsman $\mathcal{J}-$convergent sequences by $\mathcal{J}_W$.

Example 2. *Let $X = \mathbb{R}^2$ and $\{A_k\}$ be a sequence as follows:*

$$A_k = \begin{cases} \{(x,y) \in \mathbb{R}^2 : x^2 + y^2 - 2ky = 0\} & \text{if, } k \neq n^2 \\ \{(x,y) \in \mathbb{R}^2 : y = -1\} & \text{if, } k = n^2 \end{cases}$$

and

$$A = \left\{(x,y) \in \mathbb{R}^2 : y = 0\right\}.$$

The sequence $\{A_k\}$ is not Wijsman convergent to the set A. Howeverm if we choose the ideal $\mathcal{J} = \mathcal{J}_d$, then $\{A_k\}$ is Wijsman $\mathcal{J}-$convergent to set A,, where $\mathcal{J}_d = \{T \subseteq \mathbb{N} : d(T) = 0\}$, and where d is the natural density.

Definition 9. *([19]) In any metric space, let $\mathcal{J} \subseteq 2^{\mathbb{N}}$ be a non-trivial ideal and $A, A_k \subset X$. The sequence $\{A_k\}$ is said to be Wijsman $\mathcal{J}-$statistically convergent to A or $S(\mathcal{J}_W)$-convergent to A if*

$$\left\{n \in \mathbb{N} : \frac{1}{n}|\{k \leq n : |\rho(x,A_k) - \rho(x,A)| \geq \varepsilon\}| \geq \delta\right\} \in \mathcal{J}$$

for each $\varepsilon > 0$ and each $x \in X$ and $\delta > 0$, and we write $A_k \to A(S(\mathcal{J}_W))$. The class of all Wijsman $\mathcal{J}-$ statistically convergent sequences is denoted by $S(\mathcal{J}_W)$.

Recently, Hazarika and Esi [20] and Savas [21] obtained some results about asymptotically $\mathcal{J}-$statistically equivalent set sequences.

1.4. The Number α

In recent years, many concepts that are considered essential in this area has been reworked using the alpha number. In references [22,23], by using the natural density of order α, the statistical convergence of order α $(0 < \alpha < 1)$ was introduced. The new definition is not exactly parallel to that of statistical convergence. Some other applications of this concept are the $\lambda-$statistical convergence of order α by Çolak and Bektaş [24], the lacunary statistical convergence of order α by Şengül and Et [25], the weighted statistical convergence of order α and its applications by Ghosal [26], and the almost statistical convergence of order α by Et, Altın and Çolak [27]. $\mathcal{J}-$statistical convergence and $\mathcal{J}-$lacunary statistical convergence of order α were introduced by Das and Savaş in 2014 [28]. In all of these studies, n was replaced by n^{α} in the denominator in the definition of natural density, and a different direction was given.

In 2017, Savas [21] gave a new definition about Wijsman asymptotically $\mathcal{J}-$statistical equivalence of order α $(0 < \alpha \leq 1)$ as follows:

Definition 10. *([21]) In any metric space, let $A_k, B_k \subseteq X$ be any non-empty closed subsets such that $d(x, A_k) > 0$ and $d(x, B_k) > 0$ for all $x \in X$. We say that the sequences $\{A_k\}$ and $\{B_k\}$ are Wijsman asymptotically $\mathcal{J}-$ statistically equivalent of order α $(0 < \alpha \leq 1)$ to multiple L if for each $\varepsilon > 0, \delta > 0$, and $x \in X$,*

$$\left\{n \in \mathbb{N} : \frac{1}{n^\alpha}\left|\left\{k \leq n : \left|\frac{\rho(x, A_k)}{\rho(x, B_k)} - L\right| \geq \varepsilon\right\}\right| \geq \delta\right\} \in \mathcal{J}.$$

In this case, we write $\{A_k\} \overset{[S^L(\mathcal{J}_W)]^\alpha}{\sim} \{B_k\}$. It is obvious that $[S^L(\mathcal{J}_W)]^\alpha$ denotes the the set of all sequences such that $\{A_k\} \overset{[S^L(\mathcal{J}_W)]^\alpha}{\sim} \{B_k\}$.

1.5. Present Study

It should be mentioned that the generalization of the concept of $W-$asymptotically $\mathcal{J}-$statistical equivalence of order α for $\lambda = (\lambda_n)$ sequences has not been studied until now. So, this brings to mind the question of how our new results will be if we use λ and p sequences. This makes the study interesting. In this study, we searched for the answer to this question. We generalized $W-$asymptotically $\mathcal{J}_\lambda-$statistical equivalence of order α and compared the properties of this new concept with the other type of convergences without λ.

2. Main Results

Following this information, we now consider our main definitions and results. Throughout the paper, (X, ρ) is a metric space, $\mathcal{J} \subseteq 2^{\mathbb{N}}$ is an admissible ideal, $(\lambda_n) \in \Lambda$ and λ_n^α is the αth power of $(\lambda_n)^\alpha$ of λ_n, that is $\lambda^\alpha = (\lambda_n)^\alpha = (\lambda_1^\alpha, \lambda_2^\alpha, ..., \lambda_n^\alpha, ...)$, and $p = (p_k)$ is a positive real number sequence. We use the W $(Wijsman)$ symbol since our expressions are defined for set sequences.

Definition 11. *Let $A_k, B_k \subseteq X$ be non-empty closed subsets such that $d(x, A_k) > 0$ and $d(x, B_k) > 0$ for all $x \in X$. Then, the sequences $\{A_k\}$ and $\{B_k\}$ are $W-$strongly Cesáro asymptotically $\mathcal{J}-$statistically equivalent of order α $(0 < \alpha \leq 1)$ to multiple L if for each $\delta > 0$ and $x \in X$,*

$$\left\{n \in \mathbb{N} : \frac{1}{n^\alpha}\sum_{k=1}^{n}\left|\frac{\rho(x, A_k)}{\rho(x, B_k)} - L\right|^{p_k} \geq \delta\right\} \in \mathcal{J}.$$

For this situation, we write $\{A_k\} \overset{[\sigma^{L(p)}(\mathcal{J}_W)]^\alpha}{\sim} \{B_k\}$, and $\left[\sigma^{L(p)}(\mathcal{J}_W)\right]^\alpha$ denotes the set of all sequences $\{A_k\}$ and $\{B_k\}$ such that $\{A_k\} \overset{[\sigma^{L(p)}(\mathcal{J}_W)]^\alpha}{\sim} \{B_k\}$.

Now let us give our definitions with the λ sequence.

Definition 12. *Let $A_k, B_k \subseteq X$ be non-empty closed subsets such that $d(x, A_k) > 0$ and $d(x, B_k) > 0$ for all $x \in X$. Then, the sequences $\{A_k\}$ and $\{B_k\}$ are $W-$ asymptotically $\mathcal{J}_\lambda-$statistical equivalent of order α $(0 < \alpha \leq 1)$ to multiple L and denoted by $\{A_k\} \overset{[S_\lambda^L(\mathcal{J}_W)]^\alpha}{\sim} \{B_k\}$ if for each $\varepsilon > 0, \delta > 0$, and $x \in X$,*

$$\left\{n \in \mathbb{N} : \frac{1}{\lambda_n^\alpha}\left|\left\{k \in I_n : \left|\frac{\rho(x, A_k)}{\rho(x, B_k)} - L\right| \geq \varepsilon\right\}\right| \geq \delta\right\} \in \mathcal{J}.$$

We denote the set of all sequences of $\{A_k\}$ and $\{B_k\}$ such that $\{A_k\} \overset{[S_\lambda^L(\mathcal{J}_W)]^\alpha}{\sim} \{B_k\}$ by $[S_\lambda^L(\mathcal{J}_W)]^\alpha$.

Definition 13. *Let $A_k, B_k \subseteq X$ be non-empty closed subsets such that $d(x, A_k) > 0$ and $d(x, B_k) > 0$ for all $x \in X$. Then, the sequences $\{A_k\}$ and $\{B_k\}$ are $W-$ strongly asymptotically $\mathcal{J}_\lambda-$statistically equivalent of order α $(0 < \alpha \leq 1)$ to multiple L if for each $\delta > 0$ and each $x \in X$,*

$$\left\{ n \in \mathbb{N} : \frac{1}{\lambda_n^\alpha} \sum_{k \in I_n} \left| \frac{\rho(x, A_k)}{\rho(x, B_k)} - L \right|^{p_k} \geq \delta \right\} \in \mathcal{J}.$$

We denote the set of this kind of sequence by $\left[V_\lambda^{L(p)}(\mathcal{J}_W) \right]^\alpha$.

The next theorem examines the relation between Savas' definition and our second definition.

Theorem 1. *(i) If $\liminf\limits_{n\to\infty} \frac{\lambda_n^\alpha}{n^\alpha} > 0$ then $\left[S^L(\mathcal{J}_W)\right]^\alpha \subset \left[S_\lambda^L(\mathcal{J}_W)\right]^\alpha$.*
(ii) If $\liminf\limits_{n\to\infty} \frac{\lambda_n^\alpha}{n^\alpha} = 1$ then $\left[S_\lambda^L(\mathcal{J}_W)\right]^\alpha \subset \left[S^L(\mathcal{J}_W)\right]^\alpha$.

Proof. (i) Assume that $\frac{\lambda_n^\alpha}{n^\alpha} > 0$ and $\{A_k\} \overset{S^L(\mathcal{J}_W)}{\sim} \{B_k\}$. Then, there exists a $\eta > 0$ such that $\frac{\lambda_n^\alpha}{n^\alpha} \geq \eta$ for sufficiently large n. For every $\varepsilon > 0$, we have,

$$\frac{1}{n^\alpha} \left\{ k \leq n : \left| \frac{\rho(x,A_k)}{\rho(x,B_k)} - L \right| \geq \varepsilon \right\} \supseteq \frac{1}{n^\alpha} \left\{ k \in I_n : \left| \frac{\rho(x,A_k)}{\rho(x,B_k)} - L \right| \geq \varepsilon \right\}.$$

If we think about the number of elements of the sets that provide this relation,

$$\begin{aligned} \frac{1}{n^\alpha} \left| \left\{ k \leq n : \left| \frac{\rho(x,A_k)}{\rho(x,B_k)} - L \right| \geq \varepsilon \right\} \right| &\geq \frac{1}{n^\alpha} \left| \left\{ k \in I_n : \left| \frac{\rho(x,A_k)}{\rho(x,B_k)} - L \right| \geq \varepsilon \right\} \right| \\ &\geq \frac{\lambda_n^\alpha}{n^\alpha} \frac{1}{\lambda_n^\alpha} \left| \left\{ k \in I_n : \left| \frac{\rho(x,A_k)}{\rho(x,B_k)} - L \right| \geq \varepsilon \right\} \right| \\ &\geq \eta \frac{1}{\lambda_n^\alpha} \left| \left\{ k \in I_n : \left| \frac{\rho(x,A_k)}{\rho(x,B_k)} - L \right| \geq \varepsilon \right\} \right|, \end{aligned}$$

we get, for any $\delta > 0$,

$$\begin{aligned} &\left\{ n \in \mathbb{N} : \frac{1}{\lambda_n^\alpha} \left| \left\{ k \in I_n : \left| \frac{\rho(x,A_k)}{\rho(x,B_k)} - L \right| \geq \varepsilon \right\} \right| \geq \delta \right\} \\ &\subseteq \left\{ n \in \mathbb{N} : \frac{1}{n^\alpha} \left| \left\{ k \leq n : \left| \frac{\rho(x,A_k)}{\rho(x,B_k)} - L \right| \geq \varepsilon \right\} \right| \geq \delta\eta \right\} \in \mathcal{J}. \end{aligned}$$

Then, we have the proof.
(ii) Let $\delta > 0$. Since $\liminf\limits_{n\to\infty} \frac{\lambda_n^\alpha}{n^\alpha} = 1$, we have $m \in N$ such that $\left| \frac{\lambda_n^\alpha}{n^\alpha} - 1 \right| < \frac{\delta}{2}$ for all $n \geq m$. For $\varepsilon > 0$,

$$\begin{aligned} \frac{1}{n^\alpha} \left| \left\{ k \leq n : \left| \frac{\rho(x, A_k)}{\rho(x, B_k)} - L \right| \geq \varepsilon \right\} \right| &\leq \frac{n^\alpha - \lambda_n^\alpha}{n^\alpha} + \frac{1}{\lambda_n^\alpha} \left| \left\{ k \in I_n : \left| \frac{\rho(x, A_k)}{\rho(x, B_k)} - L \right| \geq \varepsilon \right\} \right| \\ &\leq 1 - (1 - \frac{1}{\delta}) + \frac{1}{\lambda_n^\alpha} \left| \left\{ k \in I_n : \left| \frac{\rho(x, A_k)}{\rho(x, B_k)} - L \right| \geq \varepsilon \right\} \right| \end{aligned}$$

for all $n \geq m$. Hence,

$$\begin{aligned} &\left\{ n \in \mathbb{N} : \frac{1}{n^\alpha} \left| \left\{ k \leq n : \left| \frac{\rho(x,A_k)}{\rho(x,B_k)} - L \right| \geq \varepsilon \right\} \right| \geq \delta \right\} \\ &\subseteq \left\{ n \in \mathbb{N} : \frac{1}{\lambda_n^\alpha} \left| \left\{ k \in I_n : \left| \frac{\rho(x,A_k)}{\rho(x,B_k)} - L \right| \geq \varepsilon \right\} \right| \geq \frac{\delta}{2} \right\} \cup \{1, 2, ...m\}. \end{aligned}$$

We know that the right side belongs to the ideal because of the theorem expression. So we have the proof. □

Now let us investigate how the $p = (p_k)$ sequence affects the previous definitions. Initially, we use the constant $p = (p) = (p, p, p, p, ...)$ sequence of positive real numbers in the following two theorems.

Theorem 2. *(i) If* $\{A_k\} \overset{\left[V_\lambda^{Lp}(\mathcal{J}_W)\right]^\alpha}{\sim} \{B_k\}$ *then* $\{A_k\} \overset{\left[S_\lambda^{L}(\mathcal{J}_W)\right]^\alpha}{\sim} \{B_k\}$.

$ii)$ If $\{A_k\},\{B_k\} \in L_\infty$ and $\{A_k\} \overset{\left[S_\lambda^{L}(\mathcal{J}_W)\right]^\alpha}{\sim} \{B_k\}$ then $\{A_k\} \overset{\left[V_\lambda^{Lp}(\mathcal{J}_W)\right]^\alpha}{\sim} \{B_k\}$.

Proof. (i) Let $\{A_k\} \overset{\left[V_\lambda^{Lp}(\mathcal{J}_W)\right]^\alpha}{\sim} \{B_k\}$ and $\varepsilon > 0$. For each $x \in X$,

$$\begin{aligned}\sum_{k\in I_n} \left|\frac{\rho(x,A_k)}{\rho(x,B_k)} - L\right|^p &\geq \sum_{\substack{k\in I_n \\ \left|\frac{\rho(x,A_k)}{\rho(x,B_k)} - L\right| \geq \varepsilon}} \left|\frac{\rho(x,A_k)}{\rho(x,B_k)} - L\right|^p \\ &\geq \varepsilon^p \left|\left\{k \in I_n : \left|\frac{\rho(x,A_k)}{\rho(x,B_k)} - L\right| \geq \varepsilon\right\}\right|\end{aligned}$$

and so,

$$\frac{1}{\varepsilon^p}\frac{1}{\lambda_n^\alpha}\sum_{k\in I_n}\left|\frac{\rho(x,A_k)}{\rho(x,B_k)} - L\right| \geq \frac{1}{\lambda_n^\alpha}\left|\left\{k \in I_n : \left|\frac{\rho(x,A_k)}{\rho(x,B_k)} - L\right| \geq \varepsilon\right\}\right|.$$

Then, for any $\delta > 0$ we have,

$$\begin{aligned}&\left\{n \in \mathbb{N} : \tfrac{1}{\lambda_n^\alpha}\left|\left\{k \in I_n : \left|\tfrac{\rho(x,A_k)}{\rho(x,B_k)} - L\right| \geq \varepsilon\right\}\right| \geq \delta\right\} \\ &\subseteq \left\{n \in \mathbb{N} : \tfrac{1}{\lambda_n^\alpha}\sum_{k\in I_n}\left|\tfrac{\rho(x,A_k)}{\rho(x,B_k)} - L\right|^p \geq \varepsilon^p\delta\right\}.\end{aligned}$$

Therefore, $\{A_k\} \overset{\left[S_\lambda^{L}(\mathcal{J}_W)\right]^\alpha}{\sim} \{B_k\}$.

(ii) Assume that $\{A_k\},\{B_k\} \in L_\infty$ and $\{A_k\} \overset{\left[S_\lambda^{L}(\mathcal{J}_W)\right]^\alpha}{\sim} \{B_k\}$. There is an M such that $\left|\frac{\rho(x,A_k)}{\rho(x,B_k)} - L\right| \leq M$ for each $x \in X$ and all k. For each $\varepsilon > 0$,

$$\begin{aligned}\frac{1}{\lambda_n^\alpha}\sum_{k\in I_n}\left|\frac{\rho(x,A_k)}{\rho(x,B_k)} - L\right|^p &= \frac{1}{\lambda_n^\alpha}\sum_{\substack{k\in I_n \\ \left|\frac{\rho(x,A_k)}{\rho(x,B_k)} - L\right| \geq \varepsilon}} \left|\frac{\rho(x,A_k)}{\rho(x,B_k)} - L\right|^p \\ &+ \frac{1}{\lambda_n^\alpha}\sum_{\substack{k\in I_n \\ \left|\frac{\rho(x,A_k)}{\rho(x,B_k)} - L\right| < \varepsilon}} \left|\frac{\rho(x,A_k)}{\rho(x,B_k)} - L\right|^p \\ &\leq \frac{1}{\lambda_n^\alpha}M^p\left|\left\{k \in I_n : \left|\frac{\rho(x,A_k)}{\rho(x,B_k)} - L\right| \geq \varepsilon\right\}\right| \\ &+ \frac{1}{\lambda_n^\alpha}\varepsilon^p\left|\left\{k \in I_n : \left|\frac{\rho(x,A_k)}{\rho(x,B_k)} - L\right| < \varepsilon\right\}\right| \\ &\leq \frac{M^p}{\lambda_n^\alpha}\left|\left\{k \in I_n : \left|\frac{\rho(x,A_k)}{\rho(x,B_k)} - L\right| \geq \varepsilon\right\}\right| + \varepsilon^p\end{aligned}$$

and then for any $\delta > 0$,

$$\left\{ n \in \mathbb{N} : \frac{1}{\lambda_n^\alpha} \sum_{k \in I_n} \left| \frac{\rho(x,A_k)}{\rho(x,B_k)} - L \right|^p \geq \varepsilon \right\}$$

$$\subseteq \left\{ n \in \mathbb{N} : \frac{1}{\lambda_n^\alpha} \left| \left\{ k \in I_n : \left| \frac{\rho(x,A_k)}{\rho(x,B_k)} - L \right| \geq \varepsilon \right\} \right| \geq \frac{\varepsilon^p}{M^p} \right\} \in \mathcal{J}.$$

□

Now let us examine the above theorems for a non-constant $p = (p_k)$ sequence of positive real numbers.

Theorem 3. *(i) Let* $p = (p_k)$ *be a positive real number sequence,* $\inf p_k = h$ *and* $\sup_k p = H$. $\{A_k\} \overset{\left[V_\lambda^{L(p)}(\mathcal{J}_W)\right]^\alpha}{\sim} \{B_k\}$ *implies* $\{A_k\} \overset{\left[S_\lambda^L(\mathcal{J}_W)\right]^\alpha}{\sim} \{B_k\}$.

(ii) Let $\{A_k\}$ and $\{B_k\}$ be bounded sequences, $\inf p_k = h$, $\sup_k p = H$ and $\varepsilon > 0$. Then, $\{A_k\} \overset{\left[S_\lambda^L(\mathcal{J}_W)\right]^\alpha}{\sim} \{B_k\}$ implies $\{A_k\} \overset{\left[V_\lambda^{L(p)}(\mathcal{J}_W)\right]^\alpha}{\sim} \{B_k\}$.

Proof. (i) Assume that $\{A_k\} \overset{\left[V_\lambda^{L(p)}(\mathcal{J}_W)\right]^\alpha}{\sim} \{B_k\}$ and $\varepsilon > 0$. Then, we can write

$$\begin{aligned}
\frac{1}{\lambda_n^\alpha} \sum_{k \in I_n} \left| \frac{\rho(x,A_k)}{\rho(x,B_k)} - L \right|^{p_k} &= \frac{1}{\lambda_n^\alpha} \sum_{\substack{k \in I_n \\ \left| \frac{\rho(x,A_k)}{\rho(x,B_k)} - L \right| \geq \varepsilon}} \left| \frac{\rho(x,A_k)}{\rho(x,B_k)} - L \right|^{p_k} + \frac{1}{\lambda_n^\alpha} \sum_{\substack{k \in I_n \\ \left| \frac{\rho(x,A_k)}{\rho(x,B_k)} - L \right| < \varepsilon}} \left| \frac{\rho(x,A_k)}{\rho(x,B_k)} - L \right|^{p_k} \\
&\geq \frac{1}{\lambda_n^\alpha} \sum_{\substack{k \in I_n \\ \left| \frac{\rho(x,A_k)}{\rho(x,B_k)} - L \right| \geq \varepsilon}} \left| \frac{\rho(x,A_k)}{\rho(x,B_k)} - L \right|^{p_k} \\
&\geq \frac{1}{\lambda_n^\alpha} \sum_{\substack{k \in I_n \\ \left| \frac{\rho(x,A_k)}{\rho(x,B_k)} - L \right| \geq \varepsilon}} (\varepsilon)^{p_k} \\
&\geq \frac{1}{\lambda_n^\alpha} \sum_{\substack{k \in I_n \\ \left| \frac{\rho(x,A_k)}{\rho(x,B_k)} - L \right| \geq \varepsilon}} \min\left\{ (\varepsilon)^h, (\varepsilon)^H \right\} \\
&\geq \frac{1}{\lambda_n^\alpha} \min\left\{ (\varepsilon)^h, (\varepsilon)^H \right\} \left\{ k \in I_n : \left| \frac{\rho(x,A_k)}{\rho(x,B_k)} - L \right| \geq \varepsilon \right\}
\end{aligned}$$

and so for $\delta > 0$, we have

$$\left\{ n \in \mathbb{N} : \left| \left\{ k \in I_n : \left| \frac{\rho(x,A_k)}{\rho(x,B_k)} - L \right| \geq \varepsilon \right\} \right| \geq \delta \right\}$$

$$\subseteq \left\{ n \in \mathbb{N} : \frac{1}{\lambda_n^\alpha} \sum_{k \in I_n} \left| \frac{\rho(x,A_k)}{\rho(x,B_k)} - L \right|^{p_k} \geq \delta \min\left\{ (\varepsilon)^h, (\varepsilon)^H \right\} \right\} \in \mathcal{J}.$$

(ii) From the theorem's statement there is an integer (M) such that $\left|\frac{\rho(x,A_k)}{\rho(x,B_k)} - L\right| \leq M$ for each $x \in X$ and all k. For each $\varepsilon > 0$,

$$\begin{aligned}
\frac{1}{\lambda_n^\alpha}\sum_{k\in I_n}\left|\frac{\rho(x,A_k)}{\rho(x,B_k)} - L\right|^{p_k} &= \frac{1}{\lambda_n^\alpha}\sum_{\substack{k\in I_n \\ \left|\frac{\rho(x,A_k)}{\rho(x,B_k)}-L\right|\geq\varepsilon}}\left|\frac{\rho(x,A_k)}{\rho(x,B_k)} - L\right|^{p_k} \\
&+ \frac{1}{\lambda_n^\alpha}\sum_{\substack{k\in I_n \\ \left|\frac{\rho(x,A_k)}{\rho(x,B_k)}-L\right|<\varepsilon}}\left|\frac{\rho(x,A_k)}{\rho(x,B_k)} - L\right|^{p_k} \\
&\leq \frac{1}{\lambda_n^\alpha}\left|\left\{k \in I_n : \left|\frac{\rho(x,A_k)}{\rho(x,B_k)} - L\right| \geq \frac{\varepsilon}{2}\right\}\right| \max\left\{M^h, M^H\right\} \\
&+ \frac{1}{\lambda_n^\alpha}\left|\left\{k \in I_n : \left|\frac{\rho(x,A_k)}{\rho(x,B_k)} - L\right| < \frac{\varepsilon}{2}\right\}\right| \frac{\max(\varepsilon)^{p_k}}{2} \\
&\leq \max\left\{M^h, M^H\right\}\frac{1}{\lambda_n^\alpha}\left|\left\{k \in I_n : \left|\frac{\rho(x,A_k)}{\rho(x,B_k)} - L\right| \geq \frac{\varepsilon}{2}\right\}\right| + \frac{\max\left\{(\varepsilon)^h,(\varepsilon)^H\right\}}{2}
\end{aligned}$$

and

$$\left\{n \in \mathbb{N} : \frac{1}{\lambda_n^\alpha}\sum_{k\in I_n}\left|\frac{\rho(x,A_k)}{\rho(x,B_k)} - L\right|^{p_k} \geq \varepsilon\right\}$$

$$\subseteq \left\{n \in \mathbb{N} : \frac{1}{\lambda_n^\alpha}\left|\left\{k \in I_n : \left|\frac{\rho(x,A_k)}{\rho(x,B_k)} - L\right| \geq \frac{\varepsilon}{2}\right\}\right| \geq \frac{2\varepsilon-\max\left\{(\varepsilon)^h,(\varepsilon)^H\right\}}{2\max\left\{M^h,M^H\right\}}\right\} \in \mathcal{J}.$$

□

Finally, in the last theorem we investigate the relationship between $\mathcal{W}-$ strongly asymptotically $\mathcal{J}_\lambda-$statistical equivalence of order α and $\mathcal{W}-$strongly Cesáro asymptotically $\mathcal{J}-$statistical equivalence of order α.

Theorem 4. *If* $\{A_k\} \overset{\left[V_\lambda^{L(p)}(\mathcal{J}_W)\right]^\alpha}{\sim} \{B_k\}$, *then* $\{A_k\} \overset{\left[\sigma^{L(p)}(\mathcal{J}_W)\right]^\alpha}{\sim} \{B_k\}$.

Proof. Now, assume that $\{A_k\} \overset{\left[V_\lambda^{L(p)}(\mathcal{J}_W)\right]^\alpha}{\sim} \{B_k\}$ and $\varepsilon > 0$.

$$\begin{aligned}
\frac{1}{n^\alpha}\sum_{k=1}^{n}\left|\frac{\rho(x,A_k)}{\rho(x,B_k)} - L\right|^{p_k} &= \frac{1}{n^\alpha}\sum_{k\in I_n}\left|\frac{\rho(x,A_k)}{\rho(x,B_k)} - L\right|^{p_k} + \frac{1}{n^\alpha}\sum_{k=1}^{n-\lambda n}\left|\frac{\rho(x,A_k)}{\rho(x,B_k)} - L\right|^{p_k} \\
&\leq \frac{1}{\lambda_n^\alpha}\sum_{k\in I_n}\left|\frac{\rho(x,A_k)}{\rho(x,B_k)} - L\right|^{p_k} + \frac{1}{\lambda_n^\alpha}\sum_{k=1}^{n-\lambda n}\left|\frac{\rho(x,A_k)}{\rho(x,B_k)} - L\right|^{p_k} \\
&\leq \frac{2}{\lambda_n^\alpha}\sum_{k\in I_n}\left|\frac{\rho(x,A_k)}{\rho(x,B_k)} - L\right|^{p_k}.
\end{aligned}$$

□

According to these operations,

$$\left\{n \in \mathbb{N} : \frac{1}{n^\alpha}\sum_{k=1}^{n}\left|\frac{\rho(x,A_k)}{\rho(x,B_k)} - L\right|^{p_k} \geq \varepsilon\right\} \subseteq \left\{n \in \mathbb{N} : \frac{1}{\lambda_n^\alpha}\sum_{k\in I_n}\left|\frac{\rho(x,A_k)}{\rho(x,B_k)} - L\right|^{p_k} \geq \frac{\varepsilon}{2}\right\} \in \mathcal{J}.$$

3. Conclusions and Future Developments

In our paper, we obtained some different results by defining the $W-$asymptotically $\mathcal{J}-$statistical equivalence of order α for $\lambda = (\lambda_n)$ sequences. Later on, we generalized our results by using a positive real number sequence $p = (p_k)$. Firstly, we compared the $W-$asymptotically $\mathcal{J}-$statistical equivalence of order α and the $W-$asymptotically $\mathcal{J}_\lambda-$statistical equivalence of order α for set sequences. These results are important to understand the role of λ. In other theorems, we investigated the relations between $W-$asymptotically $\mathcal{J}_\lambda-$statistical equivalence and $W-$strongly asymptotically $\mathcal{J}_\lambda-$statistically equivalence of order α according to whether p is constant or not. Then, we searched for the relation between $W-$strongly Cesáro asymptotically $\mathcal{J}-$statistically equivalent sequences of order α and $W-$ asymptotically $\mathcal{J}_\lambda-$statistical equivalent sequences of order α.

We know that the p sequence mentioned in this article is a sequence of positive integers. It is a matter of curiosity as to how the results will be obtained if the p sequence does not provide these conditions. On the other hand, it would be interesting to compare the results obtained using a different sequence to λ with the results in this article.

Author Contributions: Conceptualization, H.G. and N.D.; Investigation, H.G. and N.D.; Writing—original draft, H.G.

Acknowledgments: The authors are grateful to the referees and the editor for their corrections and suggestions, which have greatly improved the readability of the paper.

References

1. Fast, H. Sur la convergence statistique. *Colloq. Math.* **1951**, *2*, 241–244. [CrossRef]
2. Steinhaus, H. Sur la convergence ordinaire et la convergence asymptotique. *Colloq. Math.* **1951**, *2*, 73–74.
3. Schoenberg, I.J. The integrability of certain functions and related summability methods. *Am. Math. Mon.* **1959**, *66*, 361–375. [CrossRef]
4. Anastassiou, G.A.; Duman, O. Statistical Korovkin theory for multivariate stochastic processes. *Stoch. Anal. Appl.* **2010**, *28*, 648–661. [CrossRef]
5. Freedman, A.R.; Sember, J.; Raphael, M. Some Cesàro-type summability spaces. *Proc. Lon. Math. Soc.* **1978**, *37*, 508–520. [CrossRef]
6. Gadjiev, A.D.; Orhan, C. Some approximation theorems via statistical convergence. *Rocky Mt. J. Math.* **2002**, *32*, 129–138. [CrossRef]
7. H. I. Miller, A measure theoretical subsequence characterization of statistical convergence. *Trans. Am. Math. Soc.* **1995**, *347*, 1811–1819. [CrossRef]
8. Zygmund, A. *Trigonometric Series*; Cambridge University Press: Cambridge, UK, 1979.
9. Mursaleen, M. $\lambda-$statistical convergence. *Math. Slovaca* **2000**, *50*, 111–115.
10. Kostyrko, P.; Šalát, T.; Wilezyński, W. $\mathcal{I}-$Convergence. *Real Anal. Exch.* **2000**, *26*, 669–686.
11. Das, P.; Savaş, E.; Ghosal, S. On generalized of certain summability methods using ideals. *Appl. Math. Lett.* **2011**, *26*, 1509–1514. [CrossRef]
12. Pobyvanets, I.P. Asymptotic equivalence of some linear transformation defined by a nonnegative matrix and reduced to generalized equivalence in the sense of Cesàro and Abel. *Matematicheskaya Fizika* **1980**, *28*, 83–87.
13. Marouf, M. Asymptotic equivalence and summability. *Int. J. Math. Math. Sci.* **1993**, *16*, 755–762. [CrossRef]
14. Bilgin, T. f-Asymptotically equivalent sequences. *Acta Univ. Apulensis* **2011**, *28*, 271–278.
15. Patterson, R.F. On asymptotically statistically equivalent sequences. *Demostratio Math.* **2003**, *36*, 149–153.
16. Ümüş, H.G.; Savaş, E. On $S_\lambda^L(I)-$asymptotically statistical equivalent sequences, *Numer. Anal. Appl. Math.* **2012**, *1479*, 936–941.
17. Baronti, M.; Papini, P. *Convergence of Sequences of Sets, Methods of Functional Analysis in Approximation Theory*; Birkhauser: Basel, Switzerland, 1986; pp. 133–155.
18. Nuray, F.; Rhodes, B.E. Statistical convergence of sequences of sets. *Fasc. Math.* **2012**, *49*, 87–99.
19. Kişi, Ö.; Nuray, F. New convergence definitions for sequence of sets. *Abstr. Appl. Anal.* **2013**, *2013*. [CrossRef]
20. Hazarika, B.; Esi, A. On asymptotically Wijsman lacunary statistical convergence of set sequences in ideal context. *Filomat* **2017**, *31*, 2691–2703. [CrossRef]
21. Savas, E. Asymptotically $\mathcal{J}$—lacunary statistical equivalent of order α for sequences of sets. *J. Nonlinear Sci. Appl.* **2017**, *10*, 2860–2867. [CrossRef]

22. Hunia, S.; Das, P.; Pal, S.K. Restricting statistical convergence. *Acta Math. Hungarica* **2012**, *13*, 153–161.
23. Çolak, R. *Statistical Convergence of Order α, Modern Methods in Analysis and Its Applications*; Anamaya Publishers: New Delhi, India, 2010; pp. 121–129.
24. Çolak, R.; Bektaş, Ç.A. λ-statistical convergence of order α. *Acta Math. Sci. Ser. B Engl. Ed.* **2011**, *31*, 953–959. [CrossRef]
25. Şengül, H.; Et, M. On lacunary statistical convergence of order α. *Acta Math. Sci. Ser. B Engl. Ed.* **2014**, *34*, 473–482. [CrossRef]
26. Ghosal, S. Weighted statistical convergence of order α and its applications. *J. Egypt. Math. Soc.* **2016**, *24*, 60–67. [CrossRef]
27. Et, M.; Altın, Y.; Çolak, R. Almost statistical convergence of order α. *Acta Sci. Mar.* **2015**, *37*, 55–61. [CrossRef]
28. Das, P.; Savaş, E. On $\mathcal{I}-$statistical and $\mathcal{I}-$lacunary statistical convergence of order a. *Bull. Irani. Math. Soc.* **2014**, *40*, 459–472.

7

On the Fixed-Circle Problem and Khan Type Contractions

Nabil Mlaiki [1,*], Nihal Taş [2] and Nihal Yılmaz Özgür [2]

[1] Department of Mathematics and General Sciences, Prince Sultan University, 11586 Riyadh, Saudi Arabia
[2] Department of Mathematics, Balıkesir University, 10145 Balıkesir, Turkey; nihaltas@balikesir.edu.tr (N.T.); nihal@balikesir.edu.tr (N.Y.Ö.)
* Correspondence: nmlaiki@psu.edu.sa

Abstract: In this paper, we consider the fixed-circle problem on metric spaces and give new results on this problem. To do this, we present three types of F_C-Khan type contractions. Furthermore, we obtain some solutions to an open problem related to the common fixed-circle problem.

Keywords: fixed circle; common fixed circle; fixed-circle theorem

---

1. Introduction

Recently, the fixed-circle problem has been considered for metric and some generalized metric spaces (see [1–6] for more details). For example, in [1], some fixed-circle results were obtained using the Caristi type contraction on a metric space. Using Wardowski's technique and some classical contractive conditions, new fixed-circle theorems were proved in [5,6]. In [2,3], the fixed-circle problem was studied on an S-metric space. In [7], a new fixed-circle theorem was proved using the modified Khan type contractive condition on an S-metric space. Some generalized fixed-circle results with geometric viewpoint were obtained on S_b-metric spaces and parametric N_b-metric spaces (see [8,9] for more details, respectively). Also, it was proposed to investigate some fixed-circle theorems on extended M_b-metric spaces [10]. On the other hand, an application of the obtained fixed-circle results was given to discontinuous activation functions on metric spaces (see [1,4,11]). Hence it is important to study new fixed-circle results using different techniques.

Let (X, d) be a metric space and $C_{x_0,r} = \{x \in X : d(x, x_0) = r\}$ be any circle on X. In [5], it was given the following open problem.

Open Problem CC: What is (are) the condition(s) to make any circle $C_{x_0,r}$ as the common fixed circle for two (or more than two) self-mappings?

In this paper, we give new results to the fixed-circle problem using Khan type contractions and to the above open problem using both of Khan and Ćirić type contractions on a metric space. In Section 2, we introduce three types of F_C-Khan type contractions and obtain new fixed-circle results. In Section 3, we investigate some solutions to the above Open Problem CC. In addition, we construct some examples to support our theoretical results.

2. New Fixed-Circle Theorems

In this section, using Khan type contractions, we give new fixed-circle theorems (see [12–15] for some Khan type contractions used to obtain fixed-point theorems). At first, we recall the following definitions.

Definition 1 ([16]). *Let $\mathbb{F}$ be the family of all functions $F : (0,\infty) \to \mathbb{R}$ such that*

(F1) F is strictly increasing,

(F2) For each sequence $\{\alpha_n\}_{n=1}^{\infty}$ of positive numbers, $\lim_{n\to\infty} \alpha_n = 0$ if and only if $\lim_{n\to\infty} F(\alpha_n) = -\infty$,

(F3) There exists $k \in (0,1)$ such that $\lim_{\alpha\to 0^+} \alpha^k F(\alpha) = 0$.

Definition 2 ([16]). *Let (X,d) be a metric space. A mapping $T : X \to X$ is said to be an F-contraction on (X,d), if there exist $F \in \mathbb{F}$ and $\tau \in (0,\infty)$ such that*

$$d(Tx,Ty) > 0 \Longrightarrow \tau + F(d(Tx,Ty)) \leq F(d(x,y)),$$

for all $x,y \in X$.

Definition 3 ([15]). *Let $\mathbb{F}_k$ be the family of all increasing functions $F : (0,\infty) \to \mathbb{R}$, that is, for all $x,y \in (0,\infty)$, if $x < y$ then $F(x) \leq F(y)$.*

Definition 4 ([15]). *Let (X,d) be a metric space and $T : X \to X$ be a self-mapping. T is said to be an F-Khan-contraction if there exist $F \in \mathbb{F}_k$ and $t > 0$ such that for all $x,y \in X$ if $\max\{d(Ty,x), d(Tx,y)\} \neq 0$ then $Tx \neq Ty$ and*

$$t + F(d(Tx,Ty)) \leq F\left(\frac{d(Tx,x)d(Ty,x) + d(Ty,y)d(Tx,y)}{\max\{d(Ty,x), d(Tx,y)\}}\right),$$

and if $\max\{d(Ty,x), d(Tx,y)\} = 0$ then $Tx = Ty$.

Now we modify the definition of an F-Khan-contractive condition, which is used to obtain a fixed point theorem in [15], to get new fixed-circle results. Hence, we define the notion of an F_C-Khan type I contractive condition as follows.

Definition 5. *Let (X,d) be a metric space and $T : X \to X$ be a self-mapping. T is said to be an F_C-Khan type I contraction if there exist $F \in \mathbb{F}_k$, $t > 0$ and $x_0 \in X$ such that for all $x \in X$ if the following condition holds*

$$\max\{d(Tx_0,x_0), d(Tx,x)\} \neq 0, \tag{1}$$

then

$$t + F(d(Tx,x)) \leq F\left(h\frac{d(Tx,x)d(Tx_0,x) + d(Tx_0,x_0)d(Tx,x_0)}{\max\{d(Tx_0,x_0), d(Tx,x)\}}\right),$$

where $h \in \left[0, \frac{1}{2}\right)$ and if $\max\{d(Tx_0,x_0), d(Tx,x)\} = 0$ then $Tx = x$.

One of the consequences of this definition is the following proposition.

Proposition 1. *Let (X,d) be a metric space. If a self-mapping T on X is an F_C-Khan type I contraction with $x_0 \in X$ then we get $Tx_0 = x_0$.*

Proof. Let $Tx_0 \neq x_0$. Then using the hypothesis, we find

$$\max\{d(Tx_0,x_0), d(Tx,x)\} \neq 0$$

and

$$\begin{aligned} t + F(d(Tx_0,x_0)) &\leq F\left(h\frac{d(Tx_0,x_0)d(Tx_0,x_0) + d(Tx_0,x_0)d(Tx_0,x_0)}{d(Tx_0,x_0)}\right) \\ &= F(2hd(Tx_0,x_0)) < F(d(Tx_0,x_0)). \end{aligned}$$

This is a contradiction since $t > 0$ and so it should be $Tx_0 = x_0$. □

Consequently, the condition (1) can be replaced with $d(Tx, x) \neq 0$ and so $Tx \neq x$. Considering this, now we give a new fixed-circle theorem.

Theorem 1. *Let (X, d) be a metric space, $T : X \to X$ be a self-mapping and*

$$r = \inf \{d(Tx, x) : Tx \neq x\}. \tag{2}$$

If T is an F_C-Khan type I contraction with $x_0 \in X$ then $C_{x_0,r}$ is a fixed circle of T.

Proof. Let $x \in C_{x_0,r}$. Assume that $Tx \neq x$. Then we have $d(Tx, x) \neq 0$ and by the F_C-Khan type I contractive condition, we obtain

$$\begin{aligned} t + F(d(Tx, x)) &\leq F\left(h\frac{d(Tx, x)d(Tx_0, x) + d(Tx_0, x_0)d(Tx, x_0)}{\max\{d(Tx_0, x_0), d(Tx, x)\}}\right) \\ &= F(hr) \leq F(hd(Tx, x)) < F(d(Tx, x)), \end{aligned}$$

a contradiction since $t > 0$. Therefore, we have $Tx = x$ and so T fixes the circle $C_{x_0,r}$. □

Corollary 1. *Let (X, d) be a metric space, $T : X \to X$ be a self-mapping and r be defined as in (2). If T is an F_C-Khan type I contraction with $x_0 \in X$ then T fixes the disc $D_{x_0,r} = \{x \in X : d(x, x_0) \leq r\}$.*

We recall the following theorem.

Theorem 2 ([12]). *Let (X, d) be a metric space and $T : X \to X$ satisfy*

$$d(Tx, Ty) \leq \begin{cases} k\frac{d(x,Tx)d(x,Ty)+d(y,Ty)d(y,Tx)}{d(x,Ty)+d(y,Tx)} & \text{if } d(x, Ty) + d(y, Tx) \neq 0 \\ 0 & \text{if } d(x, Ty) + d(y, Tx) = 0 \end{cases}, \tag{3}$$

where $k \in [0, 1)$ and $x, y \in X$. Then T has a unique fixed point $x^* \in X$. Moreover, for all $x \in X$, the sequence $\{T^n x\}_{n \in \mathbb{N}}$ converges to x^*.

We modify the inequality (3) using Wardowski's technique to obtain a new fixed-point theorem. We give the following definition.

Definition 6. *Let (X, d) be a metric space and $T : X \to X$ be a self-mapping. T is said to be an F_C-Khan type II contraction if there exist $F \in \mathbb{F}_k$, $t > 0$ and $x_0 \in X$ such that for all $x \in X$ if $d(Tx_0, x_0) + d(Tx, x) \neq 0$ then $Tx \neq x$ and*

$$t + F(d(Tx, x)) \leq F\left(h\frac{d(Tx, x)d(Tx_0, x) + d(Tx_0, x_0)d(Tx, x_0)}{d(Tx_0, x_0) + d(Tx, x)}\right),$$

where $h \in \left[0, \frac{1}{2}\right)$ and if $d(Tx_0, x_0) + d(Tx, x) = 0$ then $Tx = x$.

An immediate consequence of this definition is the following result.

Proposition 2. *Let (X, d) be a metric space. If a self-mapping T on X is an F_C-Khan type II contraction then we get $Tx_0 = x_0$.*

Proof. Let $Tx_0 \neq x_0$. Then using the hypothesis, we find

$$d(Tx_0, x_0) + d(Tx, x) \neq 0$$

and

$$\begin{aligned} t + F(d(Tx_0, x_0)) &\leq F\left(h\frac{d(Tx_0, x_0)d(Tx_0, x_0) + d(Tx_0, x_0)d(Tx_0, x_0)}{2d(Tx_0, x_0)}\right) \\ &= F(hd(Tx_0, x_0)) < F(d(Tx_0, x_0)), \end{aligned}$$

which is a contradiction since $t > 0$. Hence it should be $Tx_0 = x_0$. □

Theorem 3. *Let (X, d) be a metric space, $T : X \to X$ be a self-mapping and r be defined as in (2). If T is an F_C-Khan type II contraction with $x_0 \in X$ then $C_{x_0,r}$ is a fixed circle of T.*

Proof. Let $x \in C_{x_0,r}$. Assume that $Tx \neq x$. Then using Proposition 2, we get

$$d(Tx_0, x_0) + d(Tx, x) = d(Tx, x) \neq 0.$$

By the F_C-Khan type II contractive condition, we obtain

$$\begin{aligned} t + F(d(Tx, x)) &\leq F\left(h\frac{d(Tx, x)d(Tx_0, x) + d(Tx_0, x_0)d(Tx, x_0)}{d(Tx_0, x_0) + d(Tx, x)}\right) \\ &= F(hr) \leq F(hd(Tx, x)) < F(d(Tx, x)), \end{aligned}$$

a contradiction since $t > 0$. Therefore, we have $Tx = x$ and T fixes the circle $C_{x_0,r}$. □

Corollary 2. *Let (X, d) be a metric space, $T : X \to X$ be a self-mapping and r be defined as in (2). If T is an F_C-Khan type II contraction with $x_0 \in X$ then T fixes the disc $D_{x_0,r}$.*

In the following theorem, we see that the F_C-Khan type I and F_C-Khan type II contractive conditions are equivalent.

Theorem 4. *Let (X, d) be a metric space and $T : X \to X$ be a self-mapping. T satisfies the F_C-Khan type I contractive condition if and only if T satisfies the F_C-Khan type II contractive condition.*

Proof. Let the F_C-Khan type I contractive condition be satisfied by T. Using Proposition 1 and Proposition 2, we get

$$\begin{aligned} t + F(d(Tx, x)) &\leq F\left(h\frac{d(Tx, x)d(Tx_0, x) + d(Tx_0, x_0)d(Tx, x_0)}{\max\{d(Tx_0, x_0), d(Tx, x)\}}\right) \\ &= F\left(h\frac{d(Tx, x)d(Tx_0, x)}{d(Tx, x)}\right) \\ &= F(hd(Tx_0, x)) \\ &= F\left(h\frac{d(Tx, x)d(Tx_0, x) + d(Tx_0, x_0)d(Tx, x_0)}{d(Tx_0, x_0) + d(Tx, x)}\right). \end{aligned}$$

Using the similar arguments, the converse statement is clear. Consequently, the F_C-Khan type I contractive and the F_C-Khan type II contractive conditions are equivalent. □

Remark 1. *By Theorem 4, we see that Theorem 1 and Theorem 3 are equivalent.*

Now we give an example.

Example 1. *Let $X = \mathbb{R}$ be the metric space with the usual metric $d(x,y) = |x-y|$. Let us define the self-mapping $T : \mathbb{R} \to \mathbb{R}$ as*

$$Tx = \begin{cases} x & \text{if } |x| < 6 \\ x+1 & \text{if } |x| \geq 6 \end{cases},$$

for all $x \in \mathbb{R}$. The self-mapping T is both of an F_C-Khan type I and an F_C-Khan type II contraction with $F = \ln x$, $t = \ln 2$, $x_0 = 0$ and $h = \frac{1}{3}$. Indeed, we get

$$d(Tx,x) = 1 \neq 0,$$

for all $x \in \mathbb{R}$ such that $|x| \geq 6$. Then we have

$$\begin{aligned} \ln 2 &\leq \ln\left(\frac{1}{3}|x|\right) \\ &\Longrightarrow \ln 2 + \ln 1 \leq \ln(hd(x,0)) = \ln(hd(x,x_0)) \\ &\Longrightarrow t + F(d(Tx,x)) \leq F\left(h\frac{d(Tx,x)d(Tx_0,x) + d(Tx_0,x_0)d(Tx,x_0)}{\max\{d(Tx_0,x_0), d(Tx,x)\}}\right) \end{aligned}$$

and

$$\begin{aligned} \ln 2 &\leq \ln\left(\frac{1}{3}|x|\right) \\ &\Longrightarrow \ln 2 + \ln 1 \leq \ln(hd(x,0)) = \ln(hd(x,x_0)) \\ &\Longrightarrow t + F(d(Tx,x)) \leq F\left(h\frac{d(Tx,x)d(Tx_0,x) + d(Tx_0,x_0)d(Tx,x_0)}{d(Tx_0,x_0) + d(Tx,x)}\right). \end{aligned}$$

Also we obtain

$$r = \min\{d(Tx,x) : Tx \neq x\} = 1.$$

Consequently, T fixes the circle $C_{0,1} = \{-1,1\}$ and the disc $D_{0,1} = \{x \in X : |x| \leq 1\}$. Notice that the self-mapping T has other fixed circles. The above results give us only one of these circles. Also, T has infinitely many fixed circles.

Now we consider the case if $T : X \to X$ is a self-mapping, then for all $x, y \in X$,

$$x \neq y \Longrightarrow d(Ty,x) + d(Tx,y) \neq 0.$$

Definition 7. *Let (X,d) be a metric space and $T : X \to X$ be a self-mapping. Then T is called a C-Khan type contraction if there exists $x_0 \in X$ such that*

$$d(Tx,x) \leq h\frac{d(Tx,x)d(Tx_0,x) + d(Tx_0,x_0)d(Tx,x_0)}{d(Tx_0,x) + d(Tx,x_0)}, \tag{4}$$

where $h \in [0,1)$ for all $x \in X - \{x_0\}$.

We can give the following fixed-circle result.

Theorem 5. *Let (X,d) be a metric space, $T : X \to X$ be a self-mapping and $C_{x_0,r}$ be a circle on X. If T satisfies the C-Khan type contractive condition (4) for all $x \in C_{x_0,r}$ with $Tx_0 = x_0$, then T fixes the circle $C_{x_0,r}$.*

Proof. Let $x \in C_{x_0,r}$. Suppose that $Tx \neq x$. Using the C-Khan type contractive condition with $Tx_0 = x_0$, we find

$$
\begin{aligned}
d(Tx,x) &\leq h\frac{d(Tx,x)d(Tx_0,x)+d(Tx_0,x_0)d(Tx,x_0)}{d(Tx_0,x)+d(Tx,x_0)} \\
&= \frac{hrd(Tx,x)}{r+d(Tx,x_0)} \\
&\leq \frac{hrd(Tx,x)}{r} = hd(Tx,x),
\end{aligned}
$$

which is a contradiction since $h<1$. Consequently, T fixes the circle $C_{x_0,r}$. □

Theorem 6. *Let (X,d) be a metric space, $x_0 \in X$ and $T: X \to X$ be a self-mapping. If T is a C-Khan type contraction for all $x \in X - \{x_0\}$ with $Tx_0 = x_0$, then T is the identity map I_X on X.*

Proof. Let $x \in X - \{x_0\}$ be any point. If $Tx \neq x$ then using the C-Khan type contractive condition (4) with $Tx_0 = x_0$, we find

$$
\begin{aligned}
d(Tx,x) &\leq h\frac{d(Tx,x)d(Tx_0,x)+d(Tx_0,x_0)d(Tx,x_0)}{d(Tx_0,x)+d(Tx,x_0)} \\
&= h\frac{d(Tx,x)d(x_0,x)}{d(x_0,x)+d(Tx,x_0)} \\
&\leq h\frac{d(Tx,x)d(x_0,x)+d(Tx,x)d(Tx,x_0)}{d(x_0,x)+d(Tx,x_0)} \\
&= h\frac{d(Tx,x)\left[d(x_0,x)+d(Tx,x_0)\right]}{d(x_0,x)+d(Tx,x_0)} \\
&= hd(Tx,x),
\end{aligned}
$$

which is a contradiction since $h<1$. Consequently, we have $Tx = x$ and hence T is the identity map I_X on X. □

Example 2. *Let $X = \mathbb{R}$ be the usual metric space and consider the circle $C_{0,3} = \{-3,3\}$. Let us define the self-mapping $T: \mathbb{R} \to \mathbb{R}$ as*

$$
Tx = \begin{cases} \frac{-9x+8}{2x-9} & \text{if} \quad x \in \{-3,3\} \\ 0 & \text{if} \quad x \in \mathbb{R} - \{-3,3\} \end{cases},
$$

for all $x \in \mathbb{R}$. Then the self-mapping T satisfies the C-Khan type contractive condition for all $x \in C_{0,3}$ and $T0 = 0$. Consequently, $C_{0,3}$ is a fixed circle of T.

3. Common Fixed-Circle Results

Recently, it was obtained some coincidence and common fixed-point theorems using Wardowski's technique and the Ćirić type contractions (see [17] for more details). In this section, we extend the notion of a Khan type F_C-contraction to a pair of maps to obtain a solution to the Open Problem CC. At first, we give the following definition.

Definition 8. *Let (X,d) be a metric space and $T,S: X \to X$ be two self-mappings. A pair of self-mappings (T,S) is called a Khan type $F_{T,S}$-contraction if there exist $F \in \mathbb{F}_k$, $t>0$ and $x_0 \in X$ such that for all $x \in X$ if the following condition holds*

$$
\max\{d(Tx_0,x_0),d(Sx_0,x_0)\} \neq 0,
$$

then

$$
t + F(d(Tx,Sx)) \leq F\left(h\frac{d(Tx,Sx)d(Tx,x_0)+d(Tx_0,Sx_0)d(Sx,x_0)}{\max\{d(Tx_0,x_0),d(Sx_0,x_0)\}}\right),
$$

where $h \in \left[0,\frac{1}{2}\right)$ and if $\max\{d(Tx_0,x_0),d(Sx_0,x_0)\} = 0$ then $Tx = Sx$.

An immediate consequence of this definition is the following proposition.

Proposition 3. *Let (X, d) be a metric space and $T, S : X \to X$ be two self-mappings. If the pair of self-mappings (T, S) is a Khan type $F_{T,S}$-contraction with $x_0 \in X$ then x_0 is a coincidence point of T and S, that is, $Tx_0 = Sx_0$.*

Proof. We prove this proposition under the following cases:

Case 1: Let $Tx_0 = x_0$ and $Sx_0 \neq x_0$. Then using the hypothesis, we get

$$\max\{d(Tx_0, x_0), d(Sx_0, x_0)\} = d(Sx_0, x_0) \neq 0$$

and so

$$\begin{aligned} t + F(d(Tx_0, Sx_0)) &\leq F\left(h\frac{d(Tx_0, Sx_0)d(Tx_0, x_0) + d(Tx_0, Sx_0)d(Sx_0, x_0)}{d(Sx_0, x_0)}\right) \\ &= F(hd(Tx_0, Sx_0)), \end{aligned}$$

which is a contradiction since $h \in \left[0, \frac{1}{2}\right)$ and $t > 0$.

Case 2: Let $Tx_0 \neq x_0$ and $Sx_0 = x_0$. By the similar arguments used in the proof of Case 1, we get a contradiction.

Case 3: Let $Tx_0 = x_0$ and $Sx_0 = x_0$. Then we get $Tx_0 = Sx_0$.

Case 4: Let $Tx_0 \neq x_0$, $Sx_0 \neq x_0$ and $Tx_0 \neq Sx_0$. Using the hypothesis, we obtain

$$\max\{d(Tx_0, x_0), d(Sx_0, x_0)\} \neq 0$$

and so

$$t + F(d(Tx_0, Sx_0)) \leq F\left(h\frac{d(Tx_0, Sx_0)d(Tx_0, x_0) + d(Tx_0, Sx_0)d(Sx_0, x_0)}{\max\{d(Tx_0, x_0), d(Sx_0, x_0)\}}\right). \quad (5)$$

Assume that $d(Tx_0, x_0) > d(Sx_0, x_0)$. Using the inequality (5), we get

$$\begin{aligned} t + F(d(Tx_0, Sx_0)) &\leq F\left(h\frac{d(Tx_0, Sx_0)d(Tx_0, x_0) + d(Tx_0, Sx_0)d(Sx_0, x_0)}{d(Tx_0, x_0)}\right) \\ &= F\left(hd(Tx_0, Sx_0) + h\frac{d(Tx_0, Sx_0)d(Sx_0, x_0)}{d(Tx_0, x_0)}\right) \\ &< F(2hd(Tx_0, Sx_0)) < F(d(Tx_0, Sx_0)), \end{aligned}$$

which is a contradiction. Suppose that $d(Tx_0, x_0) < d(Sx_0, x_0)$. Using the inequality (5), we find

$$t + F(d(Tx_0, Sx_0)) < F(d(Tx_0, Sx_0)),$$

which is a contradiction. Consequently, x_0 is a coincidence point of T and S, that is, $Tx_0 = Sx_0$. □

Now we use the following number given in [17] (see Definition 3.1 on page 183):

$$M(x, y) = \max\left\{d(Sx, Sy), d(Sx, Tx), d(Sy, Ty), \frac{d(Sx, Ty) + d(Sy, Tx)}{2}\right\}. \quad (6)$$

We give the following definition.

Definition 9. *Let (X, d) be a metric space and $T, S : X \to X$ be two self-mappings. A pair of self-mappings (T, S) is called a Ćirić type $F_{T,S}$-contraction if there exist $F \in \mathbb{F}_k$, $t > 0$ and $x_0 \in X$ such that for all $x \in X$*

$$d(Tx, x) > 0 \Longrightarrow t + F(d(Tx, x)) \leq F(M(x, x_0)).$$

We get the following proposition.

Proposition 4. *Let (X, d) be a metric space and $T, S : X \to X$ be two self-mappings. If the pair of self-mappings (T, S) is both a Khan type $F_{T,S}$-contraction and a Ćirić type $F_{T,S}$-contraction with $x_0 \in X$ then x_0 is a common fixed point of T and S, that is, $Tx_0 = Sx_0 = x_0$.*

Proof. By the Khan type $F_{T,S}$-contractive property and Proposition 3, we know that x_0 is a coincidence point of T and S, that is, $Tx_0 = Sx_0$. Now we prove that x_0 is a common fixed point of T and S. Let $Tx_0 \neq x_0$. Then using the Ćirić type $F_{T,S}$-contractive condition, we get

$$\begin{aligned} t + F(d(Tx_0, x_0)) &\leq F(M(x_0, x_0)) \\ &= F\left(\max\left\{\begin{array}{c} d(Sx_0, Sx_0), d(Sx_0, Tx_0), d(Sx_0, Tx_0), \\ \frac{d(Sx_0, Tx_0) + d(Sx_0, Tx_0)}{2} \end{array}\right\}\right) \\ &= F(d(Sx_0, Tx_0)) = F(0), \end{aligned}$$

which is a contradiction because of the definition of F. Therefore it should be $Tx_0 = x_0$. Consequently, x_0 is a common fixed point of T and S, that is, $Tx_0 = Sx_0 = x_0$. □

Notice that we get a coincidence point result for a pair of self-mappings using the Khan type $F_{T,S}$-contractive condition by Proposition 3. We obtain a common fixed-point result for a pair of self-mappings using the both of Khan type $F_{T,S}$-contractive condition and the Ćirić type $F_{T,S}$-contractive condition by Proposition 4.

We prove the following common fixed-circle theorem as a solution to the Open Problem CC.

Theorem 7. *Let (X, d) be a metric space, $T, S : X \to X$ be two self-mappings and r be defined as in (2). If $d(Tx, x_0) = d(Sx, x_0) = r$ for all $x \in C_{x_0,r}$ and the pair of self-mappings (T, S) is both a Khan type $F_{T,S}$-contraction and a Ćirić type $F_{T,S}$-contraction with $x_0 \in X$ then $C_{x_0,r}$ is a common fixed circle of T and S, that is, $Tx = Sx = x$ for all $x \in C_{x_0,r}$.*

Proof. Let $x \in C_{x_0,r}$. We show that x is a coincidence point of T and S. Using Proposition 4, we get

$$\max\{d(Tx_0, x_0), d(Sx_0, x_0)\} = 0$$

and so by the definition of the Khan type $F_{T,S}$-contraction we obtain

$$Tx = Sx.$$

Now we prove that $C_{x_0,r}$ is a common fixed circle of T and S. Assume that $Tx \neq x$. Using Proposition 4 and the hypothesis Ćirić type $F_{T,S}$-contractive condition, we find

$$\begin{aligned} t + F(d(Tx, x)) &\leq F(M(x, x_0)) \\ &= F\left(\max\left\{\begin{array}{c} d(Sx, Sx_0), d(Sx, Tx), d(Sx_0, Tx_0), \\ \frac{d(Sx, Tx_0) + d(Sx_0, Tx)}{2} \end{array}\right\}\right) \\ &= F\left(\max\left\{d(Sx, x_0), d(Sx, Tx), \frac{d(Sx, x_0) + d(x_0, Tx)}{2}\right\}\right) \\ &= F(\max\{r, d(Sx, Tx), r\}) = F(r), \end{aligned}$$

which contradicts with the definition of r. Consequently, we have $Tx = x$ and so $C_{x_0,r}$ is a common fixed circle of T and S. □

Corollary 3. *Let (X, d) be a metric space, $T, S : X \to X$ be two self-mappings and r be defined as in (2). If $d(Tx, x_0) = d(Sx, x_0) = r$ for all $x \in C_{x_0,r}$ and the pair of self-mappings (T, S) is both a Khan type*

$F_{T,S}$-contraction and a Ćirić type $F_{T,S}$-contraction with $x_0 \in X$ then T and S fix the disc $D_{x_0,r}$, that is, $Tx = Sx = x$ for all $x \in D_{x_0,r}$.

We give an illustrative example.

Example 3. *Let $X = [1,\infty) \cup \{-1,0\}$ be the metric space with the usual metric. Let us define the self-mappings $T : X \to X$ and $S : X \to X$ as*

$$Tx = \begin{cases} x^2 & \text{if } x \in \{0,1,3\} \\ -1 & \text{if } x = -1 \\ x+1 & \text{otherwise} \end{cases}$$

and

$$Sx = \begin{cases} \frac{1}{x} & \text{if } x \in \{-1,1\} \\ 3x & \text{if } x \in \{0,3\} \\ x+1 & \text{otherwise} \end{cases},$$

for all $x \in X$. The pair of the self-mappings (T,S) is both a Khan type $F_{T,S}$-contraction and a Ćirić type $F_{T,S}$-contraction with $F = \ln x$, $t = \ln\frac{3}{2}$ and $x_0 = 0$. Indeed, we get

$$\max\{d(T0,0), d(S0,0)\} = 0$$

and so $Tx = Sx$. Therefore, the pair (T,S) is a Khan type $F_{T,S}$-contraction. Also we get

$$d(T3,3) = 6 \neq 0,$$

for $x = 3$ and

$$d(Tx,x) = 1 \neq 0,$$

for all $x \in X \setminus \{-1,0,1,3\}$. Then we have

$$\begin{aligned} \ln\frac{3}{2} &\leq \ln 9 \\ &\Longrightarrow \ln\frac{3}{2} + \ln 6 \leq \ln 9 \\ &\Longrightarrow \ln\frac{3}{2} + \ln(d(T3,3)) \leq \ln(M(3,0)) \end{aligned}$$

and

$$\begin{aligned} \ln\frac{3}{2} &\leq \ln|x+1| \\ &\Longrightarrow \ln\frac{3}{2} + \ln 1 \leq \ln|x+1| \\ &\Longrightarrow \ln\frac{3}{2} + \ln(d(Tx,x)) \leq \ln(M(x,0)). \end{aligned}$$

Hence the pair (T,S) is a Ćirić type $F_{T,S}$-contraction. Also we obtain

$$r = \min\{d(Tx,x) : Tx \neq x\} = \min\{1,6\} = 1.$$

Consequently, T fixes the circle $C_{0,1} = \{-1,1\}$ and the disc $D_{0,1}$.

In closing, we want to bring to the reader attention the following question, under what conditions we can prove the results in [18–20] in fixed circle?

Author Contributions: All authors contributed equally in writing this article. All authors read and approved the final manuscript.

References

1. Özgür, N.Y.; Taş, N. Some fixed-circle theorems on metric spaces. *Bull. Malays. Math. Sci. Soc.* **2017**. [CrossRef]
2. Özgür, N.Y.; Taş, N.; Çelik, U. New fixed-circle results on S-metric spaces. *Bull. Math. Anal. Appl.* **2017**, *9*, 10–23.
3. Özgür, N.Y.; Taş, N. Fixed-circle problem on S-metric spaces with a geometric viewpoint. *arXiv* **2017**, arXiv:1704.08838.
4. Özgür, N.Y.; Taş, N. Some fixed-circle theorems and discontinuity at fixed circle. *AIP Conf. Proc.* **2018**. [CrossRef]
5. Taş, N.; Özgür, N.Y.; Mlaiki, N. New fixed-circle results related to F_c-contractive and F_c-expanding mappings on metric spaces. **2018**, submitted for publication.
6. Taş, N.; Özgür, N.Y.; Mlaiki, N. New types of F_c-contractions and the fixed-circle problem. *Mathematics* **2018**, *6*, 188. [CrossRef]
7. Taş, N. Various types of fixed-point theorems on S-metric spaces. *J. BAUN Inst. Sci. Technol.* **2018**. [CrossRef]
8. Özgür, N.Y.; Taş, N. Generalizations of metric spaces: From the fixed-point theory to the fixed-circle theory. In *Applications of Nonlinear Analysis*; Rassias T., Ed.; Springer: Cham, Switzerland, 2018; Volume 134.
9. Taş, N.; Özgür, N.Y. Some fixed-point results on parametric N_b-metric spaces. *Commun. Korean Math. Soc.* **2018**, *33*, 943–960.
10. Mlaiki, N.; Özgür, N.Y.; Mukheimer, A.; Taş, N. A new extension of the M_b-metric spaces. *J. Math. Anal.* **2018**, *9*, 118–133.
11. Taş, N.; Özgür, N.Y. A new contribution to discontinuity at fixed point. *arXiv* **2017**, arXiv:1705.03699.
12. Fisher, B. On a theorem of Khan. *Riv. Math. Univ. Parma* **1978**, *4*, 135–137.
13. Khan, M.S. A fixed point theorem for metric spaces. *Rend. Inst. Math. Univ. Trieste* **1976**, *8*, 69–72. [CrossRef]
14. Piri, H.; Rahrovi, S.; Kumam, P. Generalization of Khan fixed point theorem. *J. Math. Comput. Sci.* **2017**, *17*, 76–83. [CrossRef]
15. Piri, H.; Rahrovi, S.; Marasi, H.; Kumam, P. A fixed point theorem for F-Khan-contractions on complete metric spaces and application to integral equations. *J. Nonlinear Sci. Appl.* **2017**, *10*, 4564–4573. [CrossRef]
16. Wardowski, D. Fixed points of a new type of contractive mappings in complete metric spaces. *Fixed Point Theory Appl.* **2012**, *2012*, 94. [CrossRef]
17. Tomar, A.; Sharma, R. Some coincidence and common fixed point theorems concerning F-contraction and applications. *J. Int. Math. Virtual Inst.* **2018**, *8*, 181–198.
18. Kadelburg, Z.; Radenovic, S. Notes on some recent papers concerning Fcontractions in b-metric spaces. *Constr. Math. Anal.* **2018**, *1*, 108–112.
19. Satish, S.; Stojan, R.; Zoran, K. Some fixed point theorems for F-generalized contractions in 0-orbitally complete partial metric spaces. *Theory Appl. Math. Comput. Sci.* **2014**, *4*, 87–98.
20. Lukacs, A.; Kajanto, S. Fixed point therorems for various types of F-contractions in complete b-metric spaces. *Fixed Point Theory* **2018**, *19*, 321–334 [CrossRef]

8

Yukawa Potential, Panharmonic Measure and Brownian Motion

Antti Rasila [1,*] and Tommi Sottinen [2]

[1] Department of Mathematics and Systems Analysis, School of Science, Aalto University, P.O. Box 1100, FIN-00076 Aalto, Finland

[2] Department of Mathematics and Statistics, Faculty of Technology, University of Vaasa, P.O. Box 700, FIN-65101 Vaasa and Finland; tommi.sottinen@iki.fi

* Correspondence: antti.rasila@iki.fi;

Abstract: This paper continues our earlier investigation, where a walk-on-spheres (WOS) algorithm for Monte Carlo simulation of the solutions of the Yukawa and the Helmholtz partial differential equations (PDEs) was developed by using the Duffin correspondence. In this paper, we investigate the foundations behind the algorithm for the case of the Yukawa PDE. We study the panharmonic measure, which is a generalization of the harmonic measure for the Yukawa PDE. We show that there are natural stochastic definitions for the panharmonic measure in terms of the Brownian motion and that the harmonic and the panharmonic measures are all mutually equivalent. Furthermore, we calculate their Radon–Nikodym derivatives explicitly for some balls, which is a key result behind the WOS algorithm.

Keywords: potential theory; Brownian motion; Duffin correspondence; harmonic measure; Bessel functions; Monte Carlo simulation; panharmonic measure; walk-on-spheres algorithm; Yukawa equation

---

1. Introduction and Preliminaries

The harmonic measure is a fundamental tool in geometric function theory, and it has interesting applications in the study of bounded analytic functions, quasiconformal mappings and potential theory. For example, the harmonic measure has proven very useful in the study of quasidisks and related topics (see, e.g., [1–3]). Results involving the harmonic measure have been given by numerous authors since the 1930s (see [4] and references therein). In this paper, we consider the panharmonic measure, which is a natural counterpart of the classical harmonic measure, whereby the harmonic functions related are replaced with the smooth solutions to the *Yukawa equation:*

$$\Delta u(x) = \mu^2 u(x), \quad \mu^2 \geq 0. \tag{1}$$

Equation (1) first arose from the work of the Japanese physicist Hideki Yukawa in particle physics. Here, $u\colon D \to \mathbb{R}$ is a two-times differentiable function and $D \subset \mathbb{R}^n$, $n \geq 2$ is a domain. The Yukawa equation was first studied in order to describe the nuclear potential of a point charge. This model led to the concept of the Yukawa potential (also called a screened Coulomb potential), which satisfies an equation of the type given by Equation (1). The Yukawa equation also arises from certain problems related to optics (see [5]). Clearly, when $\mu = 0$, we have the Laplace equation, and, indeed, the results given in this paper reduce to the classical results.

Using the terminology of Duffin [6,7], we call a function $u\colon D \to \mathbb{R}$ *panharmonic*, or μ-*panharmonic*, in a domain D if its second derivatives are continuous and if it satisfies the Yukawa equation (Equation (1)) for all $x \in D$. The function u is called panharmonic at $x_0 \in D$ if there is a neighborhood of x_0 where u is panharmonic.

In Definition 1 of the panharmonic measure below, and in all that follows, we always assume that $n \geq 2$, although some results are also true in the dimension $n = 1$. For Definition 1, we need the notions of smallness and the regularity of a domain. These are best given by using the stochastic characterization via the Brownian motion. We refer to any of the classical textbooks [8–10] for further details.

We recall that the n-dimensional Brownian motion $W = (W(t); t \geq 0)$ starting from the point $x \in \mathbb{R}^n$ is the time-homogeneous Markov process with the Markov semigroup

$$P(t)f(x) = \mathbb{E}^x\left[f(W(t))\right]$$

given by

$$P(t) = e^{t\frac{1}{2}\Delta};$$

that is, $\frac{1}{2}\Delta$ is the generator of the Markov semigroup of the Brownian motion.

A domain $D \subset \mathbb{R}^n$ is *regular* if the Brownian motion does not dwell on its boundary; more precisely, D is (Wiener) regular if

$$\mathbb{P}^x\left[\tau_D = 0\right] = 1, \quad \text{for all } x \in \partial D,$$

where $\mathbb{P}^x$ is the probability measure under which $\mathbb{P}^x[W(0) = x] = 1$, and

$$\tau_D = \inf\{t > 0; W(t) \in D^c\}$$

is the first hitting time of the Brownian motion in the set D^c. We call a regular domain D (Wiener) *small* if a Brownian motion starting inside D eventually will leave the domain; that is, D is small if

$$\mathbb{P}^x\left[\tau_D < \infty\right] = 1, \quad \text{for all } x \in D.$$

For example, all bounded domains are small. All half-spaces are also small.

The panharmonic, or μ-panharmonic, measure is a generalization of the harmonic measure:

Definition 1. *Let $D \subset \mathbb{R}^n$ be a small regular domain, and let $\mu^2 \geq 0$. The μ-*panharmonic measure *on a boundary ∂D with a pole at $x \in D$ is the measure $H^x_\mu(D;\cdot)$ such that* any bounded *μ-panharmonic function u on $\bar{D}$ admits the representation*

$$u(x) = \int_{y\in\partial D} u(y)\, H^x_\mu(D; \mathrm{d}y). \tag{2}$$

The existence and uniqueness of a panharmonic measure is established by Theorem 1 and Corollary 2 later. Indeed, by Theorem 1, all *bounded* solutions to the Dirichlet problem $\Delta u - \mu^2 u = 0$ on a *small* regular domain with continuous and bounded boundary data are given by the panharmonic measure as in Equation (2). By Corollary 2, if $\mu^2 > 0$, then the assumption that the domain is small can be removed; that is, all bounded solutions on a regular domain are of the form given by Equation (2) if the boundary data is bounded and continuous. Of course, it is well known that there are unbounded solutions to the Laplace equation that do not admit the harmonic measure representation. The same is true for the Yukawa equation. We refer to Evans [11] for more details on the solutions of the Laplace equation.

We note that if we replace the "killing parameter" μ^2 in the Yukawa equation (Equation (1)) with a "creation parameter" $\lambda < 0$, we obtain another important partial differential equation (PDE), the

Helmholtz equation. In principle, the stochastic approaches taken in this paper can be applied to the solutions of the Helmholtz equation if the domain D is small enough compared to the parameter λ. For details, we refer to Chung and Zhao [8]. If we replace μ^2 by a (positive) function, we obtain the *Schrödinger equation*. Again, the stochastic approaches taken in this paper can be applied, in principle, to the Schrödinger equation, but the results may not be mathematically very tractable. Again, we refer to Chung and Zhao [8] for details. The rest of the paper is organized as follows: In Section 2, we show three different connections between the panharmonic measures and the Brownian motion. The first two (Theorem 1 and Corollary 1) are essentially well known. The third (Corollary 2) is new. In Section 3, we show that the panharmonic measures and the harmonic measures are all mutually equivalent (Theorem 3) and provide some corollaries; namely, we provide a domination principle for the Dirichlet problem related to the Yukawa equation (Corollary 3) and analogs of theorems of Riesz–Riesz, Makarov and Dahlberg for the panharmonic measures (Corollary 4). In Section 4, we consider the panharmonic measures on balls and prove an analogue of the Gauss mean value theorem, or the average property, for the panharmonic functions (Theorem 4), and as a corollary, we obtain the Liouville theorem for panharmonic functions (Corollary 5). Finally, in Section 5, we discuss extensions to the Schrödinger and the Helmholtz PDEs and the walk-on-spheres (WOS) simulation of PDEs.

2. Yukawa Equation and Brownian Motion

We first recall the celebrated connection between the harmonic measure and the Brownian motion first noticed by Kakutani [12] in the 1940s; the harmonic measure is the *hitting measure*:

$$H^x(D;\mathrm{d}y) = \mathbb{P}^x\left[W(\tau_D) \in \mathrm{d}y, \tau_D < \infty\right]. \tag{3}$$

Theorem 1 below is a variant of the Kakutani connection (Equation (3)). A key component of the variant is the following disintegration of the harmonic measure at the time the associated Brownian motion hits the boundary ∂D:

Lemma 1. *Let $D \subset \mathbb{R}^n$ be a regular domain, and let $x \in D$. Then*

$$H^x(D;\mathrm{d}y) = \int_{t=0}^{\infty} h^x(D;\mathrm{d}y,t)\,\mathrm{d}t,$$

where

$$h^x(D;\mathrm{d}y,t) = \mathbb{P}^x\left[W(\tau_D) \in \mathrm{d}y \mid \tau_D = t\right] \frac{\mathrm{d}\mathbb{P}^x}{\mathrm{d}t}\left[\tau_D \le t\right] \tag{4}$$

is the harmonic kernel.

Proof. First, we show the existence of the regular conditional distribution:

$$\mathfrak{p}^x(\mathrm{d}y \,|\, t) = \mathbb{P}^x\left[W(\tau_D) \in \mathrm{d}y \mid \tau_D = t\right]. \tag{5}$$

For this, we note that the random vector $(W(\tau_D), \tau_D)$ can be considered as a function from a space of continuous functions that are the Brownian trajectories equipped with the metric

$$d(f,g) = \sum_{T=1}^{\infty} 2^{-T} \left\| f\mathbf{1}_{[T-1,T)} - g\mathbf{1}_{[T-1,T)} \right\|_{\infty}.$$

For Brownian trajectories, the metric d is almost surely finite because of the independent increments of the Brownian motion and the Borel–Cantelli lemma. Additionally, with the metric d, the space of Brownian paths is a Polish space. Now, by Theorem A1.2 of [13], Polish spaces are Borel spaces. Consequently, for any fixed $x \in D$, by Theorems 6.3 and 6.4 of [13], the probability kernel (Equation (5)) exists and is measurable with respect to t. Consequently, the harmonic kernel is measurable with respect to t.

Second, we show that the distribution of the hitting time τ_D is absolutely continuous with respect to the Lebesgue measure. Let $\varepsilon > 0$ be small enough so that $B = B(x,\varepsilon) \subset D$. Then $\tau_D = \tau_B + (\tau_D - \tau_B)$. Now, the distribution of τ_B is absolutely continuous; see, for example, the section of Bessel processes in Borodin and Salminen [14]. Additionally, because of the rotation symmetry of the Brownian motion, τ_B and $\tau_D - \tau_B$ are independent. Hence, by disintegration and independence, we obtain that

$$\begin{aligned}\mathbb{P}^x[\tau_D \in \mathrm{d}t] &= \mathbb{P}^x[\tau_B + (\tau_D - \tau_B) \in \mathrm{d}t] \\ &= \int_{s=0}^{\infty} \mathbb{P}^x[t + (\tau_D - \tau_B) \in \mathrm{d}s \mid \tau_B = t]\, \mathbb{P}^x[\tau_B \in \mathrm{d}t] \\ &= \int_{s=0}^{\infty} \mathbb{P}^x[t + (\tau_D - \tau_B) \in \mathrm{d}s]\, \mathbb{P}^x[\tau_B \in \mathrm{d}t] \\ &= \varphi^x(t)\, \mathbb{P}^x[\tau_B \in \mathrm{d}t].\end{aligned}$$

Thus, the distribution of τ_D is absolutely continuous when the distribution of τ_B is absolutely continuous.

Third, we show that the formula given by Equation (4) holds. By disintegrating and conditioning, and by using the continuity of the distribution of τ_D, we obtain that

$$\begin{aligned}&\mathbb{P}^x\left[W(\tau_D) \in \mathrm{d}y, \tau_D < \infty\right] \\ &= \int_{t=0}^{\infty} \mathbb{P}^x\left[W(\tau_D) \in \mathrm{d}y, \tau_D \in \mathrm{d}t\right] \\ &= \int_{t=0}^{\infty} \mathbb{P}^x\left[W(\tau_D) \in \mathrm{d}y \mid \tau_D = t\right] \mathbb{P}^x\left[\tau_D \in \mathrm{d}t\right] \\ &= \int_{t=0}^{\infty} \mathbb{P}^x\left[W(\tau_D) \in \mathrm{d}y \mid \tau_D = t\right] \frac{\mathrm{d}\mathbb{P}^x}{\mathrm{d}t}\left[\tau_D \leq t\right] \mathrm{d}t.\end{aligned}$$

The claim follows now from the Kakutani connection (Equation (3)). □

The following Theorem 1 is a version of the Kakutani theorem [12] for the Yukawa equation. In some sense, it is a special case of the Kakutani connection to the Schrödinger equation studied extensively by Chung and Zhao [8]. However, it seems that this version with an unbounded and non-small domain D does not appear in any classical texts.

Theorem 1. *Let $D \subset \mathbb{R}^n$ be a regular domain, and let $f : \partial D \to \mathbb{R}$ be bounded and continuous.*

(i) *Then*

$$u(x) = \mathbb{E}^x\left[e^{-\frac{\mu^2}{2}\tau_D} f(W(\tau_D))\,;\, \tau_D < \infty\right] \tag{6}$$

is a solution to the Yukawa–Dirichlet problem:

$$\begin{cases} \Delta u = \mu^2 u & \text{on } D, \\ u = f & \text{on } \partial D. \end{cases}$$

(ii) *Moreover, if u is bounded and D is small, then Equation (6) is* the only *solution to the Yukawa–Dirichlet problem.*

(iii) *As a consequence, the panharmonic measure admits the representation*

$$H_\mu^x(D;\mathrm{d}y) = \int_{t=0}^{\infty} e^{-\frac{\mu^2}{2}t}\, h^x(D;\mathrm{d}y,t)\,\mathrm{d}t, \tag{7}$$

where $h^x(D;\cdot,\cdot)$ is the harmonic kernel defined in Equation (4).

Proof. The first and the second claim of Theorem 1 follow from the classical Kakutani theorem (cf., e.g., [15] Sections 4.4. and 4.6). Indeed, we note that the difficulties involving the Schrödinger equation in [15], Section 4.6 vanish, because

$$\mathbb{E}^x\left[e^{-\frac{\mu^2}{2}\tau_D}\right] \leq 1.$$

To show the third claim, we condition on $\{\tau_D = t\}$ and use the law of total probability:

$$\begin{aligned} u(x) &= \mathbb{E}^x\left[e^{-\frac{\mu^2}{2}\tau_D} f(W(\tau_D))\,;\, \tau_D < \infty\right] \\ &= \int_{y\in\partial D} f(y) \int_{t=0}^{\infty} e^{-\frac{\mu^2}{2}t}\, h^x(D;\mathrm{d}y,t)\,\mathrm{d}t \\ &= \int_{y\in\partial D} f(y)\, H^x_\mu(D;\mathrm{d}y). \end{aligned}$$

□

Remark 1. *Unfortunately, even for very simple D, the harmonic kernel (Equation (4)) is quite difficult to find. The same is true for the regular conditional distribution (Equation (5)). For smooth boundaries ∂D, one can try the following approach: If ∂D is smooth, then the harmonic kernel $h^x(D;\mathrm{d}y,t)$ is absolutely continuous with respect to the Lebesgue measure $\mathrm{d}y$. Indeed, define $p:\mathbb{R}_+\times\mathbb{R}^n\to\mathbb{R}_+$ by*

$$p(t,x) = \frac{1}{(2\pi t)^{n/2}}\exp\left(-\frac{\|x\|^2}{2t}\right). \tag{8}$$

Then p is the Brownian transition kernel*:*

$$p(t,x-y)\,\mathrm{d}y = \mathbb{P}^x\left[W(t)\in\mathrm{d}y\right],$$

and, as a result of [16], Theorem 1, the harmonic kernel can be written as

$$h^x(D;\mathrm{d}y,t) = \frac{1}{2}\frac{\partial p}{\partial \mathrm{n}_y}(D;t,x-y)\,\mathrm{d}y,$$

where n_y is the inward normal at $y\in\partial D$ and $p(D;\cdot,\cdot)$ is the transition density of a Brownian motion that is killed when it hits the boundary ∂D, which can be written as

$$p(D;t,x-y) = p(t,x-y) - \mathbb{E}^x\Big[p(t-\tau_D, W(\tau_D)-y)\,;\,\tau_D < t\Big] \tag{9}$$

as a result of [10], Equation (3), on page 34.

Consequently, for C^3 boundaries, the harmonic measure admits a Poisson kernel *representation, and therefore, as a result of the representation given by Equation (7), the panharmonic measure also admits a Poisson kernel representation:*

$$\begin{aligned} H^x_\mu(D;\mathrm{d}y) &= \int_{t=0}^{\infty} e^{-\frac{\mu^2}{2}t} h^x(D;\mathrm{d}y,t)\,\mathrm{d}t \\ &= \int_{t=0}^{\infty} e^{-\frac{\mu^2}{2}t}\frac{1}{2}\frac{\partial p}{\partial \mathrm{n}_y}(D;t,x-y)\,\mathrm{d}y\,\mathrm{d}t \\ &= \left[\frac{1}{2}\int_{t=0}^{\infty} e^{-\frac{\mu^2}{2}t}\frac{\partial p}{\partial \mathrm{n}_y}(D;t,x-y)\,\mathrm{d}t\right]\mathrm{d}y. \end{aligned}$$

Theorem 1 gives an interpretation of the panharmonic measure in terms of exponentially discounted Brownian motion. We give a second interpretation in terms of exponentially killed Brownian motion. Indeed, exponential discounting is closely related to exponential killing. The *exponentially killed Brownian motion* W_μ is

$$W_\mu(t) = W(t)\mathbf{1}_{\{Y_\mu > t\}} + \dagger \mathbf{1}_{\{Y_\mu \le t\}},$$

where $\dagger$ is a *coffin state* (by convention $f(\dagger) = 0$ for all functions f) and Y_μ is an independent exponential random variable with mean $2/\mu^2$; that is, $\mathbb{P}\left[Y_\mu > t\right] = e^{-\frac{\mu^2}{2}t}$. Let

$$\tau_D^\mu = \inf\left\{t > 0\,;\, W_\mu(t) \in D^c\right\}.$$

Then we have the following representation of the panharmonic measure:

Corollary 1. *Let $D \subset \mathbb{R}^n$ be a regular domain. Then the panharmonic measure admits the representation*

$$H_\mu^x(D; \mathrm{d}y) = \mathbb{P}^x\left[W_\mu(\tau_D^\mu) \in \mathrm{d}y\,;\, \tau_D^\mu < \infty\right]. \tag{10}$$

Proof. Let $f : \partial D \to \mathbb{R}$ be bounded. Then, by Theorem 1 and the independence of W and Y_μ,

$$\begin{aligned}
&\int_{y\in\partial D} f(y)\, H_\mu^x(D; \mathrm{d}y) \\
&= \mathbb{E}^x\left[e^{-\frac{\mu^2}{2}\tau_D} f\left(W(\tau_D)\right); \tau_D < \infty\right] \\
&= \int_{y\in\partial D} f(y) \int_{t=0}^\infty e^{-\frac{\mu^2}{2}t}\, \mathbb{P}^x\left[W(t) \in \mathrm{d}y, \tau_D \in \mathrm{d}t\right] \\
&= \int_{y\in\partial D} f(y) \int_{t=0}^\infty \mathbb{P}^x\left[Y_\mu > t\right] \mathbb{P}^x\left[W(t) \in \mathrm{d}y, \tau_D \in \mathrm{d}t\right] \\
&= \int_{y\in\partial D} f(y) \int_{t=0}^\infty \mathbb{P}^x\left[Y_\mu > t, W(t) \in \mathrm{d}y, \tau_D \in \mathrm{d}t\right] \\
&= \int_{y\in\partial D} f(y) \int_{t=0}^\infty \mathbb{P}^x\left[W_\mu(t) \in \mathrm{d}y, \tau_D^\mu \in \mathrm{d}t\right] \\
&= \mathbb{E}^x\left[f\left(W_\mu(\tau_D^\mu)\right); \tau_D^\mu < \infty\right].
\end{aligned}$$

Because f was arbitrary, the claim follows. □

The two representations, Theorem 1 and Corollary 1, for the panharmonic measures are, at least in spirit, classical. Now we give a third representation for the panharmonic measure in terms of an *escaping Brownian motion*. This representation is apparently new in spirit. The representation is due to the following *Duffin correspondence* [6]: Let $D \subset \mathbb{R}^n$ be a regular domain, and let $u : D \to \mathbb{R}$. Let $I \subset \mathbb{R}$ be any open interval that contains 0. Set $\bar{D} = D \times I$ and define $\bar{u} : \bar{D} \to \mathbb{R}$ by

$$\bar{u}(\bar{x}) = \bar{u}(x, \tilde{x}) = u(x)\cos(\mu\tilde{x}). \tag{11}$$

Theorem 2. *The function $\bar{u}$ defined by Equation (11) is harmonic on $\bar{D}$ if and only if u is μ-panharmonic on D.*

Proof. We first show that D is regular if and only if $\bar{D}$ is regular. Let $\bar{W} = (W, \tilde{W})$ be $(n+1)$-dimensional Brownian motion. Denote

$$\begin{aligned} \tau &= \inf\{t > 0;\, W(t) \in D^c\}, \\ \tilde{\tau} &= \inf\{t > 0;\, \tilde{W}(t) \in I^c\}, \\ \bar{\tau} &= \inf\{t > 0;\, \bar{W}(t) \in \bar{D}^c\}. \end{aligned}$$

We note that for $\{\tilde{\tau} = \tilde{x}\}$ to happen, $\tilde{x}$ must be an endpoint of the interval I. Then, by the independence of W and $\tilde{W}$,

$$\begin{aligned} \mathbb{P}^{x,\tilde{x}}[\bar{\tau} = 0] &= \mathbb{P}^{x,\tilde{x}}[\tau = 0, \tilde{\tau} = 0] \\ &= \mathbb{P}^{x}[\tau = 0]\mathbb{P}^{\tilde{x}}[\tilde{\tau} = 0] \\ &= \mathbb{P}^{x}[\tau = 0], \end{aligned}$$

because I is clearly regular. This shows that $\bar{D}$ is regular if and only if D is regular.

We then show that u satisfies the Laplace equation if and only if $\bar{u}$ satisfies the Yukawa equation; this is straightforward calculus:

$$\begin{aligned} \Delta_{\bar{x}}\bar{u}(\bar{x}) &= \Delta_{x,\tilde{x}}\left[u(x)\cos(\mu\tilde{x})\right] \\ &= \cos(\mu\tilde{x})\Delta_x u(x) + u(x)\frac{\mathrm{d}^2}{\mathrm{d}\tilde{x}^2}\cos(\mu\tilde{x}) \\ &= \cos(\mu\tilde{x})\Delta_x u(x) - \mu^2\cos(\mu\tilde{x})u(x) \\ &= \cos(\mu\tilde{x})\left(\Delta_x u(x) - \mu^2 u(x)\right) \\ &= 0 \end{aligned}$$

if and only if $\Delta_x u(x) = \mu^2 u(x)$. □

Let $\tilde{W}$ be a 1-dimensional standard Brownian motion that is independent of W. Then $\bar{W} = (W, \tilde{W})$ is a $(n+1)$-dimensional standard Brownian motion.

Now the idea of how to use the Duffin correspondence is clear. We start the Brownian particle $\bar{W}$ and count the boundary data on the side of the cylinder $\bar{D} = D \times I$, if the Brownian motion does not escape the cylinder from the bottom or from the top. In that case we count zero in the boundary; whence the name *escaping Brownian motion*.

Corollary 2. *Let $D \subset \mathbb{R}^n$ be a regular domain. Then the panharmonic measure admits the representation*

$$\begin{aligned} &H_\mu^x(D; \mathrm{d}y) \\ &= \mathbb{E}^{x,0}\left[\cos\left(\mu\tilde{W}(\tau_D)\right); W(\tau_D) \in \mathrm{d}y, \sup_{t \le \tau_D}|\tilde{W}(t)| < \frac{\pi}{2\mu}\right] \\ &= \int_{\tilde{y}=-\frac{\pi}{2\mu}}^{\frac{\pi}{2\mu}} \cos\left(\mu\tilde{y}\right) H^{x,0}\left(D \times \left(-\frac{\pi}{2\mu}, \frac{\pi}{2\mu}\right); \mathrm{d}y \otimes \mathrm{d}\tilde{y}\right). \end{aligned} \tag{12}$$

Here we have chosen $I = (-\frac{\pi}{2\mu}, \frac{\pi}{2\mu})$ in the Duffin correspondence.

Consequently, all bounded solutions to the Yukawa–Dirichlet problem on a regular domain with $\mu^2 > 0$ and continuous and bounded boundary data are given by the panharmonic measure.

Proof. The claim follows by combining the Kakutani connection (Equation (3)) with the Duffin correspondence (Equation (11)) by noticing that it is enough to integrate over $\partial D \times (-\pi/(2\mu), \pi/(2\mu))$, as $\cos(\mu\tilde{y}) = 0$ on the boundary $\partial(-\pi/(2\mu), \pi/(2\mu))$.

Finally, we note that for a regular domain D, the domain $\bar{D}$ is regular and small. □

Remark 2. *Equation (12) is exceptionally well suited for calculations of the panharmonic measures on upper half-spaces* $\mathbb{H}^n_+ = \{x \in \mathbb{R}^n; x_n > 0\}$. *Indeed, Duffin ([6], Theorem 5) used it to calculate the Poisson kernel representation for panharmonic measures in the dimension* $n = 2$. *Similar calculations can also be carried out for the general case,* $n \geq 2$.

3. Equivalence of Harmonic and Panharmonic Measures

The probabilistic interpretation provided by Corollary 1 implies that the harmonic measure and the panharmonic measures are equivalent. Indeed, the harmonic measure counts the Brownian particles on the boundary, and the panharmonic measures count the killed Brownian particles on the boundary. However the killing happens with independent exponential random variables. Thus, if the Brownian motion can reach the boundary with positive probability, so can the killed Brownian motion, and vice versa. Additionally, it does not matter, as far as the equivalence is concerned, what the starting point is of the Brownian motion, killed or not.

Theorem 3 below makes the heuristics above precise. As corollaries of Theorem 3, we obtain a domination principle for the Dirichlet problem related to the Yukawa equation (Corollary 3) and analogs of theorems of Riesz–Riesz, Makarov and Dahlberg for the panharmonic measures (Corollary 4).

The same arguments that give the existence of the regular conditional law (Equation (5)) in the proof of Lemma 1 also give the existence and regular measurability of the following conditional Radon–Nikodym derivative:

$$Z^x_\mu(D;y) = \mathbb{E}^x\left[e^{-\frac{\mu^2}{2}\tau_D}\,\middle|\,W(\tau_D) = y\right]. \tag{13}$$

Theorem 3. *Let* D *be a regular domain. Then all the panharmonic measures* $H^x_\mu(D;\cdot)$, $\mu \geq 0, x \in D$ *are mutually equivalent. The Radon–Nikodym derivative of* $H^x_\mu(D;\cdot)$ *with respect to* $H^x(D;\cdot)$ *is the function* $Z^x_\mu(D;\cdot)$ *given by Equation (13). Moreover* $Z^x_\mu(D;y)$ *is strictly decreasing in* μ, *and* $0 < Z^x_\mu(D;y) \leq 1$.

Remark 3. *By Corollary 1, the Radon–Nikodym derivative* $Z^x_\mu(D;\cdot)$ *in Equation (13) can be interpreted as the probability that a Brownian motion killed with intensity* $\mu^2/2$, *and that would exit the domain* D *at* $y \in \partial D$, *survives to the boundary* ∂D:

$$Z^x_\mu(D;y) = \mathbb{P}^x\left[Y_\mu > \tau_D \,|\, W(\tau_D) = y\right], \tag{14}$$

where Y_μ *is an exponentially distributed random variable with mean* $2/\mu^2$ *that is independent of the Brownian motion* W.

Proof of Theorem 3. Let $x, y \in D$, and let $D_0 \subset D$ be a subdomain of D such that $x \in D_0$ and $y \in \partial D_0$. Then, by the Markov property of the Brownian motion and the Kakutani connection (Equation (3)), we have

$$H^x(D;A) = \int_{y \in \partial D_0} H^y(D;A) H^x(D_0;\mathrm{d}y)$$

for all measurable $A \subset \partial D$. This shows that the harmonic measures $H^x(D;\cdot)$, $x \in D$ are mutually equivalent.

To see that $Z^x_\mu(D;\cdot)$ is the Radon–Nikodym derivative, we note that, by the representation given by Equation (7) and the Kakutani connection (Equation (3)),

$$\begin{aligned} H^x_\mu(D;\mathrm{d}y) &= \int_{t=0}^{\infty} e^{-\frac{\mu^2}{2}t}\, h^x(D;\mathrm{d}y,t)\,\mathrm{d}t \\ &= \int_{t=0}^{\infty} e^{-\frac{\mu^2}{2}t}\, \mathbb{P}^x\left[W(\tau_D)\in \mathrm{d}y, \tau_D\in \mathrm{d}t\right] \\ &= \int_{y\in\partial D} \mathbb{E}^x\left[e^{-\frac{\mu^2}{2}\tau_D}\,\Big|\, W(\tau_D)=y\right]\mathbb{P}^x\left[W(\tau_D)\in \mathrm{d}y\right] \\ &= \int_{y\in\partial D} Z^x_\mu(D;y)\, H^x(D;\mathrm{d}y). \end{aligned}$$

Finally, the fact that $0 < Z^x_\mu(D;\cdot) \le 1$ is clear from the representation given by Equation (13). The fact that $Z^z_\mu(D;\cdot)$ is strictly decreasing follows immediately from the representation given by Equation (14). □

From Theorem 3, we obtain immediately the following *domination principle* for the Dirichlet problem related to panharmonic functions:

Corollary 3. *Let D be a regular domain, and let $u_\mu \ge 0$ be μ-panharmonic and $u_\nu \ge 0$ be ν-panharmonic, respectively, on D with $\mu \le \nu$. Then, $u_\nu \le u_\mu$ on ∂D implies $u_\nu \le u_\mu$ on D.*

Because domains with a rectifiable boundary are regular, we obtain immediately from Theorem 3 the following analogs of the theorems of F. Riesz and M. Riesz, Makarov and Dalhberg (see [17–19], respectively).

Corollary 4. *Let $\mathcal{H}^s(D;\cdot)$ be the s-dimensional Hausdorff measure on ∂D.*

(i) *Let $D\subset\mathbb{R}^2$ be a simply connected planar domain bounded by a rectifiable curve. Then $H^x_\mu(D;\cdot)$ and $\mathcal{H}^1(D;\cdot)$ are equivalent for all $\mu\ge 0$ and $x\in D$.*
(ii) *Let $D\subset\mathbb{R}^2$ be a simply connected planar domain. If $E\subset\partial D$ and $\mathcal{H}^s(D;E)=0$ for some $s<1$, then $H^x_\mu(D;E)=0$ for all $\mu\ge 0$ and $x\in D$. Moreover, $H^x_\mu(D;\cdot)$ and $\mathcal{H}^t(D;\cdot)$ are singular for all $\mu\ge 0$ and $x\in D$ if $t>1$.*
(iii) *Let $D\subset\mathbb{R}^n$ is a bounded Lipschitz domain. Then $H^x_\mu(D;\cdot)$ and $\mathcal{H}^{n-1}(D;\cdot)$ are equivalent for all $\mu\ge 0$ and $x\in D$.*

4. The Average Property for Panharmonic Measures and Functions

By using the representation given by Equation (7), one can calculate the panharmonic measures if one can calculate the corresponding harmonic kernels; or, equivalently, one can calculate the panharmonic measures if one can calculate the corresponding harmonic measures and the Radon–Nikodym derivatives given by Equation (13).

The harmonic kernels for balls are calculated in [16]. We do not present the general formula here. Instead, we confine ourselves to the case in which the center of the ball and the pole of the panharmonic measure coincide, and give the Gauss mean value theorem, or the average property, for panharmonic measures. As a corollary, we have the Liouville theorem for the panharmonic measures.

Let $D\subset\mathbb{R}^n$ be a regular domain. For the harmonic measure, the Gauss mean value theorem states that a function $u: D\to\mathbb{R}$ is harmonic if and only if for all balls $B_n(x,r)\subset D$ we have the *average property*:

$$u(x) = \int_{y\in\partial B_n(x,r)} u(y)\,\sigma_n(r;\mathrm{d}y),$$

where

$$\sigma_n(r; \mathrm{d}y) = \frac{\Gamma(n/2)}{2\pi^{n/2}} r^{1-n}\,\mathrm{d}y$$

is the uniform probability measure on the sphere $\partial B_n(x,r)$.

For the panharmonic measures, the situation is similar to the harmonic measure: the only difference is that the uniform probability measure has to be replaced by a uniform sub-probability measure that depends on the killing parameter μ and the radius of the ball r. Indeed, we denote

$$\psi_n(\mu) = \frac{\mu^\nu}{2^\nu \Gamma(\nu+1) I_\nu(\mu)}, \quad \mu > 0, \tag{15}$$

where $\nu = (n-2)/2$, and

$$I_\nu(x) = \sum_{m=0}^{\infty} \frac{1}{m!\Gamma(m+\nu+1)} \left(\frac{x}{2}\right)^{2m+\nu}$$

is the modified Bessel function of the first kind of order ν.

Theorem 4. *Let $D \subset \mathbb{R}^n$ be a regular domain, and let $\mu > 0$. A function $u : D \to \mathbb{R}$ is μ-panharmonic if and only if it has the average property*

$$u(x) = \psi_n(\mu r) \int_{y \in \partial B_n(x,r)} u(y)\, \sigma_n(r; \mathrm{d}y).$$

for all open balls $B_n(x,r) \subset D$. Equivalently,

$$H_\mu^x\left(B_n(x,r); \mathrm{d}y\right) = \psi_n(\mu r)\, \sigma_n(r; \mathrm{d}y).$$

Remark 4. *Theorem 4 states that $\psi_n(\mu r)$ is the Radon–Nikodym derivative:*

$$\psi_n(\mu r) = Z_\mu^x\left(B_n(x,r); y\right) = \mathbb{E}^x\left[e^{-\frac{\mu^2}{2}\tau_{B_n(x,r)}} \middle| W\left(\tau_{B_n(x,r)}\right) = y\right].$$

Proof of Theorem 4. We note that we may assume $x = 0$.

Denote by τ_r^n the first hitting time of the Brownian motion W on the boundary $\partial B_n(0,r)$; that is, τ_r^n is identical in law to the first hitting time of the Bessel process with index $\nu = (n-2)/2$ reaching the level r when it starts from zero.

From the rotation symmetry of the Brownian motion, it follows that the hitting place is uniformly distributed on $\partial B_n(0,r)$ for all hitting times t. Consequently, by Theorem 1 and the independence of the hitting time τ_r^n and place $W(\tau_r^n)$,

$$\begin{aligned} H_\mu^0\left(B_n(0,r); \mathrm{d}y\right) &= \mathbb{E}^0\left[e^{-\frac{\mu^2}{2}\tau_r^n}; W(\tau_r^n) \in \mathrm{d}y\right] \\ &= \mathbb{E}^0\left[e^{-\frac{\mu^2}{2}\tau_r^n}\right] \mathbb{P}^0\left[W(\tau_r^n) \in \mathrm{d}y\right] \\ &= \mathbb{E}^0\left[e^{-\frac{\mu^2}{2}\tau_r^n}\right] \sigma_n(r; \mathrm{d}y). \end{aligned}$$

The hitting-time distributions for the Bessel process are well known. By, for example, Wendel ([20], Theorem 4),

$$\mathbb{E}^0\left[e^{-\frac{\mu^2}{2}\tau_r^n}\right] = \frac{(\mu r)^\nu}{2^\nu \Gamma(\nu+1) I_\nu(\mu r)}.$$

The claim follows from this. □

Remark 5. *The Radon–Nikodym derivative, or the "killing constant", $\psi_n(\mu)$, is rather complicated. However, some of its properties are easy to see (cf. Figure 1):*

(i) $\psi_n(\mu)$ *is continuous in* μ,
(ii) $\psi_n(\mu)$ *is strictly decreasing in* μ,
(iii) $\psi_n(\mu) \to 0$ *as* $\mu \to \infty$,
(iv) $\psi_n(\mu) \to 1$ *as* $\mu \to 0$,
(v) $\psi_n(\mu)$ *is increasing in* n.

The items (i)–(iv) are clear, because $\psi_n(\mu)$ is the probability that an exponentially killed Brownian motion starting from the origin and with killing intensity $\mu^2/2$ is not killed before it hits the boundary of the unit ball. A non-probabilistic argument for (i)–(iv) is to note that

$$\psi_n(\mu r) = \mathbb{E}^0\left[e^{-\frac{\mu^2}{2}\tau_r^n}\right]$$

and to use the monotone convergence. The item (v) is somewhat surprising: the higher the dimension n, the more likely it is for the killed Brownian motion to survive to the boundary of the unit ball. A possible intuitive explanation is that the higher the dimension, the more transitive the unit ball is, combined with the remarkable result by Ciesielski and Taylor [21] that the probability distribution for the total time spent in a ball by $(n+2)$-dimensional Brownian motion is the same as the probability distribution of the hitting time of n-dimensional Brownian motion on the boundary of the ball.

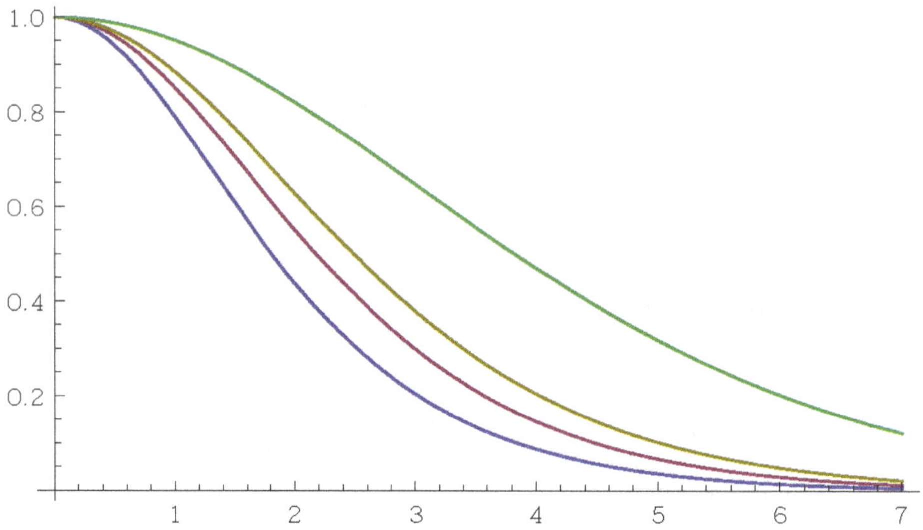

Figure 1. Function ψ_n with (from bottom to top) $n = 2, 3, 4, 10$.

Corollary 5. *Let $\mu > 0$, and let u be μ-panharmonic on the entire space $\mathbb{R}^n$. If u is bounded, then $u \equiv 0$.*

Proof. By Theorem 4,

$$\begin{aligned}
&|u(x) - u(0)| \\
&= \left|\psi_n(\mu r)\int_{\partial B_n(x,r)} u(y)\,\sigma_n(r;\mathrm{d}y) - \psi_n(\mu r)\int_{\partial B_n(0,r)} u(y)\,\sigma_n(r;\mathrm{d}y)\right| \\
&\leq \left|\psi_n(\mu r)\int_{\partial B_n(x,r)} u(y)\,\sigma_n(r;\mathrm{d}y)\right| + \left|\psi_n(\mu r)\int_{\partial B_n(0,r)} u(y)\,\sigma_n(r;\mathrm{d}y)\right| \\
&\leq 2\psi_n(\mu r)\|u\|_\infty,
\end{aligned}$$

which tends to 0 as $r \to \infty$ by property (iii) of Remark 5. This shows that u is constant. However, for a constant u, the Yukawa equation (Equation (1)) yields $0 = \mu^2 u$, which implies that $u \equiv 0$. $\square$

5. Discussion on Extensions and Simulation

The Yukawa equation (Equation (1)) is a special case of the Schrödinger equation:

$$\Delta u(x) = q(x)u(x). \tag{16}$$

The Schrödinger equation and its connection to the Brownian motion has been studied, for example, by Chung and Zhao [8]. Our investigation here can be seen as a special case. For example, analogs of Theorem 1 and Corollary 1 are known for the Schrödinger equation. However, analogs of the Duffin correspondence (Equation (11)) and Corollary 2 are not known even to exist. Moreover, the results given here cannot easily be calculated for the Schrödinger equation. The problem is that the prospective Radon–Nikodym derivate of the measure associated with the solutions of the Schrödinger equation with respect to the harmonic measure takes the form

$$Z_q^x(D;y) = \mathbb{E}^x\left[e_q(\tau_D)\big|W(\tau_D) = y\right], \tag{17}$$

where

$$e_q(t) = e^{-\frac{1}{2}\int_0^t q(W(s))\,\mathrm{d}s}$$

is the *Feynman–Kac functional*. Thus, we see that in order to calculate the Radon–Nikodym derivative, we need to know the joint density of the Feynman–Kac functional and the Brownian motion when the Brownian motion hits the boundary ∂D. If q is constant, that is, we have either the Yukawa equation or the Helmholtz equation, then it is enough to know the joint distribution of the hitting time and place of the Brownian motion on the boundary ∂D. These distributions are well studied (see, e.g., [14,16,21–23]), but few joint distributions involving the Feynman–Kac functionals are known.

In addition to the Yukawa equation, the other important special case of the Schrödinger equation (Equation (16)) is the Helmholtz equation:

$$\Delta u(x) = -\lambda u(x), \quad \lambda \geq 0. \tag{18}$$

It is possible to also provide a Duffin correspondence for the Helmholtz equation. Indeed, for example, setting

$$\bar{u}(\bar{x}) = \bar{u}(x, \tilde{x}) = u(x)\cosh(\lambda\tilde{x})$$

provides a correspondence (see [24] for details). Thus, our results extend in a straightforward manner to the Helmholtz equation (Equation (18)) for domains that are small enough with respect to the creation parameter λ that the associated Feynman–Kac functional is finite:

$$\mathbb{E}^x\left[e^{\frac{\lambda}{2}\tau_D}\right] < \infty. \tag{19}$$

Finally, we note that Theorem 1, Corollary 1 and Corollary 2 give three different ways to simulate the panharmonic measures. Indeed, in [24,25], the classical WOS algorithm given by Muller [26] was extended for the Yukawa PDE, and also for the Helmholtz PDE, by using the results mentioned above.

Author Contributions: Antti Rasila was responsible for the general idea of connecting the Yukawa equation to the harmonic measure and the Brownian motion through the Duffin correspondence, connections to classical analysis and potential theory, and computer experiments with Mathematica. Tommi Sottinen was responsible for the proofs and methodologies making use of stochastic analysis. Much of the paper was shaped in discussions between the authors.

Acknowledgments: T. Sottinen was partially funded by the Finnish Cultural Foundation (National Foundations' Professor Pool).

Reference

1. Bishop, C.J.; Jones, P.W. Harmonic measure and arclength. *Ann. Math.* **1990**, *132*, 511–547. [CrossRef]
2. Gehring, F.W.; Hag, K. *The Ubiquitous Quasidisk*; No. 184.; American Mathematical Society: Providence, RI, USA, 2012.
3. Krzyż, J.G. Quasicircles and harmonic measure. *Ann. Acad. Sci. Fenn.* **1987**, *12*, 19–24. [CrossRef]
4. Garnett, J.B.; Marshall, D.E. *Harmonic Measure*; Cambridge University Press: Cambridge, UK, 2005.
5. Harrach, B. On uniqueness in diffuse optical tomography. *Inverse Probl.* **2009**, *25*, 1–14. [CrossRef]
6. Duffin, R.J. Yukawan potential theory. *J. Math. Anal. Appl.* **1971**, *35*, 105–130. [CrossRef]
7. Duffin, R.J. Hilbert transforms in Yukawan potential theory. *Proc. Nat. Acad. Sci. USA* **1972**, *69*, 3677–3679. [CrossRef] [PubMed]
8. Chung, K.L.; Zhao, Z. *From Brownian Motion to Schrödinger's Equation*, 2nd ed.; Springer: Berlin/ Heidelberg, Germany, 2001.
9. Doob, J.L. *Classical Potential Theory and Its Probabilistic Counterpart*; Grundlehren der Mathematischen Wissenschaften 262; Springer: New York, NY, USA, 1984.
10. Port, S.C.; Stone, C.J. *Brownian Motion and Classical Potential Theory*; Academic Press: New York, NY, USA, 1978.
11. Evans, L.C. *Partial Differential Equations (Graduate Studies in Mathematics)*, 2nd ed.; American Mathematical Society: Providence, RI, USA, 2010.
12. Kakutani, S. On Brownian motion in n-space. *Proc. Imp. Acad.* **1944**, *20*, 648–652. [CrossRef]
13. Kallenberg, O. *Foundations of Modern Probability*, 2nd ed.; Springer: New York, NY, USA, 2002.
14. Borodin, A.; Salminen, P. *Handbook of Brownian Motion—Facts and Formulae*, 2nd ed.; Birkhäuser: Basel, Switzerland, 2002.
15. Durrett, R. *Stochastic Calculus: A Practical Introduction*; CRC Press: Boca Raton, FL, USA, 1996.
16. Hsu, P. Brownian exit distribution of a ball. In *Seminar on Stochastic Processes 1985*; Birkhäuser: Boston, MA, USA, 1986.
17. Dahlberg, B. Estimates of harmonic measure. *Arch. Rat. Mech. Anal.* **1977**, *65*, 275–288. [CrossRef]
18. Makarov, N.G. On the Distortion of Boundary Sets Under Conformal Maps. *Proc. Lond. Math. Soc.* **1985**, *52*, 369–384. [CrossRef]
19. Riesz, F.; Riesz, M. Über die Randwerte einer analytischen Funktion. In *Quatrième Congrès des Mathématiciens Scandinaves*; Stockholm, Sweden, 1916; pp. 27–44.
20. Wendel, J.G. Hitting Spheres with Brownian Motion. *Ann. Probab.* **1980**, *8*, 164–169. [CrossRef]
21. Ciesielski, Z.; Taylor, S.J. First passage times and sojourn times for Brownian motion in space and the exact Hausdorff measure of the sample path. *Trans. Am. Math. Soc.* **1962**, *103*, 434–450. [CrossRef]
22. Kent, T. Eigenvalue expansion for diffusion hitting times. *Z. Wahrscheinlichkeitstheorie Ver. Gebiete* **1980,** *52*, 309–319. [CrossRef]
23. Lévy, P. La mesure de Hausdorff de la courbe du mouvement brownien. *Giorn. Ist. Ital. Attuari* **1953**, *16*, 1–37.
24. Yang, X.; Rasila, A.; Sottinen, T. Walk On Spheres Algorithm for Helmholtz and Yukawa Equations via Duffin Correspondence. *Methodol. Comput. Appl. Probab.* **2017**, *19*, 589–602. [CrossRef]
25. Yang, X.; Rasila, A.; Sottinen, T. Efficient simulation of Schrödinger equation with piecewise constant positive potential. *arXiv* **2015**, arXiv:1512.01306.
26. Muller, M.E. Some continuous Monte Carlo methods for the Dirichlet problem. *Ann. Math. Stat.* **1956**, *27*, 569–589. [CrossRef]

9

Approximation Properties of an Extended Family of the Szász–Mirakjan Beta-Type Operators

Hari Mohan Srivastava [1,2,3,*], Gürhan İçöz [4] and Bayram Çekim [4]

1 Department of Mathematics and Statistics, University of Victoria, Victoria, BC V8W 3R4, Canada
2 Department of Medical Research, China Medical University Hospital, China Medical University, Taichung 40402, Taiwan
3 Department of Mathematics and Informatics, Azerbaijan University, 71 Jeyhun Hajibeyli Street, AZ1007 Baku, Azerbaijan
4 Department of Mathematics, Gazi University, Ankara TR-06500, Turkey; gurhanicoz@gazi.edu.tr (G.İ.); bayramcekim@gazi.edu.tr (B.Ç.)
* Correspondence: harimsri@math.uvic.ca

Abstract: Approximation and some other basic properties of various linear and nonlinear operators are potentially useful in many different areas of researches in the mathematical, physical, and engineering sciences. Motivated essentially by this aspect of approximation theory, our present study systematically investigates the approximation and other associated properties of a class of the Szász-Mirakjan-type operators, which are introduced here by using an extension of the familiar Beta function. We propose to establish moments of these extended Szász-Mirakjan Beta-type operators and estimate various convergence results with the help of the second modulus of smoothness and the classical modulus of continuity. We also investigate convergence via functions which belong to the Lipschitz class. Finally, we prove a Voronovskaja-type approximation theorem for the extended Szász-Mirakjan Beta-type operators.

Keywords: gamma and beta functions; Szász-Mirakjan operators; Szász-Mirakjan Beta type operators; extended Gamma and Beta functions; confluent hypergeometric function; Modulus of smoothness; modulus of continuity; Lipschitz class; local approximation; Voronovskaja type approximation theorem

---

1. Introduction, Definitions and Preliminaries

In approximation theory and its related fields, approximation and other basic properties of various linear and nonlinear operators are investigated, because mainly of the potential for their usefulness in many areas of researches in the mathematical, physical, and engineering sciences. Our study in this article is motivated essentially by the demonstrated applications of such results as those associated with various approximation operators. With this objective in view, we begin by providing the following definitions and other (chiefly historical) background material related to our presentation here.

For a given continuous function, $f \in C[0,\infty)$, and for $x \in [0,\infty)$, Otto Szász [1] defined a family of operators in the year 1950, which we recall here as follows,

$$S_n(f;x) = e^{-nx} \sum_{k=0}^{\infty} \frac{(nx)^k}{k!} f\left(\frac{k}{n}\right) \qquad (n \in \mathbb{N} := \{1,2,3,\cdots\}), \tag{1}$$

This family of operators was considered earlier in 1941 by G. M. Mirakjan (see [2]). There are several integral and other modifications, variations, and basic (or q-) extensions of the Szász-Mirakjan-type operators. These include the Bézier, Kantorovich, Durrmeyer, and other types of modifications and extensions of the Szász-Mirakjan operators (see, for details, [3–15]). In particular, Gupta and Noor [6] introduced an integral modification of the Szász-Mirakjan operators in Equation (1) by considering a weight function in terms of the Beta basis functions as given below.

$$T_n(f;x) = \sum_{k=1}^{\infty} s_{n,k}(x) \int_0^{\infty} b_{n,k}(t)\ f(t)\ dt + s_{n,0}(x)\, f(0), \tag{2}$$

where

$$s_{n,k}(x) = e^{-nx}\,\frac{(nx)^k}{k!},$$

$$b_{n,k}(t) = \frac{1}{B(k,n+1)}\,\frac{t^{k-1}}{(1+t)^{n+k+1}} = \frac{(n+1)_k}{(k-1)!}\,\frac{t^{k-1}}{(1+t)^{n+k+1}},$$

$$B(\alpha,\beta) = \frac{\Gamma(\alpha)\Gamma(\beta)}{\Gamma(\alpha+\beta)} = B(\beta,\alpha),$$

and $(\lambda)_\ell$ $(\ell \in \mathbb{N}_0 := \mathbb{N} \cup \{0\})$ represents the Pochhammer symbol given by

$$(\lambda)_\ell := \frac{\Gamma(\lambda+\ell)}{\Gamma(\lambda)} = \begin{cases} 1 & (\ell = 0) \\ \lambda(\lambda+1)(\lambda+2)\cdots(\lambda+\ell-1) & (\ell \in \mathbb{N}) \end{cases}$$

in terms of the classical (Euler's) Gamma function $\Gamma(z)$ and the classical Beta function $B(\alpha,\beta)$.

Gupta and Noor [6] observed that the operators in Equation (2) reproduce not only the constant function, but linear functions as well. Owing to this valuable property of the operators in Equation (2), many authors investigated the different approximation properties of the summation-integral operators in Equation (2) (see, for example, [16–18]). Gupta and Noor [6] also derived some direct results for the operators T_n, a pointwise rate of convergence, a Voronovskaja-type asymptotic formula, and an error estimate in simultaneous approximation.

In recent years, some extensions of such well-known special functions as, for example, the classical Gamma and Beta functions, have been considered by several authors. For example, in 1994, Chaudhry and Zubair [19] introduced the following extension of the Gamma function,

$$\Gamma_p(x) := \int_0^{\infty} t^{x-1} \exp\left(-t - \frac{p}{t}\right) dt \qquad (\Re(p) > 0). \tag{3}$$

Subsequently, in 1997, Chaudhry et al. [20] presented the following extension of Euler's Beta function,

$$B_p(x,y) := \int_0^1 t^{x-1}\,(1-t)^{y-1} \exp\left(-\frac{p}{t\,(1-t)}\right) dt \tag{4}$$

$$(\Re(p) > 0;\ \Re(x) > 0;\ \Re(y) > 0).$$

Obviously, each of the definitions in Equations (3) and (4) also remains valid when $p = 0$, in which case we have the following relationships,

$$\Gamma_0(x) = \Gamma(x) \qquad \text{and} \qquad B_0(x,y) = B(x,y).$$

Özergin et al. [21] considered the following generalizations of the Gamma and Beta functions,

$$\Gamma_p^{(\alpha,\beta)}(x) := \int_0^\infty t^{x-1}\, {}_1F_1\left(\alpha;\beta;-t-\frac{p}{t}\right) dt$$

$$(\Re(\alpha)>0;\ \Re(\beta)>0;\ \Re(p)>0;\ \Re(x)>0)$$

and

$$B_p^{(\alpha,\beta)}(x,y) := \int_0^1 t^{x-1}\,(1-t)^{y-1}\, {}_1F_1\left(\alpha;\beta;-\frac{p}{t(1-t)}\right) dt$$

$$(\Re(\alpha)>0;\ \Re(\beta)>0;\ \Re(p)>0;\ \Re(x)>0;\ \Re(y)>0),$$

respectively. Here, as usual, ${}_1F_1$ denotes the (Kummer's) confluent hypergeometric function. It is obvious that

$$\Gamma_p^{(\alpha,\alpha)}(x) = \Gamma_p(x) \qquad \text{and} \qquad \Gamma_0^{(\alpha,\alpha)}(x) = \Gamma(x)$$

and that

$$B_p^{(\alpha,\alpha)}(x,y) = B_p(x,y) \qquad \text{and} \qquad B_0^{(\alpha,\alpha)}(x,y) = B(x,y).$$

Finding different integral representations of the generalized Beta function is important and useful for later use. It is also useful to discuss the relationships between the classical Gamma and Beta functions and their generalizations. In fact, by definition, it is easily seen that

$$\int_0^\infty \Gamma_p(x)\, dp = \Gamma(x+1) \qquad (\Re(x)>-1).$$

and that (see, for example, [21])

$$\int_0^\infty B_p(x,y)\, dp = B(x+1,y+1) \tag{5}$$

$$(\Re(x)>-1;\ \Re(y)>-1).$$

Note that various further extensions and generalizations of the classical Gamma and Beta functions, as well as their corresponding hypergeometric and related functions, were introduced and studied by, among others, Lin et al. [22] and Srivastava et al. [23].

We now introduce the following generalization of Szász-Mirakjan Beta-type operators via the above extension of the Beta function as follows,

$$\begin{aligned} S_n^*(f,x) &= \sum_{k=0}^\infty e^{-nx}\frac{(nx)^k}{k!}\frac{1}{B(k+2,n+1)} \\ &\quad \cdot \int_0^\infty\int_0^\infty \frac{t^k}{(1+t)^{n+k+1}}\exp\left(-\frac{p(1+t)^2}{t}\right) f(t)\, dt\, dp \end{aligned} \tag{6}$$

for $x \in [0,\infty)$, and for a function $f \in C_\nu[0,\infty)$, provided that the double integral in Equation (6) is convergent when $n>\nu$. Here, and in what follows, we have

$$C_\nu[0,\infty) := \{f : f \in C[0,\infty) \quad \text{and} \quad |f(t)| \leqq M(1+t)^\nu \quad (M>0;\ \nu>0)\}.$$

We note that, by setting $t = \frac{u}{1+u}$ in Equation (4), we get

$$B_p(x,y) = \int_0^\infty \frac{u^{x-1}}{(1+u)^{x+y}} \exp\left(-\frac{p(1+u)^2}{u}\right) du. \tag{7}$$

So, if we take in consideration Equations (5) and (7) in the definition Equation (6), then we can say that the operators, S_n^*, are a generalization of the operators, T_n, given by Equation (2).

In this article, we investigate the moments of the general Szász-Mirakjan Beta-type operators S_n^* and find the rate of convergence with the help of the classical and second moduli of continuity. We also derive a Voronovskaja-type approximation theorem associated with these general operators, S_n^*.

2. A Set of Auxiliary Results

In this section, we give the moments of Szász-Mirakjan Beta-type operators, S_n^*, defined by Equation (6). We first recall for the S_n that

$$S_n(1,x) = 1, \quad S_n(t,x) = x \quad \text{and} \quad S_n\left(t^2,x\right) = x^2 + \frac{x}{n}, \tag{8}$$

just as in [1].

Lemma 1. *The moments of the Szász-Mirakjan Beta-type operators, S_n^*, defined by* (6) *are given by*

$$S_n^*(1,x) = 1, \tag{9}$$

$$S_n^*(t,x) = x + \frac{2}{n} \tag{10}$$

and

$$S_n^*(t^2,x) = \frac{n}{n-1}x^2 + \frac{6}{n-1}x + \frac{6}{n(n-1)}. \tag{11}$$

Proof. By using the known formulas in Equation (8), we find from the definition (6) that

$$\begin{aligned} S_n^*(1,x) &= \sum_{k=0}^{\infty} e^{-nx}\frac{(nx)^k}{k!}\frac{1}{B(k+2,n+1)} \\ &\quad \cdot \int_0^\infty \int_0^\infty \frac{t^k}{(1+t)^{n+k+1}} \exp\left(-\frac{p(1+t)^2}{t}\right) dt\, dp \\ &= \sum_{k=0}^{\infty} e^{-nx}\frac{(nx)^k}{k!}\frac{1}{B(k+2,n+1)} \int_0^\infty B_p(k+1,n)\, dp \\ &= \sum_{k=0}^{\infty} e^{-nx}\frac{(nx)^k}{k!}\frac{B(k+2,n+1)}{B(k+2,n+1)} \\ &= S_n(1,x) = 1. \end{aligned}$$

For $n > 1$, we have

$$\begin{aligned}
S_n^*(t,x) &= \sum_{k=0}^{\infty} e^{-nx} \frac{(nx)^k}{k!} \frac{1}{B(k+2,n+1)} \\
&\quad \cdot \int_0^{\infty} \int_0^{\infty} \frac{t^{k+1}}{(1+t)^{n+k+1}} \exp\left(-\frac{p(1+t)^2}{t}\right) dt\, dp \\
&= \sum_{k=0}^{\infty} e^{-nx} \frac{(nx)^k}{k!} \frac{1}{B(k+2,n+1)} \\
&\quad \cdot \int_0^{\infty} B_p(k+2,n-1)\, dp \\
&= \sum_{k=0}^{\infty} e^{-nx} \frac{(nx)^k}{k!} \frac{B(k+3,n)}{B(k+2,n+1)} \\
&= \sum_{k=0}^{\infty} e^{-nx} \frac{(nx)^k}{k!} \frac{\Gamma(k+3)\Gamma(n)}{\Gamma(n+k+3)} \frac{\Gamma(n+k+3)}{\Gamma(k+2)\Gamma(n+1)} \\
&= \sum_{k=0}^{\infty} e^{-nx} \frac{(nx)^k}{k!} \frac{k+2}{n} \\
&= S_n(t,x) + \frac{2}{n} S_n(1,x) = x + \frac{2}{n}
\end{aligned}$$

and, for $n > 2$, we find that

$$\begin{aligned}
S_n^*(t^2,x) &= \sum_{k=0}^{\infty} e^{-nx} \frac{(nx)^k}{k!} \frac{1}{B(k+2,n+1)} \\
&\quad \cdot \int_0^{\infty} \int_0^{\infty} \frac{t^{k+2}}{(1+t)^{n+k+1}} \exp\left(-\frac{p(1+t)^2}{t}\right) dt\, dp \\
&= \sum_{k=0}^{\infty} e^{-nx} \frac{(nx)^k}{k!} \frac{1}{B(k+2,n+1)} \\
&\quad \cdot \int_{p=0}^{\infty} B_p(k+3,n-2)\, dp \\
&= \sum_{k=0}^{\infty} e^{-nx} \frac{(nx)^k}{k!} \frac{B(k+4,n-1)}{B(k+2,n+1)} \\
&= \sum_{k=0}^{\infty} e^{-nx} \frac{(nx)^k}{k!} \frac{\Gamma(k+4)\Gamma(n-1)}{\Gamma(n+k+3)} \frac{\Gamma(n+k+3)}{\Gamma(k+2)\Gamma(n+1)} \\
&= \sum_{k=0}^{\infty} e^{-nx} \frac{(nx)^k}{k!} \frac{(k+3)\,(k+2)}{n(n-1)} \\
&= \frac{n}{n-1} S_n(t^2,x) + \frac{5}{n-1} S_n(t,x) + \frac{6}{n(n-1)} S_n(1,x) \\
&= \frac{n}{n-1} x^2 + \frac{6}{n-1} x + \frac{6}{n(n-1)}.
\end{aligned}$$

The proof of Lemma 1 is thus completed. □

Lemma 2. *The central moments of the Szász-Mirakjan Beta-type operators, S_n^*, defined by Equation* (6) *are given by*

$$S_n^*(t-x,x) = \frac{2}{n} \tag{12}$$

and

$$S_n^*\left((t-x)^2, x\right) = \frac{1}{n-1}\,x^2 + \frac{2(n+2)}{n(n-1)}\,x + \frac{6}{n(n-1)} =: \varepsilon_n(x). \tag{13}$$

Proof. The assertions (12) and (13) of Lemma 2 follow easily from those of Lemma 1, so we omit the details involved. □

3. Local Approximation

Let $C_{\mathcal{B}}[0,\infty)$ be the set of all real-valued continuous and bounded functions f on $[0,\infty)$, which is endowed with the norm given by

$$\|f\| = \sup_{x\in[0,\infty)} |f(x)|.$$

Then Peetre's K-functional is defined by

$$K_2(f;\delta) = \inf\left\{\|f-g\| + \delta\,\|g''\| : g \in C_{\mathcal{B}}^2[0,\infty)\right\},$$

where

$$C_{\mathcal{B}}^2[0,\infty) := \{g : g \in C_{\mathcal{B}}[0,\infty) \quad \text{and} \quad g', g'' \in C_{\mathcal{B}}[0,\infty)\}.$$

There exists a positive constant $C > 0$ such that (see, for example, [24])

$$K_2(f,\delta) \leqq C\omega_2\left(f, \sqrt{\delta}\right), \tag{14}$$

where $\delta > 0$ and ω_2 denotes the second-order modulus of smoothness for $f \in C_{\mathcal{B}}[0,\infty)$, which is defined by

$$\omega_2\left(f; \sqrt{\delta}\right) = \sup_{0<h\leqq\delta}\ \sup_{x\in[0,\infty)} |f(x+2h) - 2f(x+h) + f(x)|.$$

The usual modulus of continuity for $f \in C_{\mathcal{B}}[0,\infty)$ is given by

$$\omega(f;\delta) = \sup_{0<h\leqq\delta}\ \sup_{x\in[0,\infty)} |f(x+h) - f(x)|.$$

Lemma 3 below provides an auxiliary inequality which is useful in proving our next theorem (see Theorem 1).

Lemma 3. *For all $g \in C_{\mathcal{B}}^2[0,\infty)$, it is asserted that*

$$\left|\widetilde{S}_n^*(g,x) - g(x)\right| \leqq \delta_n(x)\,\|g''(x)\|, \tag{15}$$

where

$$\delta_n(x) = \frac{1}{n-1}\,x^2 + \frac{2\,(n+2)}{n\,(n-1)}\,x + \frac{2\,(5n-2)}{n^2(n-1)} \tag{16}$$

and

$$\widetilde{S}_n^*(f,x) = S_n^*(f,x) + f(x) - f\left(x + \frac{2}{n}\right) \tag{17}$$

for $f \in C_{\mathcal{B}}[0,\infty)$.

Proof. First of all, we find from (17) that

$$\widetilde{S}_n^*(t-x,x) = S_n^*(t-x,x) - \frac{2}{n} = \frac{2}{n} - \frac{2}{n} = 0. \tag{18}$$

Now, by using the Taylor's formula, we have

$$g(t) - g(x) = (t-x)g'(x) + \int_x^t (t-u)g''(u)\, du,$$

which, in view of Equation (18), yields

$$\begin{aligned}\widetilde{S}_n^*(g,x) - g(x) &= \widetilde{S}_n^*(t-x,x)g'(x) + \widetilde{S}_n^*\left(\int_x^t (t-u)g''(u)\, du, x\right) \\ &= S_n^*\left(\int_x^t (t-u)g''(u)du, x\right) - \int_x^{x+\frac{2}{n}} \left(x+\frac{2}{n}-u\right) g''(u)\, du.\end{aligned}$$

On the other hand, as

$$\begin{aligned}\left|\int_x^t (t-u)g''(u)\, du\right| &\leqq \int_x^t |t-u| \cdot |g''(u)|\, du \\ &\leqq \|g''\| \int_x^t |t-u|\, du \\ &\leqq (t-x)^2 \|g''\|\end{aligned}$$

and

$$\left|\int_x^{x+\frac{2}{n}} \left(x+\frac{2}{n}-u\right) g''(u)\, du\right| \leqq \left(\frac{2}{n}\right)^2 \|g''\|,$$

we conclude that

$$\begin{aligned}\left|\widetilde{S}_n^*(g,x) - g(x)\right| &\leqq S_n^*((t-x)^2, x)\|g''\| + \frac{4}{n^2}\|g''\| \\ &= \left(\frac{1}{n-1}x^2 + \frac{2(n+2)}{n(n-1)}x + \frac{6}{n(n-1)} + \frac{4}{n^2}\right)\|g''\| \\ &= \left(\frac{1}{n-1}x^2 + \frac{2(n+2)}{n(n-1)}x + \frac{2(5n-2)}{n^2(n-1)}\right)\|g''\| = \delta_n(x)\|g''\|.\end{aligned}$$

This is the result asserted by Lemma 3. □

We now state and prove our main results in this section.

Theorem 1. *Let $f \in C_B[0,\infty)$. Then, for every $x \in [0,\infty)$, there exists a constant $L > 0$, such that*

$$|S_n^*(f,x) - f(x)| \leqq L\omega_2\left(f; \sqrt{\delta_n(x)}\right) + \omega\left(f; \frac{2}{n}\right),$$

where $\omega_2(f;\delta)$ is the second-order modulus of smoothness, $\omega(f;\delta)$ is the usual modulus of continuity, and $\delta_n(x)$ is given by Equation (16).

Proof. We observe from Equation (17) that

$$\begin{aligned}|S_n^*(f,x) - f(x)| &\leqq \left|\widetilde{S}_n^*(f,x) - f(x)\right| + \left|f(x) - f\left(x+\frac{2}{n}\right)\right| \\ &\leqq \left|\widetilde{S}_n^*(f-g,x) - (f-g)(x)\right| + \left|f(x) - f\left(x+\frac{2}{n}\right)\right| \\ &\quad + \left|\widetilde{S}_n^*(g,x) - g(x)\right|\end{aligned}$$

for $g \in C_B^2[0,\infty)$. Thus, by applying Lemma 3 for $g \in C_B^2[0,\infty)$, we get

$$|S_n^*(f,x) - f(x)| \leqq 4\|f-g\| + \delta_n(x)\|g''\| + \omega\left(f;\frac{2}{n}\right),$$

which, by taking the infimum on the right-hand side over all $g \in C_B^2[0,\infty)$ and using (14), yields

$$\begin{aligned}|S_n^*(f,x) - f(x)| &\leqq 4K_2(f;\delta_n(x)) + \omega\left(f;\frac{2}{n}\right)\\ &\leqq L\omega_2\left(f;\sqrt{\delta_n(x)}\right) + \omega\left(f;\frac{2}{n}\right).\end{aligned}$$

where $L = 4M > 0$. This evidently completes the demonstration of Theorem 1. □

Theorem 2. *Let E be any bounded subset of the interval $[0,\infty)$, and suppose that $0 < \alpha \leqq 1$. If $f \in C_B[0,\infty)$ is locally $\mathrm{Lip}_M(\alpha)$, that is, if the following inequality holds true,*

$$|f(y) - f(x)| \leqq M|y-x|^\alpha \qquad (y \in E;\ x \in [0,\infty)),$$

then, for each $x \in [0,\infty)$,

$$|S_n^*(f,x) - f(x)| \leqq M([\varepsilon_n(x)]^{\frac{\alpha}{2}} + 2[d(x,E)]^\alpha), \tag{19}$$

$\varepsilon_n(x)$ is given by Equation (13), M is a constant depending on α and f, and $d(x,E)$ is the distance between x and E defined as follows:

$$d(x,E) = \inf\{\,|y-x| : y \in E\}.$$

Proof. Let $\overline{E}$ denote the closure of E in $[0,\infty)$. Then there exists a point $x_0 \in \overline{E}$ such that

$$|x - x_0| = d(x,E).$$

By the above-mentioned definition of $\mathrm{Lip}_M(\alpha)$, we get

$$\begin{aligned}|S_n^*(f,x) - f(x)| &\leqq S_n^*(\,|f(y)-f(x)|\,,x)\\ &\leqq S_n^*(\,|f(y)-f(x_0)|\,,x) + S_n^*(|f(x)-f(x_0)|\,,x)\\ &\leqq M\{S_n^*(|y-x_0|^\alpha,x) + |x-x_0|^\alpha\}\\ &\leqq M\{S_n^*(|y-x|^\alpha + |x-x_0|^\alpha,x) + |x-x_0|^\alpha\}\\ &\leqq M\{S_n^*(|y-x_0|^\alpha,x) + 2|x-x_0|^\alpha\}.\end{aligned}$$

Now, if we use the Hölder inequality with

$$p = \frac{2}{\alpha} \qquad \text{and} \qquad q = \frac{2}{2-\alpha},$$

we find that

$$\begin{aligned}|S_n^*(f,x) - f(x)| &\leqq M\left(\left[S_n^*(\,(y-x_0)^2,x)\right]^{\frac{\alpha}{2}} + 2[d(x,E)]^\alpha\right)\\ &= M\left([\varepsilon_n(x)]^{\frac{\alpha}{2}} + 2[d(x,E)]^\alpha\right).\end{aligned}$$

We have thus completed our demonstration of the result asserted by Theorem 2. □

4. A Voronovskaja-Type Approximation Theorem

By applying Equations (5) to (7), as well as Lemma 1, we first prove the following result.

Lemma 4. *It is asserted that*

$$\begin{aligned} S_n^*(t^3,x) = & \frac{n^2}{(n-1)(n-2)}\,x^3 + \frac{12n}{(n-1)(n-2)}\,x^2 \\ & + \frac{36}{(n-1)(n-2)}x + \frac{24}{n(n-1)(n-2)} \end{aligned} \tag{20}$$

and

$$\begin{aligned} S_n^*(t^4,x) = & \frac{n^3}{(n-1)(n-2)(n-3)}\,x^4 + \frac{20n^2}{(n-1)(n-2)(n-3)}\,x^3 \\ & + \frac{120n}{(n-1)(n-2)(n-3)}\,x^2 + \frac{240}{(n-1)(n-2)(n-3)}\,x \\ & + \frac{120}{n(n-1)(n-2)(n-3)}. \end{aligned} \tag{21}$$

Furthermore, the following result holds true,

$$\begin{aligned} S_n^*(\,(t-x)^4\,,x) = & \frac{3(n+6)}{(n-1)(n-2)(n-3)}\,x^4 + \frac{4(3n^2+32n+12)}{n(n-1)(n-2)(n-3)}\,x^3 \\ & + \frac{12(n^2+21n+18)}{n(n-1)(n-2)(n-3)}\,x^2 + \frac{144(n+2)}{n(n-1)(n-2)(n-3)}\,x \\ & + \frac{120}{n(n-1)(n-2)(n-3)}. \end{aligned} \tag{22}$$

Proof. We begin by recalling the following moments of the Szász-Mirakjan operators,

$$S_n\left(t^3,x\right) = x^3 + \frac{3x^2}{n} + \frac{x}{n^2} \tag{23}$$

and

$$S_n\left(t^4,x\right) = x^4 + \frac{6x^3}{n} + \frac{7x^2}{n^2} + \frac{x}{n^3}. \tag{24}$$

Using Equations (8) and the above formulas (23) and (24), we thus find for $n > 3$ that

$$\begin{aligned}
S_n^*(t^3,x) &= \sum_{k=0}^{\infty} e^{-nx}\frac{(nx)^k}{k!}\frac{1}{B(k+2,n+1)} \\
&\quad \cdot \int_0^{\infty}\int_0^{\infty}\frac{t^{k+3}}{(1+t)^{n+k+1}}\exp\left(-\frac{p(1+t)^2}{t}\right)dt\,dp \\
&= \sum_{k=0}^{\infty} e^{-nx}\frac{(nx)^k}{k!}\frac{1}{B(k+2,n+1)}\int_0^{\infty}B_p(k+4,n-3)\,dp \\
&= \sum_{k=0}^{\infty} e^{-nx}\frac{(nx)^k}{k!}\frac{B(k+5,n-2)}{B(k+2,n+1)} \\
&= \sum_{k=0}^{\infty} e^{-nx}\frac{(nx)^k}{k!}\frac{\Gamma(k+5)\Gamma(n-2)}{\Gamma(n+k+3)}\frac{\Gamma(n+k+3)}{\Gamma(k+2)\Gamma(n+1)} \\
&= \sum_{k=0}^{\infty} e^{-nx}\frac{(nx)^k}{k!}\frac{(k+4)\,(k+3)\,(k+2)}{n(n-1)(n-2)} \\
&= \frac{n^2}{(n-1)(n-2)}S_n(t^3,x)+\frac{9n}{(n-1)(n-2)}S_n(t^2,x) \\
&\quad +\frac{26}{(n-1)(n-2)}S_n(t,x)+\frac{24}{n(n-1)(n-2)}S_n(1,x) \\
&= \frac{n^2}{(n-1)(n-2)}\left(x^3+\frac{3x^2}{n}+\frac{x}{n^2}\right)+\frac{9n}{(n-1)(n-2)}\left(x^2+\frac{x}{n}\right) \\
&\quad +\frac{26}{(n-1)(n-2)}x+\frac{24}{n(n-1)(n-2)} \\
&= \frac{n^2}{(n-1)(n-2)}x^3+\frac{12n}{(n-1)(n-2)}x^2+\frac{36}{(n-1)(n-2)}x \\
&\quad +\frac{24}{n(n-1)(n-2)}.
\end{aligned}$$

On the other hand, for $n>4$, we find that

$$\begin{aligned}
S_n^*(t^4,x) &= \sum_{k=0}^{\infty} e^{-nx}\frac{(nx)^k}{k!}\frac{1}{B(k+2,n+1)} \\
&\quad \cdot \int_0^{\infty}\int_0^{\infty}\frac{t^{k+4}}{(1+t)^{n+k+1}}\exp\left(-\frac{p(1+t)^2}{t}\right)dt\,dp \\
&= \sum_{k=0}^{\infty} e^{-nx}\frac{(nx)^k}{k!}\frac{1}{B(k+2,n+1)}\int_0^{\infty}B_p(k+5,n-4)\,dp \\
&= \sum_{k=0}^{\infty} e^{-nx}\frac{(nx)^k}{k!}\frac{B(k+6,n-3)}{B(k+2,n+1)} \\
&= \sum_{k=0}^{\infty} e^{-nx}\frac{(nx)^k}{k!}\frac{\Gamma(k+6)\Gamma(n-3)}{\Gamma(n+k+3)}\frac{\Gamma(n+k+3)}{\Gamma(k+2)\Gamma(n+1)} \\
&= \sum_{k=0}^{\infty} e^{-nx}\frac{(nx)^k}{k!}\frac{(k+5)(k+4)\,(k+3)\,(k+2)}{n(n-1)(n-2)(n-3)},
\end{aligned}$$

that is, that

$$\begin{aligned} S_n^*(t^4,x) &= \frac{n^3}{(n-1)(n-2)(n-3)}\, S_n(t^4,x) + \frac{14n^2}{(n-1)(n-2)(n-3)}\, S_n(t^3,x) \\ &\quad + \frac{71n}{(n-1)(n-2)(n-3)}\, S_n(t^2,x) + \frac{154}{(n-1)(n-2)(n-3)}\, S_n(t,x) \\ &\quad + \frac{120}{n(n-1)(n-2)(n-3)}\, S_n(1,x) \\ &= \frac{n^3}{(n-1)(n-2)(n-3)}\, x^4 + \frac{20n^2}{(n-1)(n-2)(n-3)}\, x^3 \\ &\quad + \frac{120n}{(n-1)(n-2)(n-3)}\, x^2 + \frac{240}{(n-1)(n-2)(n-3)}\, x \\ &\quad + \frac{120}{n(n-1)(n-2)(n-3)}, \end{aligned}$$

which, together, complete the proof of Lemma 4. □

Theorem 3. *Let $f, f', f'' \in C_\nu[0,\infty)$ for $\nu \geqq 4$. Then, the following Voronovskaja-type approximation result holds true,*

$$\lim_{n\to\infty} \{n\,[S_n^*(f,x) - f(x)]\,\} = 2f'(x) + \left(\frac{x^2}{2} + x\right) f''(x). \tag{25}$$

Proof. By Taylor's expansion of $f\,(t)$ at the point $t = x$, we have

$$f(t) = f(x) + f'\,(x)\,(t-x) + \frac{1}{2}\, f''(x)(t-x)^2 + \Psi(t,x)(t-x)^2, \tag{26}$$

where $\Psi(t,x)$ is remainder term, $\Psi(\cdot,x) \in C_\nu[0,\infty)$ and $\Psi(t,x) \to 0$ as $t \to x$.

Applying the Szász-Mirakjan Beta-type operators S_n^* to Equation (26) and, using Lemma 2, we obtain

$$\begin{aligned} S_n^*(f,x) - f(x) &= f'(x)\; S_n^*(t-x,x) + \frac{1}{2}\, f''(x)\; S_n^*(\,(t-x)^2\,,x) \\ &\quad + S_n^*(\Psi(t,x)\,(t-x)^2\,,x) \\ &= \frac{2}{n}\, f'(x) + \frac{1}{2}\left(\frac{1}{n-1}\, x^2 + \frac{2(n+2)}{n(n-1)}\, x + \frac{6}{n(n-1)}\right) f''(x) \\ &\quad + S_n^*(\Psi(t,x)\,(t-x)^2\,,x). \end{aligned} \tag{27}$$

We now apply the Cauchy-Schwarz inequality to the third term on the right-hand side of Equation (27). We thus find that

$$n\left|S_n^*(\Psi(t,x)\,(t-x)^2\,,x)\right| \leqq \sqrt{n^2 S_n^*(\,(t-x)^4\,,x)} \cdot \sqrt{S_n^*(\,[\Psi(t,x)]^2\!,x)}.$$

Let

$$\eta(t,x) := \left[\Psi(t,x)\right]^2.$$

In this case, we observe that $\eta(x,x) = 0$ and also that $\eta(\cdot,x) \in C_\nu[0,\infty)$. Then, it follows that

$$\lim_{n\to\infty} \{S_n^*(\,[\Psi(t,x)]^2\!,x)\} = \lim_{n\to\infty} \{S_n^*(\eta(t,x),x)\} = \eta(x,x) = 0$$

uniformly with respect to $x \in [0, b]$ $(b > 0)$ and the following limit,

$$\lim_{n \to \infty} \{n^2 S_n^*((t-x)^4, x)\}$$

is finite. Consequently, we have

$$\lim_{n \to \infty} \{n S_n^*(\Psi(t,x)(t-x)^2, x)\} = 0.$$

Thus, in the limit when $n \to \infty$ in Equation (27), we obtain

$$\lim_{n \to \infty} \{n[S_n^*(f,x) - f(x)]\} = 2f'(x) + \left(\frac{x^2}{2} + x\right) f''(x).$$

The proof of Theorem 3 is thus completed. □

5. Concluding Remarks and Observations

We find it worthwhile to reiterate the fact that, in approximation theory and related fields, the approximation and some other basic properties of various linear and nonlinear operators are investigated because mainly of the potential for their usefulness in many areas of researches in the mathematical, physical, and engineering sciences. This article has been motivated essentially by the demonstrated applications of such results as those associated with various approximation operators.

In our present investigation, we have systematically studied a number of approximation properties of a class of the Szász-Mirakjan Beta-type operators, which we have introduced here by using an extension of the familiar Beta function $B(\alpha, \beta)$. We have established the moments of these extended Szász-Mirakjan Beta-type operators and estimated several convergence results with the help of the second modulus of smoothness and the classical modulus of continuity. We have also investigated convergence via functions belonging to the Lipschitz class. Finally, we have proved a Voronovskaja-type approximation theorem for the general Szász-Mirakjan Beta-type operators.

Using the other substantially more general forms of the classical Beta function $B(\alpha, \beta)$, which we have indicated in Section 1 of this article (see, for example, [22,23]), one can analogously develop further extensions and generalizations of the various results which we have presented here. In many of these suggested areas of further researches on the subject of this article, some other, possibly deeper, mathematical analytic tools and techniques will have to be called for.

Author Contributions: All three authors contributed equally to this investigation.

References

1. Szász, O. Generalizations of S. Bernstein's polynomials to the infinite interval. *J. Res. Nat. Bur. Stand.* **1950**, *45*, 239–245. [CrossRef]
2. Mirakjan, G.M. Approximation des fonctions continues au moyen de polynômes de la forme $e^{-nx} \sum_{k=0}^{m_n} C_{k,n} x^k$. *C. R. (Doklady) Acad. Sci. URSS (New Ser.)* **1941**, *31*, 201–205.
3. Duman, O.; Özarslan, M.A. Szász-Mirakjan-type operators providing a better error estimation. *Appl. Math. Lett.* **2007**, *20*, 1184–1188. [CrossRef]
4. Gal, S.G. Approximation with an arbitrary order by generalized Szász-Mirakjan operators. *Stud. Univ. Babeş-Bolyai Math.* **2014**, *59*, 77–81.
5. Gupta, V.; Mahmudov, N. Approximation properties of the q-Szász-Mirakjan-Beta operators. *Indian J. Ind. Appl. Math.* **2012**, *3*, 41–53.
6. Gupta, V.; Noor, M.A. Convergence of derivatives for certain mixed Szász-Beta operators. *J. Math. Anal. Appl.* **2006**, *321*, 1–9. [CrossRef]

7. İçöz, G.; Mohapatra, R.N. Approximation properties by q-Durrmeyer-Stancu operators. *Anal. Theory Appl.* **2013**, *29*, 373–383.
8. Gupta, V.; Srivastava, H.M. A general family of the Srivastava-Gupta operators preserving linear functions. *Eur. J. Pure Appl. Math.* **2018**, *11*, 575–579. [CrossRef]
9. Gupta, V.; Srivastava, G.S.; Sahai, A. On simultaneous approximation by Szász-beta operators. *Soochow J. Math.* **1995**, *21*, 1–11.
10. Păltănea, R. Modified Szász-Mirakjan operators of integral form. *Carpathian J. Math.* **2008**, *24*, 378–385.
11. Srivastava, H.M.; Mursaleen, M.; Alotaibi, A.M.; Nasiruzzaman, M.; Al-Abied, A.A.H. Some approximation results involving the q-Szász-Mirakjan-Kantorovich type operators via Dunkl's generalization. *Math. Methods Appl. Sci.* **2017**, *40*, 5437–5452. [CrossRef]
12. Srivastava, H.M.; Özger, F.; Mohiuddine, S.A. Construction of Stancu-type Bernstein operators based on Bézier bases with shape parameter λ. *Symmetry* **2019**, *11*, 316. [CrossRef]
13. Srivastava, H.M.; Zeng, X.-M. Approximation by means of the Szász-Bézier integral operators. *Int. J. Pure Appl. Math.* **2004**, *14*, 283–294.
14. Xie, L.-S.; Xie, T.-F. Approximation theorems for localized Szász-Mirakjan operators. *J. Approx. Theory* **2008**, *152*, 125–134. [CrossRef]
15. Zeng, X.-M. Approximation of absolutely continuous functions by Stancu Beta operators. *Ukrainian Math. J.* **2012**, *63*, 1787–1794. [CrossRef]
16. Duman, O.; Özarslan, M.A.; Aktuğlu, H. Better error estimates for Szász-Mirakjan-Beta operators. *J. Comput. Anal. Appl.* **2008**, *10*, 53–59.
17. Özarslan, M.A.; Aktuğlu, H. A-Statistical approximation of generalized Szász-Mirakjan-Beta operators. *Appl. Math. Lett.* **2011**, *24*, 1785–1790. [CrossRef]
18. Qi, Q.-L.; Zhang, Y.-P. Pointwise approximation for certain mixed Szász-Beta operators. In *Further Progress in Analysis, Proceedings of the Sixth International Conference* (*ISAAC* 2002) *on Clifford Algebras and Their Applications in Mathematical Physics, Tennessee Technological University, Cookeville, TN, USA, 20–25 May 2002*; World Scientific Publishing Company: Singapore, 2009; pp. 152–164.
19. Chaudhry, M.A.; Zubair, S.M. Generalized incomplete gamma functions with applications. *J. Comput. Appl. Math.* **1994**, *55*, 99–124. [CrossRef]
20. Chaudhry, M.A.; Qadir, A.; Rafique, M.; Zubair, S.M. Extension of Euler's beta function. *J. Comput. Appl. Math.* **1997**, *78*, 19–32. [CrossRef]
21. Özergin, E.; Özarslan, M.A.; Altın, A. Extension of gamma, beta and hypergeometric functions. *J. Comput. Appl. Math.* **2011**, *235*, 4601–4610. [CrossRef]
22. Lin, S.-D.; Srivastava, H.M.; Yao, J.-C. Some classes of generating relations associated with a family of the generalized Gauss type hypergeometric functions. *Appl. Math. Inform. Sci.* **2015**, *9*, 1731–1738.
23. Srivastava, H.M.; Parmar, R.K.; Chopra, P. A class of extended fractional derivative operators and associated generating relations involving hypergeometric functions. *Axioms* **2012**, *1*, 238–258. [CrossRef]
24. DeVore, R.A.; Lorentz, G.G. *Constructive Approximation*; Springer: Berlin/Heidelberg, Germany; New York, NY, USA, 1993.

10

Interior Regularity Estimates for a Degenerate Elliptic Equation with Mixed Boundary Conditions

Jean-Daniel Djida [1,*] and Arran Fernandez [2]

1 Departamento de Análise Matemática, Universidade de Santiago de Compostela, 15782 Santiago de Compostela, Spain
2 Department of Applied Mathematics & Theoretical Physics, University of Cambridge, Cambridge CB3 0WA, UK; af454@hermes.cam.ac.uk
* Correspondence: jeandaniel.djida@usc.es

Abstract: The Marchaud fractional derivative can be obtained as a Dirichlet-to–Neumann map via an extension problem to the upper half space. In this paper we prove interior Schauder regularity estimates for a degenerate elliptic equation with mixed Dirichlet–Neumann boundary conditions. The degenerate elliptic equation arises from the Bernardis–Reyes–Stinga–Torrea extension of the Dirichlet problem for the Marchaud fractional derivative.

Keywords: marchaud fractional derivative; interior regularity; schauder estimate; extension problem; fractional order weighted Sobolev spaces

---

1. Introduction

In the last years there has been a growing interest in the study of fractional elliptic equations involving the right fractional Marchaud derivative $(\mathscr{D}_{right})^{\alpha}$, such as equations of the form

$$(\mathscr{D}_{right})^{\alpha}v = f \quad \text{in}\,\Omega, \quad v = 0 \quad \text{in} \quad [b,\infty), \tag{1}$$

where without loss of generality $\Omega := [a,b) \subset \mathbb{R}$, with $a < b$ and $0 < \alpha < 1$.

Fractional diffusion problems of type (1) arise for example in the modelling of neuronal transmission in Purkinje cells, whose malfunctioning is known to be related to the lack of voluntary coordination and the appearance of tremors [1]. Further motivation comes from various experimental results which showed anomalous diffusion of fractional type, see for example [2,3] and references therein.

The right fractional Marchaud derivative of a function $w : \mathbb{R} \to \mathbb{R}$ is defined via Fourier transforms as

$$\widehat{(\mathscr{D}_{right})^{\alpha}w}(\xi) = (\pm i\xi)^{\alpha}\,\widehat{w}(\xi), \tag{2}$$

and it can also be expressed by the pointwise formula

$$(\mathscr{D}_{right})^{\alpha}v(x) = \frac{c_{\alpha}}{\Gamma(-\alpha)}\int_{x}^{\infty}\frac{v(y)-v(x)}{(y-x)^{1+\alpha}}\,dy, \tag{3}$$

where c_{α} is a positive normalization constant. We observe from (3) that the right fractional Marchaud derivative is a nonlocal operator. Nonlocal operators have the peculiarity of taking memory effects into account and capturing long-range interactions, i.e., events that happen far away in time or space. Further discussion of the difference between local integro-differential operators and nonlocal or fractional ones can be found in [4] and references therein. In this context, the nonlocality of the fractional Marchaud derivative prevents us from applying local PDE techniques to treat nonlinear problems for $(\mathscr{D}_{right})^{\alpha}$. To overcome this difficulty, Bernardis, Reyes, Stinga and Torrea showed in [5] that the right fractional Marchaud derivative can be determined as an operator that maps a Dirichlet boundary condition to a Neumann-type condition via an extension problem. Similar extension properties have been found for the fractional Laplacian by Caffarelli and Silvestre in [6].

To be more precise, consider the function $\mathcal{U} : \mathbb{R}^2_+ := \mathbb{R} \times [0, \infty) \to \mathbb{R}$ that solves the boundary value problem

$$\begin{cases} \mathcal{M}_\alpha \mathcal{U}(t,x) = 0 & \text{for } (t,x) \in \mathbb{R}^2_+, \\ \lim_{t\to 0} \mathcal{N}_\alpha \mathcal{U}(t,x) = f(x) & \text{for } x \in \Omega, \\ \mathcal{U}(t,x) = 0 & \text{for } x \in \mathbb{R} \setminus \Omega, \end{cases} \tag{4}$$

Then we have [5]:

$$\lim_{t\to 0} \mathcal{N}_\alpha \mathcal{U}(t,x) = c_\alpha (\mathscr{D}_{right})^\alpha v(x),$$

where $c_\alpha := \dfrac{4^{\alpha-1/2}\Gamma(\alpha)}{\Gamma(1-\alpha)}$ is a positive multiplicative constant depending only on $\alpha \in (0,1)$. Here the differential operators $\mathcal{M}_\alpha$ and $\mathcal{N}_\alpha$ are given respectively by:

$$\mathcal{M}_\alpha \mathcal{U} := -(\mathscr{D}_{right})\mathcal{U} + \frac{1-2\alpha}{t}\mathcal{U}_t + \mathcal{U}_{tt}; \tag{5}$$

$$\mathcal{N}_\alpha \mathcal{U} := -t^{1-2\alpha}\mathcal{U}_t. \tag{6}$$

We use the notation $(\mathscr{D}_{right})$ for the derivative *from the right* at the point $x \in \mathbb{R}$, that is:

$$(\mathscr{D}_{right})v(x) = \lim_{t\to 0^+} \frac{v(x) - v(x+t)}{t}, \tag{7}$$

for good enough functions v. Observe that $(\mathscr{D}_{right})$ equals the negative of the lateral derivative $\left(\frac{d}{dx^+}\right)$ as usually defined in calculus [5].

This characterization of $(\mathscr{D}_{right})^\alpha v$ via the local (degenerate) PDE (5) was used for the first time in [5] to get maximum principles. To solve (4), Stinga and Torrea noted that (5) can be thought of as the harmonic extension of v into $2 - 2\alpha$ extra dimensions (see [5]). From there, they established the fundamental solution and, using a conjugate equation, a Poisson formula for $\mathcal{U}$. Furthermore, taking advantage of the general theory of degenerate elliptic equations developed by Fabes, Jerison, Kenig and Serapioni in 1982–83, they proved comparison principles for $\mathcal{U}$ (and thus for v).

The aim of this paper is to prove an interior Schauder estimate for the problem (4), involving any fractional power of the derivative $(\mathscr{D}_{right})^\alpha$ as an operator that maps a Dirichlet condition to a Neumann-type condition via an extension problem as in [5].

A significant contribution of the above extension problem is to provide a way of applying classical analysis methods to partial differential equations containing one-sided Marchaud derivative operators. By means of such extension techniques, a series of important results, such as comparison principles, Harnack inequalities, and regularity estimates for solutions to degenerate elliptic equations involving the fractional Laplacian, have been studied by many authors, for example [6–17]. The same analysis was done for the one-sided fractional derivative operator in the sense of Marchaud ([5] Theorem 1.1 and Corollary 1.2).

In view of these results, we immediately observe that interior regularity and boundary regularity for the degenerate elliptic equation with mixed boundary conditions involving the one-sided Marchaud derivative is missing in the literature. Indeed, from the pioneering work of [5,7,18] on the analogue extension problem for nonlocal operators that map Dirichlet to Neumann, one can reduce a nonlocal problem involving fractional derivatives to a local one by keeping their qualitative properties. Using this technique, one can study interior and boundary regularity. Hence the raison d'être for this work.

In order to underline what makes the difference between the extension problems introduced by Bernardis, Reyes, Stinga and Torrea [5], and the one introduced by Bucur and Ferrari [18], we point out that the extension problem established in [18] is based on a time-dependent initial condition, which leads to a heat conduction problem. Indeed, considering the function $\varphi : \mathbb{R} \to \mathbb{R}$ of one variable, formally representing the time variable, their approach relies on constructing a parabolic local operator by adding an extra variable, say the space variable, on the positive half-line, and working on the following problem in the half-plane $[0,\infty) \times \mathbb{R}$:

$$\begin{cases} \partial_t \mathcal{U} & = \Delta \mathcal{U} \qquad (t,x) \in (0,\infty) \times \mathbb{R}, \\ \mathcal{U}(0,t) & = \varphi(t) \qquad t \in \mathbb{R}. \end{cases} \tag{8}$$

The problem (8) is not the usual Cauchy problem for the heat operator, but a heat conduction problem.

In view of the type of problem we are interested in here, we choose to deal with the Bernardis–Reyes–Stinga–Torrea extension problem [5]. Our main result, which will be proved in Section 3 below, is as follows. We note that this result can be proved only using extension techniques.

Theorem 1. *Let $\alpha \in (0,1)$ and let $\mathcal{U} \in L^\infty(\mathbb{R}_2^+) \cap H^1(t^\mu; \mathcal{B}_2^+)$ be a weak solution to*

$$\begin{cases} \mathcal{M}_\alpha \mathcal{U} = 0 & \text{in } \mathcal{B}_2^+, \\ \lim_{t\to 0} \mathcal{N}_\alpha \mathcal{U}(t,\cdot) = f & \text{on } B_2, \\ \mathcal{U} = 0 & \text{on } \mathbb{R} \setminus \Omega. \end{cases}$$

(a) *For $1 < p < \infty$, if $f \in L^p(B_1, w)$ and $\gamma \in (0, \min(1,\alpha))$ is such that $0 < \alpha - \gamma - \frac{1}{p} < 1$, then $\mathcal{U} \in C^{0,\alpha-\gamma-\frac{1}{p}}\left(\overline{\mathcal{B}_1^+}, w\right)$. Moreover,*

$$\| \mathcal{U} \|_{C^{\alpha-\gamma-\frac{1}{p}}(\overline{\mathcal{B}_1^+}, w)} \leq C \left(\| \mathcal{U} \|_{L^p(\mathbb{R}_2^+, w)} + \| f \|_{L^p(\mathbb{R}, w)} \right),$$

where C is a positive constant depending only on α, γ, and p.

(b) *If $f \in L^\infty(B_1)$ and $\gamma \in (0, \min(1,\alpha))$, then $\mathcal{U} \in C^{\alpha-\gamma}\left(\overline{\mathcal{B}_1^+}\right)$. Moreover,*

$$\|\mathcal{U}\|_{C^{\alpha-\gamma}\left(\overline{\mathcal{B}_1^+}\right)} \leq C \left(\|\mathcal{U}\|_{\mathbb{R}_+^2} + \|f\|_{L^\infty(\mathbb{R})} \right),$$

where C is a positive constant depending only on α and γ.

The paper is organised as follows. In Section 2, we give some notations and definitions of function spaces and their associated norms which will be needed in this work. We also provide some preliminary results and finally state our main result. In Section 3, we prove an intermediate result and provide the proof of the regularity estimate up to the boundary for the degenerate Equation (4) with the Neumann boundary condition stated in Theorem 1. Finally we end with the conclusion in Section 4.

2. Notations and Preliminary Results

In this section we introduce some notations, definitions, and preliminary results used throughout the paper.

Here and in the following, we consider $\alpha \in (0,1)$, $\mathbb{R}_+^2 := \mathbb{R}_+ \times \mathbb{R} = \{z = (t,x) : t > 0\}$ and $\Omega \subset \mathbb{R}$ a bounded Lipschitz domain. For an open set Ω, an integer $k \geq 1$, and a real number $\lambda \in (0,1]$, the Hölder spaces $C^{k,\lambda}(\Omega)$ are defined as the subspaces of $C^k(\Omega)$ consisting of functions whose k-th order derivatives are uniformly Hölder continuous with exponent λ in Ω.

Furthermore, we introduce the following notation for intervals, boxes, and balls:

$$\begin{aligned} B_r(x_0) &:= \{x \in \mathbb{R} : |x - x_0| < r\}, \\ \mathcal{B}_r^+(x_0) &:= [0, r) \times B_r(x_0), \\ \mathcal{B}_r(z_0) &:= \{z = (t, x) \in \mathbb{R} \times \mathbb{R} : |z - z_0| < r\}, \end{aligned} \tag{9}$$

We consider the function space

$$\mathcal{L}^1_\alpha := \left\{ \varphi : \mathbb{R} \to \mathbb{R} \quad : \quad \|\varphi\|_{\mathcal{L}^1_\alpha} := \int_{\mathbb{R}} \frac{|\varphi(x)|}{1 + |x|^{1+\alpha}} \, dx < \infty \right\}.$$

For $\Omega \subset \mathbb{R}$ an open set, we say $v : \Omega \to \mathbb{R}$ is in $C^{0,\gamma}(\Omega)$, i.e., Hölder continuous with exponent $\gamma \in (0, 1)$, if

$$\|v\|_{C^{0,\gamma}(\Omega)} := \sup_{x \neq y} \frac{|v(x) - v(y)|}{|x - y|^\gamma} < +\infty.$$

We recall the following definition of Sobolev spaces.

Definition 1 (Sobolev spaces). *For any real number α, the αth Sobolev space on $\mathbb{R}$ is defined to be*

$$H^\alpha(\mathbb{R}) := \{u \in \mathcal{S}'(\mathbb{R}) : \hat{u} \in L^2_{loc}(\mathbb{R}), ||u||_{H^\alpha} < \infty\},$$

where the Sobolev norm $|| \cdot ||_{H^\alpha}$ is defined by

$$||u||_{H^\alpha} := \left(\int_{\mathbb{R}} |\hat{u}(\lambda)|^2 \left(1 + |\lambda|^2\right)^\alpha \mathrm{d}\lambda \right)^{1/2}.$$

For a general domain $X \subset \mathbb{R}$, the αth Sobolev space on X is defined to be

$$H^\alpha_{loc}(X) := \{u \in \mathcal{D}'(X) : u\phi \in H^\alpha(\mathbb{R}) \text{ for all } \phi \in \mathcal{D}(X)\}.$$

Let $a \in \mathbb{R}$ and $\alpha \in (0, 1)$ be two arbitrary parameters. We define the functional space

$$C^{1,\alpha}_a := \left\{ f : \mathbb{R} \to \mathbb{R} \quad : \quad \text{for any } x > a,\ f \in AC([a, x]) \text{ and } f'(\cdot)(x - \cdot)^{-\alpha} \in L^1((a, x)) \right\}. \tag{10}$$

We denote here by $AC(I)$ the space of absolutely continuous functions on I.

Definition 2 (Caputo derivative). *The Caputo derivative of $v \in C^{1,\alpha}_a$ with initial point $a \in \mathbb{R}$ at the point $x > a$ is given by*

$$D^\alpha_a v(x) := \frac{1}{\Gamma(1 - \alpha)} \int_a^x v'(y)(x - y)^{-\alpha} \, dy. \tag{11}$$

Definition 3. *The right Marchaud derivative of a well defined function v is given by*

$$(\mathscr{D}_{right})^\alpha v(x) = \lim_{\delta \to 0^+} \frac{C}{\Gamma(-\alpha)} \int_{x+\delta}^\infty \frac{v(y) - v(x)}{(y - x)^{1+\alpha}} \, dy, \tag{12}$$

with C_α a positive normalisation constant.

Remark 1. *Notice that the one-sided nonlocal derivative in the sense of Marchaud can also be obtained by extending the Caputo derivative. Indeed, by making an integration by parts of Equation (11), we obtain an equivalent definition [19,20] as follows:*

$$D^{\alpha}v(x) = C(\alpha)\int_{x}^{\infty}\frac{v(y)-v(x)}{(y-x)^{1+\alpha}}\,dy,$$

for all $x < 0$, so that $v(x) = v(0)$, where $C(\alpha)$ is a constant depending on α. Indeed, for sufficiently regular functions v, we have:

$$\begin{aligned} D^{\alpha}v(x) &= \int_0^x \frac{v'(y)}{(x-y)^{\alpha}}\,dy = \int_0^x \left(\frac{d}{dy}\frac{(v(y)-v(x))}{(x-y)^{\alpha}} - \alpha\,\frac{(v(y)-v(x))}{(x-y)^{1+\alpha}}\right)dy \\ &= \frac{v(x)-v(0)}{x^{\alpha}} - \lim_{y\to x}\frac{v(x)-v(y)}{(x-y)^{\alpha}} - \alpha\int_0^x \frac{(v(y)-v(x))}{(x-y)^{1+\alpha}}\,dy \\ &= \frac{v(x)-u(0)}{x^{\alpha}} - v'(y)\lim_{y\to x}(x-y)^{1-\alpha} - \alpha\int_0^x \frac{v(y)-v(x)}{(x-y)^{1+\alpha}}\,dy \\ &= \frac{v(x)-v(0)}{x^{\alpha}} + \alpha\int_0^x \frac{v(x)-v(y)}{(x-y)^{1+\alpha}}\,dy. \end{aligned} \tag{13}$$

Hence, we take the convention that $v(x) = v(0)$ for any $x \le 0$. With this extension, one has that, for any $x > 0$,

$$\alpha\int_{-\infty}^{0}\frac{v(x)-v(y)}{(x-y)^{1+\alpha}}\,dy = \alpha\int_{-\infty}^{0}\frac{v(x)-u(0)}{(x-y)^{1+\alpha}}\,dy = \frac{v(x)-v(0)}{x^{\alpha}}.$$

So one can write (13) *as*

$$D^{\alpha}v(x) = C(\alpha)\int_{-\infty}^{x}\frac{v(x)-v(y)}{(x-y)^{1+\alpha}}\,dy.$$

This type of formula also relates the Caputo derivative to the so-called Marchaud derivative [20,21]. Therefore the results obtained in this paper could also be applied for the extended Caputo derivative.

Note that the integral in (12) is absolutely convergent for functions in the Schwartz class $\mathcal{S}$. Furthermore one should notice that the nonlocal operators $(\mathscr{D}_{right})^{\alpha}$ and $(\mathscr{D}_{right})^{-\alpha}$ depend on the values of v on the whole half line (x, ∞).

We recall that the inverse of the right fractional Marchaud derivative $(\mathscr{D}_{right})^{-\alpha}$ is defined as

$$(\mathscr{D}_{right})^{-\alpha}v(x) := \int_{\mathbb{R}}\frac{v(y)}{|x-y|^{1-\alpha}}\,dy = \mathcal{I}_{\alpha} * v(x) \tag{14}$$

where the Riesz potential (see [7,21]) is defined as

$$\mathcal{I}_{\alpha} = C_{\alpha}|x-y|^{\alpha-1} \qquad \text{for } \alpha < 1, \tag{15}$$

with the constant $C_{\alpha} = \frac{1}{\pi}\Gamma(1-\alpha)\sin\frac{\pi\alpha}{2}$.

From [5], we have that for $u \in \mathcal{S}$, $(\mathscr{D}_{right})^{\alpha}u \in \mathcal{S}_{\alpha}$, where

$$\mathcal{S}_{\alpha} := \left\{ f \in C^{\infty}(\mathbb{R}) : (1+|x|^{1+\alpha})f^{k}(x) \in L^{\infty}(\mathbb{R}), \text{ for each } k \ge 0 \right\}.$$

The topology in $\mathcal{S}_{\alpha}$ is given by the family of seminorms $[f]_k := \sup_{x\in\mathbb{R}}\left|(1+|x|^{1+\alpha})f^{(k)}(x)\right|$, for $k \ge 0$. Let $\mathcal{S}'_{\alpha}$ be the dual space of $\mathcal{S}_{\alpha}$; then $(\mathscr{D}_{right})^{\alpha}$ defines a continuous operator from $\mathcal{S}'_{\alpha}$ into $\mathcal{S}'$.

2.1. Weighted Spaces

Weighted spaces of smooth functions play an important role in the context of partial differential equations (PDEs). They are widely used, for instance, to treat PDEs with degenerate coefficients or domains with a nonsmooth geometry (see e.g., [22–25]), as is the case here. For evolution equations, power weights in time play an important role in order to obtain results for rough initial data (see [26,27]). This subsection dedicated to weighted spaces is motivated by the appearance of the Muckenhoupt weight $w := t^{1-2\alpha}$ which appears in (5) and (6). For general literature on weighted function spaces we refer to [23–25,28–31] and references therein.

In a general framework, a function $w : \mathbb{R}^d \to [0, \infty)$, for an integer $d \geq 1$, is called a *weight* if w is locally integrable and the zero set $\{x : w(x) = 0\}$ has Lebesgue measure zero. For $p \in [1, \infty]$ we denote by A_p the *Muckenhoupt class* of weights. In the case $p \in (1, \infty)$, we say that $w \in A_p$ if

$$\sup_{\mathcal{B} \text{ cubes in } \mathbb{R}^d} \left(\frac{1}{|\mathcal{B}|} \int_{\mathcal{B}} w(x)\, dx \right) \left(\frac{1}{|\mathcal{B}|} \int_{\mathcal{B}} w(x)^{-\frac{1}{p-1}}\, dx \right)^{p-1} < \infty.$$

In the case $p = 1$, we say that $w : \mathbb{R} \to [0, \infty)$ belongs to A_1 if there exists some constant C such that

$$\frac{1}{|\mathcal{B}|} \int_{\mathcal{B}} w(y)\, dy \leq C w(x)$$

for all $x \in \mathcal{B}$ and all balls $\mathcal{B} \subset \mathbb{R}^d$. In the case $p = \infty$, we define $A_\infty = \bigcup_{1 \leq p < \infty} A_p$. Note that, for functions with support contained in $(-\infty, 0)$ or $(0, \infty)$, the class of weights is denoted by A_p^+ or A_p^- respectively. We refer to [27,30,32] for the general properties of these classes.

Example 1. *Problem (4) is a weighted—singular or degenerate, depending on the value of $\alpha \in (0,1)$—elliptic equation on $\mathbb{R}^2_+$ with mixed boundary conditions. The weight $w := t^{1-2\alpha}$ belongs to the Muckenhoupt class A_2^+, i.e., there exists a constant C such that for any $\mathcal{B} \subset \mathbb{R}^2_+$,*

$$\left(\frac{1}{|\mathcal{B}|} \int_{\mathcal{B}} |t|^{1-2\alpha} dt\, dx \right) \left(\frac{1}{|\mathcal{B}|} \int_{\mathcal{B}} |t|^{2\alpha-1} dt\, dx \right) \leq C.$$

For this reason, when working with one-sided weights, we can assume without loss of generality that $\Omega := [a, b) = \mathbb{R}$ (see e.g., [30] for more details).

Next, for a strongly measurable function f and a number $p \in [1, \infty)$, we define the weighted L^p norm by

$$\|f\|_{L^p(\mathbb{R}^d, w)} := \left(\int_{\mathbb{R}^d} \|f(x)\|^p w(x)\, dx \right)^{1/p},$$

and we define the weighted L^p space to be the following Banach space:

$$L^p(\mathbb{R}^d, w) := \{f \text{ strongly measurable} : \|f\|_{L^p(\mathbb{R}^d, w)} < \infty\}.$$

Definition 4 (see [8]). *Given $\alpha \in (0,1)$, $\mu = 1 - 2\alpha \in (-1, 1)$, and an open set $\mathcal{B} \subset \overline{\mathbb{R}^2_+}$, we denote*

$$L^2(t^\mu; \mathcal{B}) := \left\{ \mathcal{U} : \mathbb{R}^2_+ \to \mathbb{R},\ \int_{\mathcal{B}} t^\mu |\mathcal{U}|^2 dt\, dx < +\infty \right\},$$

endowed with the norm

$$\|\mathcal{U}\|_{L^2(t^\mu; \mathcal{B})} := \left(\int_{\mathcal{B}} t^\mu |\mathcal{U}|^2 dt\, dx \right)^{1/2}.$$

We also denote

$$H^1(t^\mu; \mathcal{B}) := \left\{ \mathcal{U} \in L^2(t^\mu; \mathcal{B}) : \nabla \mathcal{U} \in L^2(t^\mu; \mathcal{B}) \right\},$$

with the induced norm

$$\|\mathcal{U}\|_{H^1(t^\mu; \mathcal{B})} := \left(\int_{\mathcal{B}} t^\mu \left(|\mathcal{U}|^2 + |\nabla \mathcal{U}|^2 \right) dt\, dx \right)^{1/2}.$$

Using the variable $(t,x) \in \mathbb{R}^2_+$, the space $H^\alpha(\mathbb{R})$ coincides with the trace on $\partial\mathbb{R}^2_+$ of

$$\dot{H}^1(t^\mu;\mathcal{B}) := \left\{ \mathcal{U} \in L^2_{loc}(\mathbb{R}^2_+) : \int_{\mathbb{R}^2_+} t^\mu \left(\mathcal{U}^2 + |\nabla \mathcal{U}|^2 \right) dt\, dx < +\infty \right\}.$$

In other words [8,9], for any given function $\mathcal{U} \in \dot{H}^1(t^\mu;\mathcal{B}) \cap \mathcal{C}\left(\overline{\mathbb{R}^2_+}\right)$, we have $v := \mathcal{U}|_{\partial\mathbb{R}^2_+} \in H^\alpha(\mathbb{R})$, and there exists a constant $C = C(\alpha) > 0$ such that

$$\| v \|_{H^\alpha(\mathbb{R})} \leq C \| \mathcal{U} \|_{\dot{H}^1(t^\mu;\mathcal{B})} .$$

So by a density argument, every $\mathcal{U} \in \dot{H}^1(t^\mu;\mathcal{B}) \cap \mathcal{C}\left(\overline{\mathbb{R}^2_+}\right)$ has a well defined trace $v \in H^\alpha(\mathbb{R})$. Conversely, any $v \in H^\alpha(\mathbb{R})$ is the trace (restriction to $t = 0$) of a function $\mathcal{U} \in \dot{H}^1(t^\mu;\mathcal{B}) \cap \mathcal{C}\left(\overline{\mathbb{R}^2_+}\right)$.

Definition 5. *We say that a function $\mathcal{U} \in \dot{H}^1(t^\mu;\mathcal{B})$ is a weak solution of* (4) *if*

$$\int_{\mathcal{B}} t^\mu \nabla\mathcal{U}(t,x) \nabla\Psi(t,x) dt\, dx - c_\alpha^{-1} \int_\Omega f(x) \mathrm{Tr}(\Psi)(x)\, dx = 0, \tag{16}$$

where f is as in (1), $\mathrm{Tr}(\Psi)$ *denotes the trace* $\Psi\big|_{\{0\}\times\mathbb{R}}$, *and* $\Psi \in \dot{H}^1(t^\mu;\mathcal{B}) \cap \mathcal{C}\left(\overline{\mathbb{R}^2_+}\right)$ *is an arbitrary test function.*

2.2. The Extension Problem

In the next statement we recall the results obtained from [5] which show that the fractional derivatives on the line are Dirichlet-to–Neumann operators for an extension degenerate PDE problem in $\mathbb{R} \times (0,\infty)$, where the data f have been taken in the more general setting: more precisely a weighted $L^p(w)$ space, where w satisfies the one-sided version A_p^+ (see [30]) of the familiar A_p condition of Muckenhoupt.

Fix $0 < \alpha < 1$. Given a semigroup $\{T_t\}_{t\geq 0}$ acting on real functions, the *generalized Poisson integral* of f is given by

$$P_t^\alpha f(x) = \frac{t^{2\alpha}}{4^\alpha \Gamma(\alpha)} \int_0^\infty e^{-t^2/(4s)} T_s f(x) \frac{ds}{s^{1+\alpha}}, \quad x \in \mathbb{R}; \tag{17}$$

see ([5] (1.9)) for more details.

By considering the semigroup of translations $T_s f(x) = f(x+s)$, $s \geq 0$, we find

$$P_t^\alpha f(x) = f * P_t^\alpha(x) := \int_{\mathbb{R}} f(s) P_t^\alpha(x-s)\, ds,$$

where

$$P_t^\alpha(x) := \frac{t^{2\alpha} e^{t^2/4x}}{4^\alpha \Gamma(\alpha)(-x)^{1+\alpha}} \chi_{(-\infty,0)}(x). \tag{18}$$

Since the kernel P_t^α is increasing and integrable in $(-\infty,0)$, it is well known that the function

$$P_*^\alpha f(x) := \sup_{t>0} |f| * P_t^\alpha(x) = \int_{\mathbb{R}} |f(t)| P_t^\alpha(x-t)\, dt,$$

is pointwise controlled by the usual Hardy–Littlewood maximal operator. However, since the support of P_t^α is $(-\infty,0)$, a sharper control can be obtained by using the one-sided Hardy–Littlewood maximal operator. This control and the behavior of P_*^α in weighted L^p-spaces will be used in the results of this paper. We revise briefly recall the two fundamental theorems from [5].

Theorem 2 ([5]). *Consider the semigroup of translations $T_t f(x) = f(x+t)$, $t \geq 0$, initially acting on functions $f \in \mathcal{S}$. Let $P_t^\alpha f$, $0 < \alpha < 1$, be as in* (17). *Then:*

1. *For $1 \le p \le \infty$, P_t^α is a bounded linear operator from $L^p(\mathbb{R})$ into itself and $\|P_t^\alpha f\|_{L^p(\mathbb{R})} \le \|f\|_{L^p(\mathbb{R})}$.*
2. *When $f \in \mathcal{S}$, the Fourier transform of $P_t^\alpha f$ is given by*

$$\widehat{P_t^\alpha f}(\xi) = \frac{2^{1-\alpha}}{\Gamma(\alpha)} (-it\xi^{1/2})^\alpha \|_\alpha(-it\xi^{1/2})\widehat{f}(\xi), \quad \xi \in \mathbb{R},$$

where $\|_\nu(z)$ is the modified Bessel function of the third kind or Macdonald's function, which is defined for arbitrary ν and $z \in \mathbb{C}$, see ([33] Chapter 5). In particular,

$$\widehat{P_t f}(\xi) = e^{-t(-i\xi)^{1/2}}\widehat{f}(\xi).$$

3. *The maximal operator P_*^α defined by $P_*^\alpha f(x) = \sup_{t>0} |P_t^\alpha f(x)|$ is bounded from $L^p(\mathbb{R}, w)$ into itself, for $w \in A_p^+, 1 < p < \infty$, and from $L^1(\mathbb{R}, w)$ into weak-$L^1(\mathbb{R}, w)$, for $w \in A_1^+$.*
4. *Let $f \in L^p(w)$, for $w \in A_p^+$, $1 \le p < \infty$. The function $\mathcal{U}(x,t) \equiv P_t^\alpha f(x)$ is a classical solution to the extension problem (4).*

Theorem 3 (Extension problem). *Let $f \in L^p(w)$, $w \in A_p^+$, $1 < p < \infty$. Then the function*

$$\mathcal{U}(x,t) := \frac{t^{2\alpha}}{4^\alpha \Gamma(\alpha)} \int_0^\infty e^{-t^2/(4\tau)} T_\tau f(x) \frac{d\tau}{\tau^{1+\alpha}}, \quad x \in \mathbb{R},\ t > 0,$$

is a classical solution to the extension problem

$$\begin{cases} -(\mathscr{D}_{right})\mathcal{U} + \frac{1-2\alpha}{t}\mathcal{U}_t + \mathcal{U}_{tt} = 0, & \text{in } \mathbb{R} \times (0,\infty), \\ \lim_{t\to 0^+} \mathcal{U}(x,t) = f(x), & \text{a.e. and in } L^p(w). \end{cases}$$

Moreover, for $c_\alpha := \dfrac{4^{\alpha-1/2}\Gamma(\alpha)}{\Gamma(1-\alpha)} > 0$, *we have*

$$-c_\alpha \lim_{t\to 0^+} t^{1-2\alpha}\mathcal{U}_t(x,t) = (\mathscr{D}_{right})^\alpha f(x) \quad \text{in the distributional sense.}$$

Remark 2. *This parallel result regarding the extension problem in the case of the Marchaud fractional time derivative has been derived as well in [18,20].*

3. Regularity Estimate up to the Boundary for the Degenerate Equation with the Neumann Boundary Condition

In this section, we prove the interior regularity estimate up to the boundary for the degenerate equation with the Neumann boundary condition associated to problem (4). Namely we provide the proof of Theorem 1. But before we get into that, it is necessary to explain the main ideas in the proof of interior regularity provided by Theorem 1. The proof of Theorem 1 is inspired by [5,7,8,34]. The method for this proof differs substantially from interior regularity methods for second-order equations, but is similar to the proof for the fractional Laplacian. Recall that for second-order equations, one first shows that D^2u is bounded, and then the estimate for equations with bounded measurable coefficients implies a $C^{2,\sigma}$ estimate for $\sigma \in (0, \min(1,\alpha))$. This is also true for the boundary regularity for solutions to fully nonlinear equations [35].

We shall start by the regularity property of the problem (1). We show in Proposition 1 that the solution of the problem (1) is of class $C^{0,\sigma}$. To the best of the authors' knowledge, the proofs available in the literature are those dealing with the case of the fractional Laplacian (see for instance [7,36] (Proposition 2.1.9)). With this result in hand, and by making an appropriate change of variables, we will use this result and estimate to prove our main theorem.

We start by recalling the following lemma from [37], which gives a Liouville-type theorem for (1) in the case $f = 0$.

Lemma 1. *Let $u \in C(\mathbb{R})$ be a function satisfying $(\mathscr{D}_{right})^{\alpha} u = 0$ in $\mathbb{R}_+$, $u = 0$ in $\mathbb{R}_-$, and $|u(x)| \leq C(1 + |x|^{\gamma})$ for some $\gamma < \alpha$. Then $u(x) = kx^{\alpha}$.*

The proof of this lemma relies on similar reasoning as the proof of ([37] (Theorem 2.2.3)) for the Caputo density function.

In the case where we have a non-vanishing right hand side ($f \neq 0$) as in (1), we state the following Liouville-type theorem for the one-sided Marchaud derivative.

Proposition 1. *Let $\alpha \in (0,1)$ and let $u \in \mathcal{L}^1_{\alpha} \cap L^{\infty}_{loc}(B_1)$ be the solution to*

$$(\mathscr{D}_{right})^{\alpha} u = f \quad in \quad B_1.$$

(a) *For $1 < p < \infty$, if $f \in L^p(B_1, w)$ and $r \in (0,1)$ and $\gamma \in (0, \min(1, \alpha))$ is such that $0 < \gamma - \frac{1}{p} < 1$, then $u \in C^{0,\gamma - \frac{1}{p}}(B_1, w)$ and there exists a constant $C := C(\alpha, r, \gamma, p) > 0$ such that*

$$\|u\|_{C^{0,\gamma-\frac{1}{p}}(B_r,w)} \leq C\left(\|u\|_{L^p(B_r,w)} + \|f\|_{L^p(B_1,w)}\right). \tag{19}$$

(b) *If $f \in L^{\infty}(B_1)$ and $r \in (0,1)$ and $\gamma \in (0, \min(1, \alpha))$, then $u \in C^{0,\gamma}(B_1)$ and there exists a constant $C := C(\alpha, r, \gamma) > 0$ such that*

$$\|u\|_{C^{0,\gamma}(B_r)} \leq C\left(\|u\|_{L^{\infty}(B_r)} + \|f\|_{L^{\infty}(B_1)}\right). \tag{20}$$

Proof. We will show that u has the corresponding regularity in a neighbourhood of the origin. We split the proof into two parts, as follows.

Proof of (a): $f \in L^p(B_1, w)$. Let $\eta \in C_c^{\infty}(\mathbb{R})$ be a smooth cutoff function such that $\eta = 1$ on B_r, $\eta = 0$ on $\mathbb{R} \setminus B_1$, and $0 \leq \eta \leq 1$ on $\mathbb{R}$. Consider the Riesz potential as defined in (15). Then the function

$$v(x) := \int_{\mathbb{R}} \mathcal{I}_{\alpha}(x,y)(\eta f)(y)dy, \quad \text{for all} \quad x \in \mathbb{R},$$

satisfies

$$(\mathscr{D}_{right})^{\alpha} v(x) = \eta(x) f(x) \quad \text{for all } x \in \mathbb{R}_+. \tag{21}$$

We first estimate the L^p norm of v for $\alpha < 1$. Since the kernel $(\mathscr{D}_{right})^{-\alpha}$ is positive and $\eta \geq 0$ is a smooth function with compact support in B_r, we write $v = (\mathscr{D}_{right})^{-\alpha}(\eta f) = (\mathscr{D}_{right})^{1-\alpha} \circ (\mathscr{D}_{right})^{-1}(\eta f)$. We note that, by using a similar argument as for the Poisson equation for the fractional Laplacian, we find that $(\mathscr{D}_{right})^{-1}(\eta f)$ is an element of $C^{1,\gamma}$ with norm depending only on $\|f\|_{C^{0,\gamma}}$.

Since ηf is compactly supported, we get

$$\|v\|_{C^{0,\gamma}(\mathbb{R})} \leq C_{\alpha,\gamma} \|\eta f\|_{L^p(B,w^p)} + C\|v\|_{L^p(\mathbb{R},w^p)} \leq C_{\alpha,\gamma,B'} \|f\|_{L^p(B,w^p)}.$$

For $\alpha < 1$ and $\gamma \in (0, \min(\alpha, 1))$ and $x, y \in B_r$, we have

$$\begin{aligned} v(x) - v(y) &= \int_{\mathbb{R}} \left(\mathcal{I}_{\alpha}(x,z) - \mathcal{I}_{\alpha}(y,z)\right)(\eta f)(z)dz \\ &= C_{1,\alpha} \int_{\mathbb{R}} \left(|x - z|^{\alpha - 1} - |y - z|^{\alpha - 1}\right)(\eta f)(z)\, dz. \end{aligned}$$

Next we consider the following inequalities [38], valid for $\gamma \in (0, \min(1, \alpha))$ and $m \in \mathbb{R}$ with $m + \gamma > 0$ and for every $x, y, z \in B_r$:

$$\left|\left(|x-z|^{-m} - |y-z|^{-m}\right)\right| \leq \frac{|m|}{m+\gamma}|x-y|^{\gamma}\left(|x-z|^{-(m+\gamma)} + |y-z|^{-(m+\gamma)}\right). \tag{22}$$

For $m = 1 - \alpha$, and for $1 < p < \infty$, we can write

$$\begin{aligned}|v(x) - v(y)| &< \frac{(1-\alpha)}{(1-\alpha+\gamma)}\int_{\mathbb{R}} |x-y|^{\gamma}\left(|x-z|^{\alpha-1-\gamma} + |y-z|^{\alpha-1-\gamma}\right)|(\eta f)(z)|dz \\ &\leq C_{\alpha,\gamma}|x-y|^{\gamma-\frac{1}{p}}\int_{\mathbb{R}} |x-y|^{\frac{1}{p}}|x-z|^{\alpha-1-\gamma}|(\eta f)(z)|dz,\end{aligned}$$

since $\dfrac{|y - z|}{|x - z|} > 1$. Using the fact that the support of η is always contained in the ball of radius 2 centred at $x \in B_r$, we have that

$$\begin{aligned}|v(x) - v(y)| &\leq C(1,\alpha,\gamma)|x-y|^{\gamma}\int_{\mathbb{R}} |x-y|^{\alpha-1-\gamma+\frac{1}{p}}|(\eta f)(y)|dy, \\ &\leq C(1,\alpha,\gamma)|x-y|^{\gamma-\frac{1}{p}}\left(\int_{\mathbb{R}} w^{-1}\,|x-y|^{p(\alpha-1-\gamma)+1}\,w\,|(\eta f)(y)|^{p}\,dy\right)^{1/p} \\ &\leq C(1,\alpha,\gamma)|x-y|^{\gamma-\frac{1}{p}}\left(\int_{B(x,2)} w\,|(\eta f)(y)|^{p}\,dy\right)^{\frac{1}{p}}\left(\int_{B(x,2)}\frac{w^{-\frac{1}{p-1}}}{|x-y|^{\frac{p(-\alpha+1+\gamma)-1}{p-1}}}dy\right)^{\frac{p-1}{p}} \\ &\leq C(1,\alpha,\gamma)|x-y|^{\gamma-\frac{1}{p}}\left(\int_{B_1} w\,|f(y)|^{p}\,dy\right)^{\frac{1}{p}}\left(\int_{B_1} w^{-\frac{1}{p-1}}\,|y|^{\frac{p(\alpha-1-\gamma)+1}{p-1}}\,dy\right)^{\frac{p-1}{p}} \\ &\leq 2C(1,\alpha,\gamma)|x-y|^{\gamma-\frac{1}{p}}\,\|f\|_{L^p(B_1,w)}\left(\int_{B_1} w^{-\frac{1}{p-1}}\,|y|^{\frac{p(\alpha-1-\gamma)+1}{p-1}}\,dy\right)^{\frac{p-1}{p}},\end{aligned}$$

up to relabelling of the positive constant $C(\alpha, \gamma)$ that depends on α and γ. Replacing $w(y)$ by its value $|y|^{1-2\alpha}$ and using the polar coordinates $y = rx$, $r > 0$, we get that

$$\begin{aligned}\int_{B_1} w(y)^{-\frac{1}{p-1}}\,|y|^{\frac{p(\alpha-1-\gamma)+1}{p-1}}\,dy &:= \int_{S_1}\int_0^1 |r|^{\frac{1}{p-1}(2\alpha+p(\alpha-1-\gamma))}\,dr\,d\sigma(x) \leq \frac{(p-1)}{p(\alpha-1-\gamma)+p-1)}|S_1| \\ &\leq C(p,\gamma,\alpha).\end{aligned}$$

Then,

$$|v(x) - v(y)| \leq C(p,\gamma,\alpha)|x-y|^{\gamma-\frac{1}{p}}\,\|f\|_{L^p(B_1,w)}.$$

Hence, we conclude that

$$\|v\|_{C^{0,\gamma-\frac{1}{p}}(B_r)} \leq C_\alpha\|f\|_{L^p(B_1,w)}, \tag{23}$$

for every $\gamma \in (0, \min(1, \alpha))$.

Next, by change of variables, the function $\xi := u - v$ satisfies $(\mathscr{D}_{right})^{\alpha}\xi = 0$ in B_r by (21). Then, thanks to the derivative estimate, for every $r' \in (0, r)$,

$$\|\nabla\xi\|_{L^p(B_{r'},w)} \leq C_{\alpha,r'}\|\xi\|_{L^p(B_1,w)} \leq C_{\alpha,r'}\left(\|u\|_{L^p(B_r,w)} + \|v\|_{L^p(B_r,w)}\right).$$

The difference function $\xi = u - v$ is smooth in B_1 and is bounded. From this observation, together with (23), we have that

$$\|u\|_{C^{0,\gamma-\frac{1}{p}}(B_{r'})} = \|\xi + v\|_{C^{0,\gamma}(B_{r'})} \leq C_{\alpha,r'}\left(\|u\|_{L^p(B_r,w)} + \|f\|_{L^p(B_1,w)}\right),$$

for every $\gamma \in (0, \min(1, \alpha))$ with $0 < \gamma - \frac{1}{p} < 1$, as required.

Proof of (b): $f \in L^\infty(B_1)$. The proof in this case is similar to the previous one. We consider as above a smooth cutoff function $\eta \in C_c^\infty(\mathbb{R})$ such that $\eta = 1$ on B_r, $\eta = 0$ on $\mathbb{R} \setminus B_1$, and $0 \le \eta \le 1$ on $\mathbb{R}$. Then we consider the Riesz potential as defined in (15), so that we can estimate the L^∞ norm of v for $\alpha < 1$. Since the kernel $\mathcal{I}_\alpha$ is positive and $\eta \ge 0$ is a smooth function with compact support in B_r, we get

$$\|v\|_{C^{0,\gamma}(\mathbb{R})} \le C_{\alpha,\gamma} \|\eta f\|_{L^\infty(\mathbb{R})} + C\|v\|_{L^\infty(\mathbb{R})} \le C_{\alpha,\gamma,B'} \|f\|_{L^\infty(B)}.$$

Next, by using the inequality stated in (22), we get that for $\gamma \in (0, \min(\alpha, 1))$ and $x, y \in B_r$,

$$\begin{aligned} |v(x) - v(y)| &\le C_{\alpha,\gamma} |x-y|^\gamma \int_{\mathbb{R}} |x-y|^{\alpha-1-\gamma} \eta(y) |f(y)| dy \\ &\le C_{\alpha,\gamma} |x-y|^\gamma \|f\|_{L^\infty(B_1)} \int_{B_2} |x-y|^{\alpha-1-\gamma} dy \\ &\le C_{\alpha,\gamma} \|f\|_{L^\infty(B_1)} |x-y|^\gamma. \end{aligned}$$

Hence, we conclude that

$$\|v\|_{C^{0,\gamma}(B_r)} \le C_\alpha \|f\|_{L^\infty(B_1)}, \tag{24}$$

for every $\gamma \in (0, \min(1, \alpha))$.

Next, by change of variables, the function $\xi := u - v$ satisfies $(\mathscr{D}_{right})^\alpha \xi = 0$ in B_r by (21). Therefore, thanks to ([7] (Corollary 1.13)), we have the derivative estimate for every $r' \in (0, r)$:

$$\|\nabla w\|_{L^\infty(B_r')} \le C_{\alpha,r'} \|w\|_{L^\infty(B_1)} \le C_{\alpha,r'} (\|u\|_{L^\infty(B_r)} + \|v\|_{L^\infty(B_r)}).$$

The difference function $\xi = u - v$ is smooth in B_1 and is bounded. From this, together with (24), we conclude that

$$\|u\|_{C^{0,\gamma}(B_{r'})} = \|w + v\|_{C^{0,\gamma}(B_{r'})} \le C_{\alpha,r'} \left(\|u\|_{L^\infty(B_r)} + \|f\|_{L^\infty(B_1)} \right),$$

for every $\gamma \in (0, \min(1, \alpha))$. □

Now we are in a position to state and prove our main result on the interior Schauder estimate for the solution function $\mathcal{U}$ on the set $\overline{\mathcal{B}_1^+}$.

Proof of the Main Result: Theorem 1

Proof. Again, we present the two parts of the proof separately.

Proof of (a): $f \in L^p(\mathbb{R}, w)$. We choose a cut-off function $\eta \in C_c^\infty(B_2)$ such that $\eta \equiv 1$ on B_2 and $0 \le \eta \le 1$ on $\mathbb{R}$. Let $\overline{v}$ be the unique solution to the equation

$$(\mathscr{D}_{right})^\alpha \overline{v} = \overline{f} \qquad \text{in} \quad \mathbb{R},$$

where $\overline{f} := \eta f$. Making use of the previous result Proposition 1, we know that $\overline{v} \in C^{\alpha-\gamma-\frac{1}{p}}(\mathbb{R}, w)$ and

$$\|\overline{v}\|_{C^{\alpha-\gamma-\frac{1}{p}}(\mathbb{R},w)} \le C \left(\|\overline{v}\|_{L^p(\mathbb{R},w)} + \left\|\overline{f}\right\|_{L^p(\mathbb{R},w)} \right),$$

where $C > 0$ is a constant that depends only on α, γ, and p.

The next step is to consider the Bernardis–Reyes–Stinga–Torrea extension $\overline{\mathcal{U}}$ of $\overline{v}$, i.e., the function

$$\overline{\mathcal{U}}(t, \cdot) = P_t^\alpha(t, \cdot) \star \overline{v},$$

which satisfies the equations

$$\begin{cases} \mathcal{M}_\alpha \overline{\mathcal{U}}(t,x) = 0 & \text{in } \mathbb{R}^2_+, \\ \lim_{t\to 0} \mathcal{N}_\alpha \overline{\mathcal{U}}(t,x) = (\mathscr{D}_{right})^\alpha \overline{v}(x) = \overline{f}(x) & \text{on } \mathbb{R}. \end{cases}$$

By a change of variables, we have

$$\overline{\mathcal{U}}(t,x) = \left(P_t^\alpha(t,\cdot) \star \overline{v}\right)(x) = \int_{\mathbb{R}} \overline{v}(x-ty)H_\alpha(y)dy, \tag{25}$$

where

$$H_\alpha(y) = P_t^\alpha(1,y) = \frac{C_{1,\alpha} e^{-1/(4(-y))}}{(-y)^{1+\alpha}} \chi_{(-\infty,0)}(y). \tag{26}$$

Then, if we set $z_1 = (t_1,x_1), z_2 = (t_2,x_2) \in \overline{\mathbb{R}^2_+}$, we have the estimate

$$\begin{aligned} |\overline{\mathcal{U}}(z_1) - \overline{\mathcal{U}}(z_2)| &\le |z_1 - z_2|^{\alpha-\gamma-\frac{1}{p}} \parallel \overline{v} \parallel_{C^{\alpha-\gamma-\frac{1}{p}}(\mathbb{R},w)} \int_{\mathbb{R}} \max\{|y|^{\alpha-\gamma-\frac{1}{p}}, 1\} H_\alpha(y)dy, \\ &\le C|z_1 - z_2|^{\alpha-\gamma-\frac{1}{p}} \left(\|\overline{v}\|_{L^p(\mathbb{R},w)} + \left\|\overline{f}\right\|_{L^p(\mathbb{R},w)} \right). \end{aligned}$$

By direct computation from (25), and using Theorem 2, we have:

$$\|\overline{\mathcal{U}}\|_{L^p(\mathbb{R}^2_+,w)} \le \|\overline{v}\|_{L^p(\mathbb{R},w)} \le C \left\|\overline{f}\right\|_{L^p(\mathbb{R},w)}.$$

Therefore,

$$\|\overline{\mathcal{U}}\|_{C^{\alpha-\gamma-\frac{1}{p}}\left(\mathbb{R}^2_+\right)} \le C \left(\|\overline{V}\|_{L^\infty(\mathbb{R}^2_+)} + \left\|\overline{f}\right\|_{L^\infty(\mathbb{R})} \right), \tag{27}$$

for a positive constant $C > 0$ depending only on α, p and γ.

Next we put $\widetilde{\mathcal{U}} = \mathcal{U} - \overline{\mathcal{U}}$, so that $\widetilde{\mathcal{U}}$ satisfies

$$\begin{cases} \mathcal{M}_\alpha \widetilde{\mathcal{U}} = 0 & \text{in } \mathcal{B}_2^+, \\ \lim_{t\to 0} \mathcal{N}_\alpha \widetilde{\mathcal{U}}(t,\cdot) = (1-\eta)f = 0 & \text{on } B_2. \end{cases}$$

Considering the even reflection $\widetilde{Z}$ of $\widetilde{\mathcal{U}}$ in the variable t, as described in ([37] (Lemma 4.1)), we have that

$$\mathcal{M}_\alpha \widetilde{Z} = 0 \text{ in } \mathcal{B}_2.$$

From the definition (7) of $\left(\mathscr{D}_{right}\right)$, and using ([5] (Corollary 1.13)) or ([39] (Corollary 1.5)), we have that for $x \in B_1$ and $t \in (-1,1)$ fixed,

$$\left|(\mathscr{D}_{right})\widetilde{Z}(t,x)\right| \le C \parallel \widetilde{Z}(t,\cdot) \parallel_{L^p(B_2,w)}. \tag{28}$$

Next, from the fact that

$$(\mathscr{D}_{right})\widetilde{Z} = \widetilde{Z}_{tt} + \frac{\mu}{|t|}\widetilde{Z}_t$$

and from the inequality (28), we obtain

$$\left|\widetilde{Z}_{tt} + \frac{\mu}{|t|}\widetilde{Z}_t\right| \le C \parallel \widetilde{Z} \parallel_{L^p(\mathcal{B}_2,w)}.$$

Therefore

$$\left|\left(|t|^\mu \widetilde{Z}_t\right)_t\right| \le C|t|^\mu \parallel \widetilde{Z} \parallel_{L^p(\mathcal{B}_2,w)}$$

Hence

$$\left|\widetilde{Z}_{tt}\right| \leq C \parallel \widetilde{Z} \parallel_{L^p(\mathcal{B}_2,w)} .$$

For any point $(t,x) \in \overline{\mathcal{B}}_1$, we have, by (28), that

$$\left|(\mathscr{D}_{right})\widetilde{Z}(t,x)\right| + \left|\widetilde{Z}_{tt}(t,x)\right| \leq C \parallel \widetilde{Z} \parallel_{L^p(\mathcal{B}_2,w)},$$

which implies that

$$\widetilde{Z} \in C^{1-\gamma-\frac{1}{p}}(\overline{\mathcal{B}}_1, w).$$

Thus, we have that $\widetilde{\mathcal{U}} \in C^{1-\gamma-\frac{1}{p}}(\overline{\mathcal{B}_1^+}, w)$ such that

$$\begin{aligned}\parallel \widetilde{\mathcal{U}} \parallel_{C^{1-\gamma-\frac{1}{p}}(\overline{\mathcal{B}_1^+},w)} &\leq C\left(\parallel \mathcal{U} \parallel_{L^p(\mathcal{B}_2^+,w)} + \parallel \overline{v} \parallel_{L^p(\mathbb{R})}\right) \\ &\leq C\left(\parallel \mathcal{U} \parallel_{L^p(\mathcal{B}_2^+,w)} + \parallel f \parallel_{L^p(B_2,w)}\right).\end{aligned}$$

We finally obtain

$$\begin{aligned}\parallel \mathcal{U} \parallel_{C^{\alpha-\gamma-\frac{1}{p}}(\overline{\mathcal{B}_1^+},w)} &\leq C\left(\parallel \widetilde{\mathcal{U}} \parallel_{C^{1-\gamma-\frac{1}{p}}(\overline{\mathcal{B}_1^+},w)} + \parallel \overline{\mathcal{U}} \parallel_{C^{\alpha-\gamma-\frac{1}{p}}(\overline{\mathbb{R}^2_+},w)}\right) \\ &\leq C\left(\parallel \mathcal{U} \parallel_{L^p(\mathcal{B}_2^+,w)} + \parallel f \parallel_{L^p(B_2),w}\right),\end{aligned}$$

since $\widetilde{\mathcal{U}} = \mathcal{U} - \overline{\mathcal{U}}$. This ends the proof of the first case.

Proof of (b): $f \in L^\infty(\mathbb{R})$. The proof here is similar to the first case above. By considering the same cut-off function $\eta \in C_c^\infty(B_2)$ with $\eta \equiv 1$ on B_2 and $0 \leq \eta \leq 1$ on $\mathbb{R}$, we let $\overline{v}$ be the unique solution to the equation

$$(\mathscr{D}_{right})^\alpha \overline{v} = \overline{f} \qquad \text{in} \quad \mathbb{R},$$

where $\overline{f} := \eta f$. Making use of the previous result Proposition 1, we have that $\overline{v} \in C^{\alpha-\gamma}(\mathbb{R})$ and

$$\|\overline{v}\|_{C^{\alpha-\gamma}(\mathbb{R})} \leq C\left(\|\overline{v}\|_{L^\infty(\mathbb{R})} + \left\|\overline{f}\right\|_{L^\infty(\mathbb{R})}\right),$$

where $C > 0$ is a constant that depends only on α and γ.

The next step is to consider the Bernardis–Reyes–Stinga–Torrea extension $\overline{\mathcal{U}}$ of $\overline{v}$, i.e.,

$$\overline{\mathcal{U}}(t,\cdot) = P_t^\alpha(t,\cdot) \star \overline{v},$$

which satisfies the equation

$$\begin{cases} \mathcal{M}_\alpha \overline{\mathcal{U}} = 0 & \text{in } \mathbb{R}^2_+, \\ \lim_{t\to 0} \mathcal{N}_\alpha \overline{\mathcal{U}}(t,x) = (\mathscr{D}_{right})^\alpha \overline{v}(x) = \overline{f}(x) & \text{on } \mathbb{R}. \end{cases}$$

Proceeding as in the previous case, it follows that

$$\|\overline{\mathcal{U}}\|_{C^{\alpha-\gamma}\left(\overline{\mathbb{R}^2}\right)} \leq C\left(\|\overline{\mathcal{U}}\|_{L^\infty(\mathbb{R}^2)} + \left\|\overline{f}\right\|_{L^\infty(\mathbb{R})}\right), \tag{29}$$

for a positive constant $C > 0$ depending only on α and γ.

Next we put $\widetilde{\mathcal{U}} = \mathcal{U} - \overline{\mathcal{U}}$, so that $\widetilde{\mathcal{U}}$ satisfies

$$\begin{cases} \mathcal{M}_\alpha \widetilde{\mathcal{U}} = 0 & \text{in } \mathcal{B}_2^+, \\ \lim_{t\to 0} \mathcal{N}_\alpha \widetilde{\mathcal{U}}(t,\cdot) = (1-\eta)f = 0 & \text{on } B_2. \end{cases}$$

We finally obtain

$$\| \mathcal{U} \|_{C^{\alpha-\gamma}(\overline{\mathcal{B}_1^+})} \le C \left(\| \widetilde{\mathcal{U}} \|_{C^{1-\gamma}(\overline{\mathcal{B}_1^+})} + \| \overline{\mathcal{U}} \|_{C^{\alpha-\gamma}(\overline{\mathbb{R}_+^2})} \right) \le C \left(\| \mathcal{U} \|_{L^\infty(\mathcal{B}_2^+)} + \| f \|_{L^\infty(B_2)} \right),$$

which ends the proof. □

4. Conclusions

Regularity theorems are an important result in the theory of PDEs, and their fractional counterparts also play a significance role in the study of problems involving nonlocal behaviour. As already observed in various papers [6,8,9,34,40–42] in the theory of fractional nonlocal PDEs, it is possible to find the qualitative behaviour of a solution. In this paper we have shown that the degenerate elliptic equation with mixed boundary conditions for a problem with fractional Marchaud derivative admits an interior regularity estimate. The current work fits in with some results obtained in the case of fractional Laplacians with Caffareli–Silvestre extensions. We stress that the types of regularity results proved herein form only a small subset of many possible versions of regularity theorems which can only be obtained using extension techniques.

Author Contributions: J.-D.D. and A.F. contributed equally to this work.

References

1. Dipierro, S.; Valdinoci, E. A Simple Mathematical Model Inspired by the Purkinje Cells: From Delayed Travelling Waves to Fractional Diffusion. *Bull. Math. Biol.* **2018**, *80*, 1849–1870. [CrossRef] [PubMed]
2. Marinov, T.; Santamaria, F. Modeling the effects of anomalous diffusion on synaptic plasticity. *BMC Neurosci.* **2013**, *14* (Suppl. 1), 343. [CrossRef]
3. Marinov, T.; Santamaria, F. Computational modeling of diffusion in the cerebellum. *Prog. Mol. Biol. Trans. Sci.* **2014**, *123*, 169–189.
4. Abatangelo, N.E.; Valdinoci, E. Getting acquainted with the fractional Laplacian. *arXiv* **2017**, arXiv:1710.11567.
5. Bernardis, A.; Martín-Reyes, F.J.; Stinga, P.R.; Torrea, J.L. Maximum principles, extension problem and inversion for nonlocal one-sided equations. *J. Differ. Equ.* **2016**, *260*, 6333–6362. [CrossRef]
6. Caffarelli, L.; Silvestre, L. An extension problem related to the fractional Laplacian. *Commun. Part. Differ. Equ.* **2007**, *32*, 1245–1260. [CrossRef]
7. Stinga, P.R.; Torrea, J.L. Problem and Harnack's Inequality for Some Fractional Operators. *Commun. Part. Differ. Equ.* **2010**, *35*, 2092–2122. [CrossRef]
8. Niang, A. Boundary regularity for a degenerate elliptic equation with mixed boundary conditions. *arXiv* **2018**, arXiv:1803.10641.
9. Cabré, X.; Sire, Y. Nonlinear equations for fractional Laplacians I: Regularity, maximum principles, and Hamiltonian estimates. *Ann. Inst. Henri Poincaré Anal. Non Linéaire* **2014**, *31*, 23–53. [CrossRef]
10. Björn, J. Regularity at infinity fot a mixed problem for degenerate elliptic operators in a half-cynlider. *Math. Scand.* **1997**, *81*, 101–126. [CrossRef]
11. Fabes, E.; Jerison, D.; Kenig, C. The Wiener test for degenerate elliptic equations. *Ann. Inst. Fourier (Grenoble)* **1982**, *32*, 151–182. [CrossRef]
12. Fabes, E.; Kenig, C.; Serapioni, R. The local regularity of solutions of degenerate elliptic equations. *Commun. Part. Differ. Equ.* **1982**, *7*, 77–116. [CrossRef]
13. Jin, T.; Li, Y.Y.; Xiong, J. On a fractional nirenberg problem Part I: Blow up analysis and compactness solutions. *J. Eur. Math. Soc.* **2014**, *16*, 1111–1171. [CrossRef]

14. Kassmann, M.; Madych, W.R. Difference quotients and elliptic mixed boundary value problems of second order. *Indiana Univ. Math. J.* **2007**, *56*, 1047–1082. [CrossRef]
15. Levendorskii, S. Degenerate Elliptic Equations. In *Mathematics and Its Applications*; Springer: Dordrecht, The Netherlands, 1993; Volume 258, 436p, ISBN 978-94-017-1215-6.
16. Savaré, G. Regularity and perturbation results for mixed second order elliptic problems. *Commun. Part. Differ. Equ.* **1997**, *22*, 869–899. [CrossRef]
17. Zaremba, S. Sur un problème mixte relatif à l'équation de Laplace. In *Bulletin International de L'Académie des Sciences de Cracovie. Classe des Sciences Mathématiques et Naturelles*; Serie A: Sciences Mathématiques; Impr. de l'Université: Cracovie, Poland, 1910; pp. 313–344.
18. Bucur, C.; Ferrari, F. An extension problem for the fractional derivative defined by Marchaud. *Fract. Calc. Appl. Anal.* **2016**, *19*, 867–887. [CrossRef]
19. Allen, M.; Caffarelli, L.; Vasseur, A. A parabolic problem with a fractional time derivative. *Arch. Ration. Mech. Anal.* **2016**, *221*, 603–630. [CrossRef]
20. Ferrari, F. Weyl and Marchaud derivatives: A forgotten history. *Mathematics* **2018**, *6*, 6. [CrossRef]
21. Samko, S.G.; Kilbas, A.A.; Marichev, O.I. *Fractional. Integrals and Derivatives, Theory and Applications*; Translated from the 1987 Russian Original; Gordon and Breach Science Publishers: Yverdon, Switzerland, 1993.
22. Kufner, A. *Weighted Sobolev Spaces*; Wiley: New York, NY, USA, 1985.
23. Maz'ya, V.G. *Sobolev Spaces with Applications to Elliptic Partial Differential Equations*; Volume 342 of Grundlehren der Mathematischen Wissenschaften [Fundamental Principles of Mathematical Sciences]; Springer: Heidelberg, Germany, 2011.
24. Triebel, H. *Interpolation Theory, Function Spaces, Differential Operators*, 2nd ed.; Johann Ambrosius Barth: Heidelberg, Germany, 1995.
25. Meyries, M.; Veraar, M. Sharp embedding results for spaces of smooth functions with power weights. *arXiv* **2011**, arXiv:1112.5388.
26. Köhne, M.; Prüss, J.; Wilke, M. On quasilinear parabolic evolution equations in weighted L_p-spaces. *J. Evol. Equ.* **2010**, *10*, 443–463.
27. Lunardi, A. *Analytic Semigroups and Optimal Regularity in Parabolic Problems*; Progress in Nonlinear Differential Equations and Their Applications; Birkhäuser Verlag: Basel, Switzerland, 1995; Volume 16.
28. Triebel, H. Theory of Function Spaces. In *Monographs in Mathematics*; Birkhäuser Verlag: Basel, Switzerland, 1983; Volume 78.
29. Coifman, R.R.; Fefferman, C. Weighted norm inequalities for maximal functions and singular integrals. *Stud. Math.* **1974**, *51*, 241–250. [CrossRef]
30. Muckenhoupt, B. Weighted norm inequalities for the Hardy maximal function. *Trans. Am. Math. Soc.* **1972**, *165*, 207–226. [CrossRef]
31. Opic, B.; Kufner, A. *Hardy-Type Inequalities*; Wiley: New York, NY, USA, 1990.
32. Stein, E.M. *Harmonic Analysis: Real-Variable Methods, Orthogonality, and Oscillatory Integrals*; Volume 43 of Princeton Mathematical Series, Monographs in Harmonic Analysis, III; Princeton University Press: Princeton, NJ, USA, 1993.
33. Lebedev, N.N. *Special Functions and Their Applications*; Prentice-Hall, Inc.: Englewood Cliffs, NJ, USA, 1965.
34. Fall, M.M. Regularity estimates for nonlocal Schrödinger equations. *arXiv* **2017**, arXiv:1711.02206.
35. Caffarelli, L.; Cabré, X. *Fully Nonlinear Elliptic Equations*; American Mathematical Society: Providence, RI, USA, 1995; Volume 43.
36. Silvestre, L. Regularity of the obstacle problem for a fractional power of the Laplace operator. *Commun. Pure Appl. Math.* **2007**, *60*, 67–112, . [CrossRef]
37. Bucur, C. Some nonlocal operators and effects due to nonlocality. *arXiv* **2017**, arXiv:1705.00953.
38. Fall, M.M.; Weth, T. Monotonicity and nonexistence results for some fractional elliptic problems in the half space. *arXiv* **2013**, arXiv:1309.7230.
39. Caffarelli, L.A.; Salsa, S.; Silvestre, L. Regularity estimates for the solution and the free boundary of the obstacle problem for the fractional Laplacian. *Invent. Math.* **2008**, *171*, 425–461. [CrossRef]
40. Silvestre, L. Regularity of the obstacle problem for a fractional power of the Laplace operator. *Comm. Pure Appl. Math.* **2007**, *60*, 67–112. [CrossRef]

41. Dipierro, S.; Soave, N.; Valdinoci, E. On fractional elliptic equations in Lipschitz sets and epigraphs: Regularity, monotonicity and rigidity results. *Math. Ann.* **2017**, *369*, 1283–1326. [CrossRef]
42. Fernandez, A. An elliptic regularity theorem for fractional partial differential operators. *Comput. Appl. Math.* **2018**, 1–12. [CrossRef]

11

Fixed Point Theorems through Modified ω-Distance and Application to Nontrivial Equations

Tariq Qawasmeh [1], Abdalla Tallafha [1] and Wasfi Shatanawi [2,3,*]

1 Department of mathematics, School of Science, University of Jordan, Amman 11942, Jordan; jorqaw@yahoo.com or tar9160251@fgs.ju.edu.jo (T.Q.); a.tallafha@ju.edu.jo (A.T.)
2 Department of Mathematics and general courses, Prince Sultan University, Riyadh 11586, Saudi Arabia
3 Department of Medical Research, China Medical University Hospital, China Medical University, Taichung 40402, Taiwan
* Correspondence: wshatanawi@psu.edu.sa or wshatanawi@yahoo.com

Abstract: In this manuscript, we utilize the concept of modified ω-distance mapping, which was introduced by Alegre and Marin [Alegre, C.; Marin, J. Modified ω-distance on quasi metric spaces and fixed point theorems on complete quasi metric spaces. *Topol. Appl.* **2016**, *203*, 120–129] in 2016 to introduce the notions of (ω, φ)-Suzuki contraction and generalized (ω, φ)-Suzuki contraction. We employ these notions to prove some fixed point results. Moreover, we introduce an example to show the novelty of our results. Furthermore, we introduce some applications for our results.

Keywords: quasi metric space; Suzuki contractions; fixed point theorems; modified ω-distance; almost perfect functions

---

1. Introduction and Preliminaries

Constructing new contractions and formulating new fixed point theorems are very important subjects in mathematics since active researchers employ the existence and uniqueness of the fixed point to solve some integral equations, differential equations, etc.

Banach was the first pioneer mathematician who constructed and formulated the first fixed point theorem, which was called after him as the Banach contraction principle [1].

Suzuki [2] introduced a new contraction and generalized the Banach contraction principle.

In the rest of this paper, the letter d refers to a metric on a set B and f_1 refers to self-mappings on B.

One of the important contractions is the Kannan contraction [3]:

$$d(f_1 l_1, f_1 l_2) \leq \alpha[d(l_1, f_1 l_1) + d(l_2, f_1 l_2)] \text{ for all } l_1, l_2 \in B,$$

where $\alpha \in [0, \frac{1}{2})$.

Moreover, Kannan proved that if f_1 satisfies Kannan contraction, then f_1 has a unique fixed point.

In 1931, Wilson [4] generalized the notion of metric spaces to a new notion called quasi metric spaces.

Definition 1. *We call $q : B\times B \rightarrow [0, \infty)$ a quasi metric if q satisfies:*

(i) $q(l_1, l_2) = 0 \Longleftrightarrow l_1 = l_2$
and:

(ii) $q(l_1, l_2) \leq q(l_1, l_3) + q(l_3, l_2)$ *for all* $l_1, l_2, l_3 \in B$.

(B, q) *is called a quasi metric space.*

From now on, by (B,q), we mean a quasi metric space.

Defining $q_m : B \times B \to [0,+\infty)$ via

$$q_m(l_1,l_2) = \max\{q(l_1,l_2), q(l_2,l_1)\},$$

we generate a metric on B.

Recall the following definitions.

Definition 2. *[5,6] The sequence (l_t) converges to $l \in B$ if $\lim_{t\to\infty} q(l_t,l) = \lim_{n\to\infty} q(l,l_t) = 0$.*

Definition 3. *[6] Let (l_t) be a sequence in (B,q). Then, we say that:*

(i) (l_t) is left-Cauchy if for any $\varepsilon > 0$, there exists $n_0 \in \mathbb{N}$ such that $q(l_t,l_m) < \varepsilon\ \forall\ t \geq m > n_0$.

(ii) (l_t) is right-Cauchy if for any $\varepsilon > 0$, there exists $n_0 \in \mathbb{N}$ such that $q(l_t,l_m) < \varepsilon\ \forall\ m \geq t > n_0$.

Definition 4. *[5,6] We say that (l_t) is Cauchy if for any $\varepsilon > 0$, there exists $n_0 \in \mathbb{N}$ such that $q(l_t,l_m) \leq \epsilon\ \forall\ t,m > n_0$.*

We note that (l_t) in (B,q) is Cauchy if and only if (l_t) is right and left Cauchy.

Definition 5. *[5,6] We say that (B,q) is complete if every Cauchy sequence in B is convergent.*

For some theorems in quasi-metric space, see [5–9].

Alegre and Marin [10] introduced the concept of modified ω-distance mappings on (B,d).

Definition 6. *[10] A modified ω-distance (shortened as $m\omega$-distance) on (B,q) is a function $p : B\times B \to [0,\infty)$, which satisfies:*

(W1) $p(l_1,l_2) \leq p(l_1,l_3) + p(l_3,l_2)$ for all $l_1,l_2,l_3 \in B$;

(W2) $p(l,.) : B \to [0,\infty)$ is lower semi-continuous for all $l \in B$; and

(mW3) for each $\varepsilon > 0$, there exist $v > 0$ such that if $p(l_1,l_2) \leq v$ and $p(l_2,l_3) \leq v$, then $q(l_1,l_3) \leq \epsilon$ for all $l_1,l_2,l_3 \in B$.

Definition 7. *[10] We call an $m\omega$-distance function a p strong $m\omega$-distance if p is lower semi-continuous on its second coordinate.*

Remark 1. *[10] If q is a quasi metric on B, then q is $m\omega$-distance.*

Lemma 1. *[11] Let (α_t), (β_t) be two sequences of nonnegative real numbers converging to zero. Assume that p is $m\omega$-distance. Then, we have the following:*

(i) If $p(l_t,l_m) \leq \alpha_t$ for any $t,m \in \mathbb{N}$ with $m \geq t$, then (l_t) is right Cauchy in (B,q).

(ii) If $p(l_t,l_m) \leq \beta_m$ for any $t,m \in \mathbb{N}$ with $t \geq m$, then (l_t) is left Cauchy in (B,q).

Remark 2. *[11] The above lemma implies that if $\lim_{m,t\to\infty} p(l_t,l_m) = 0$, then (l_t) is Cauchy in (B,q).*

For some works on ω-distance, we ask the readers to see [11–13].

Abodayeh et al. [14] generalized the definition of altering the distance function [15] to the concept of the almost perfect function.

Definition 8. *We call a non-decreasing function $\varphi : [0,\infty) \to [0,\infty)$ almost perfect if φ satisfies:*

(i) $\varphi(l) = 0$ if and only if $l = 0$.

(ii) If (l_t) is a sequence in $[0,\infty)$ such that $\lim_{t\to\infty} \varphi(l_t) = 0$, then $\lim_{t\to\infty} l_t = 0$.

2. Main Results

We begin our work with the following definition:

Definition 9. *Let $\varphi : \mathbb{R}^+ \to \mathbb{R}^+$ be an almost perfect function and p be modified ω-distance on B. We say that p is bounded with respect to φ if there exists an integer $A > 0$ such that:*

$$\varphi p(l,e) \leq A \text{ for all } l, e \in B.$$

Definition 10. *Equip (B,q) with an $m\omega$-distance mapping p. Then, we call that $f_1 : B \to B$ an (ω, φ)-Suzuki contraction if there are an almost perfect function φ and a constant $k \in [0,1)$ such that for all $l, e \in X$ and $t \in \mathbb{N}$, we have:*

$$(1-k)p\left(l, {f_1}^t l\right) \leq p(l,e) \Longrightarrow \varphi p(f_1 l, f_1 e) \leq k \varphi p(l,e),$$

and:

$$(1-k)p\left({f_1}^t l, l\right) \leq p(e,l) \Longrightarrow \varphi p(f_1 e, f_1 l) \leq k \varphi p(e,l).$$

Now, we introduce and prove our first result.

Theorem 1. *Equip (B,q) with an $m\omega$-distance mapping p. Let p be bounded with respect to the almost perfect function φ and f_1 be an (ω, φ)-Suzuki contraction mapping. Suppose that:*

(i) f_1 *is continuous,*
or
(ii) *if $u^* \in B$ and $u^* \neq f_1 u^*$, then:*

$$\inf\left\{p\left(e, u^*\right) + p\left(f_1 e, u^*\right) : e \in B\right\} > 0. \tag{1}$$

Then, f_1 has a unique fixed point in B.

Proof. By starting with $l_0 \in B$, we produce a sequence (l_t) in B inductively by putting $l_{t+1} = f_1 l_t$ for all $t \in \mathbb{N} \cup \{0\}$. Given $m, t \in \mathbb{N} \cup \{0\}$ with $m > t$, then $m = t + s$ for some $s \in \mathbb{N}$. From the definition, we have:

$$\begin{aligned}(1-k)p(l_{t-1}, l_{m-1}) &= (1-k)p(l_{t-1}, l_{t+s-1}) \\ &\leq p(l_{t-1}, l_{t+s-1}).\end{aligned}$$

Therefore, we get that:

$$\begin{aligned}\varphi p(l_t, l_m) &= \varphi p(f_1 l_{t-1}, {f_1}^s l_{t-1}) \\ &= \varphi p(f_1 l_{t-1}, f_1 l_{t+s-1}) \\ &\leq k \varphi p(l_{t-1}, l_{t+s-1}).\end{aligned} \tag{2}$$

Repeating (2) t-times, we get that:

$$\varphi p(l_t, l_m) \leq k^t \varphi p(l_0, l_s). \tag{3}$$

Since (B,p) is bounded with respect to φ, then we have:

$$\varphi p(l_t, l_m) \leq k^t A \text{ for some integer } A > 0. \tag{4}$$

By letting $t, m \to \infty$, we get that:

$$\lim_{t,m \to \infty} \varphi p(l_t, l_m) = 0. \tag{5}$$

By the definition of φ, we get that:

$$\lim_{t,m\to\infty} p(l_t, l_m) = 0. \tag{6}$$

Since $m > t$, Lemma 1 implies that (l_t) is right Cauchy. Now, suppose that $t, m \in \mathbb{N} \cup \{0\}$ with $t > m$. Then, $t = m + q$ for some $q \in \mathbb{N}$. We note that:

$$(1-k)p(l_{t-1}, l_{m-1}) \quad \leq p(l_{m+q-1}, l_{m-1}).$$

Therefore, we get that:

$$\begin{aligned} \varphi p(l_t, l_m) &= \varphi p(f_1 l_{t-1}, f_1 x_{m-1}) \\ &\leq \cdots \leq k^m \varphi p(l_q, l_0) \end{aligned} \tag{7}$$

$$\varphi p(l_t, l_m) \leq k^m \varphi p(l_q, l_0). \tag{8}$$

Since (B, p) is bounded with respect to φ, we get that:

$$\varphi p(l_n, l_m) \leq k^m A \text{ for some integer } A > 0. \tag{9}$$

By letting $t, m \to \infty$, we have:

$$\lim_{t,m\to\infty} \varphi p(l_n, l_m) = 0. \tag{10}$$

Therefore,

$$\lim_{t,m\to\infty} p(l_t, l_m) = 0. \tag{11}$$

Since $t > m$, Lemma 1 implies that (l_t) is left Cauchy. Therefore, we deduce that (l_t) is Cauchy. The completeness of (B, q) implies that there exists an element $l^* \in B$ such that $l_t \to l^*$. If f_1 is continuous, then $l_{t+1} = f_1 l_t$ converges to $f_1 l^*$. The uniqueness of the limit ensures that $f_1 l^* = l^*$. Let $\epsilon > 0$. Since $\lim_{t,m\to\infty} p(l_t, l_m) = 0$, we choose $k_0 \in \mathbb{N}$ such that $p(l_t, l_m) \leq \frac{\epsilon}{2}$ for all $l, m \geq k_0$. The lower semi continuity of p implies that:

$$p(l_t, l^*) \leq \lim_{j\to\infty} \inf p(l_t, l_j) \leq \frac{\epsilon}{2} \text{ for all } n \geq k_0.$$

Assume that $l^* \neq f_1 l^*$. Then, by (1), we have:

$$\begin{aligned} \inf\{p(e, l^*) + p(f_1 e, l^*) : e \in B\} &\leq \inf\{p(l_t, l^*) + p(f_1 l_t, l^*) : t \in \mathbb{N}\} \\ &= \inf\{p(l_t, l^*) + p(l_{t+1}, l^*) : t \in \mathbb{N}\} \leq \epsilon, \end{aligned}$$

a contradiction. Therefore, $l^* = f_1 l^*$. Now, assume that $z^* \in B$ is a fixed point of f_1. Therefore:

$$(1-k)p(z^*, {f_1}^t z^*) = (1-k)p(z^*, z^*) \leq p(z^*, z^*).$$

Thus,

$$\varphi p(z^*, z^*) = \varphi p(f_1 z^*, f_1 z^*) \leq k \varphi p(z^*, z^*).$$

Since $k < 1$ and φ is an almost perfect function, we conclude that $p(z^*, z^*) = 0$. Assume that there exists $v^* \in B$ such that $v^* = f_1 v^*$. Since $p(z^*, z^*) = 0$, we have:

$$(1-k)p(z^*, f_1^t z^*) = (1-k)p(z^*, z^*) \leq p(z^*, v^*).$$

Therefore,

$$\varphi p(z^*, v^*) = \varphi p(f_1 z^*, f_1 v^*) \leq k \varphi p(z^*, v^*).$$

Thus, we have $\varphi p(z^*, v^*) = 0$, and so, $p(z^*, v^*) = 0$. Hence, by (mW3), we have $q(z^*, v^*) = 0$. Thus, $v^* = z^*$. Therefore, the fixed point of f_1 is unique. □

Corollary 1. *Equip (B, q) with an mω-distance mapping p. Assume p is bounded with respect to φ. Assume for all $e, l \in B$, we have:*

$$\varphi p(f_1 e, f_1 l) \leq k\varphi(p(e,l)), \text{ where } k \in [0,1). \tag{12}$$

Furthermore, assume that:

(i) *f_1 is continuous,*
or
(ii) *if $u^* \in B$ and $u^* \neq f_1 u^*$, then:*

$$\inf\{p(e,u^*) + p(f_1 e, u^*) : e \in B\} > 0.$$

Then, f_1 has a unique fixed point in B.

By taking the almost perfect function φ in Corollary 1 as follows: $\varphi(e) = e$, we get the following result:

Corollary 2. *Equip (B, q) with an mω-distance mapping p. Assume there exists $A > 0$ such that $p(e,l) \leq A$ for all $e, l \in B$. Furthermore, assume that there exists $k \in [0,1)$ such that for all $e, l \in B$, we have:*

$$p(f_1 e, f_1 l) \leq kp(e,l), \text{ where } k \in [0,1).$$

Furthermore, assume that:

(i) *f_1 is continuous,*
or
(ii) *if $u^* \in B$ and $u^* \neq f_1 u^*$, then:*

$$\inf\{p(e,u^*) + p(f_1 e, u^*) : e \in B\} > 0.$$

Then, f_1 has a unique fixed point in B.

Example 1. *Let $B = \{0, 1, 2, \cdots, n\}$, where $n \in \mathbb{N}$. Define $p, q : B \times B \to [0, +\infty)$ as follows:*

$$q(e,l) = \begin{cases} 0 & \text{if } e = l; \\ 3e + l & \text{if } e \neq l, \end{cases}$$

and:

$$p(e,l) = \begin{cases} 0 & \text{if } e = l; \\ \frac{1}{2}(3e + l) & \text{if } e \neq l. \end{cases}$$

Furthermore, define $f_1 : B \to B$ by:

$$f_1 e = \begin{cases} 0 & \text{if } e = 0, 1; \\ 1 & \text{if } e = 2, 3, \cdots, n, \end{cases}$$

and $\varphi : \mathbb{R}^+ \to \mathbb{R}^+$ by:

$$\varphi(l) = \begin{cases} 3^l - 1 & \text{if } l \in [0, n]; \\ 3^l & \text{if } l > n. \end{cases}$$

Then,

1. *φ is an almost perfect function.*
2. *p is an $m\omega$-distance function on q.*
3. *q is a quasi metric on B.*
4. *(B, q) is complete.*
5. *f_1 satisfies (ω, φ)-Suzuki contraction with $k = \frac{1}{\sqrt{3}}$, i.e., $\forall e, l \in B$, $j \in \mathbb{N}$, we have:*

$$\left(1 - \frac{1}{\sqrt{3}}\right) p\left(e, f_1^j e\right) \leq p(e, l) \Longrightarrow \varphi p(f_1 e, f_1 l) \leq k \varphi p(e, l),$$

and:

$$\left(1 - \frac{1}{\sqrt{3}}\right) p\left(f_1^j e, e\right) \leq p(l, e) \Longrightarrow \varphi p(f_1 l, f_1 e) \leq k \varphi p(l, e).$$

Proof. The proofs of (1), (2), and (3) are obvious. To show that q is complete, let (l_t) be a Cauchy sequence in B. Then, for each $t, m \in \mathbb{N}$, we have:

$$\lim_{m,t \to \infty} q(l_t, l_m) = 0.$$

Therefore, we deduce that $l_t = l_m$ for all $t, m \in \{0, 1, 2, \cdots\}$, but possible for finitely many. Thus, (l_t) converges in B. Hence, (B, q) is complete. To prove (5), given $e, l \in B$, we divide our proof into the following cases: Case (1): $e = 0$. Here, we have:

$$\left(1 - \frac{1}{\sqrt{3}}\right) p(0, 0) = \left(1 - \frac{1}{\sqrt{3}}\right) p(e, f_1^j e) \leq p(0, l) \text{ where } l = 0, 1, \cdots, n.$$

If $l \in \{0, 1\}$, then:

$$\varphi p(f_1 0, f_1 l) = \varphi p(0, 0) = 0 \leq \left(\frac{1}{\sqrt{3}}\right) \varphi p(0, l).$$

If $l \in \{2, 3, \cdots, n\}$, then:

$$\varphi p(f_1 0, f_1 l) = \varphi p(0, 1) = \varphi\left(\frac{1}{2}\right) = 3^{\frac{1}{2}} - 1.$$

Therefore,

$$\varphi p(0, l) = \varphi\left(\frac{l}{2}\right) = 3^{\frac{l}{2}} - 1.$$

$$\varphi p(f_1 0, f_1 l) = 3^{\frac{1}{2}} - 1 \leq \left(\frac{1}{\sqrt{3}}\right)\left(3^{\frac{l}{2}} - 1\right).$$

Case (2): $e = 1$. Here:

$$\left(1 - \frac{1}{\sqrt{3}}\right) p(e, 0) = \left(1 - \frac{1}{\sqrt{3}}\right) p(1, f_1 1) \leq p(1, l) \text{ where } l = 0, 2, 3, \cdots, n.$$

If $l = 0$, then we have $\varphi p(f1, fl) = 0$. Therefore,

$$\varphi p(f1, fl) = 0 \leq \left(\frac{1}{\sqrt{3}}\right)\left(3^{\frac{3}{2}} - 1\right).$$

If $l = 2, 3, \cdots, n$, then:

$$\varphi p(f_1 1, f_1 l) = \varphi p(0, 1) = \varphi\left(\frac{1}{2}\right) = 3^{\frac{1}{2}} - 1.$$

Now,

$$\varphi p(1,l) = \varphi\left(\frac{3+l}{2}\right) = 3^{\frac{3+l}{2}} - 1.$$

Thus,

$$\varphi p(f_1 1, f_1 l) = 3^{\frac{1}{2}} - 1 \leq \left(\frac{1}{\sqrt{3}}\right)\varphi\left(\frac{3+l}{2}\right) = \left(\frac{1}{\sqrt{3}}\right)\left(3^{\frac{3+l}{2}} - 1\right).$$

Case (3): $e \in \{2, 3, \cdots, n\}$. Here,

$$\left(1 - \frac{1}{\sqrt{3}}\right)p(e,1) = \left(1 - \frac{1}{\sqrt{3}}\right)p(e, f_1 e) \leq p(e,l) \text{ where } l = 1, 2, \cdots, n.$$

If $l = 1$, then:

$$\varphi p(fe, f1) = \varphi p(1,0) = \varphi\left(\frac{3}{2}\right) = 3^{\frac{3}{2}} - 1.$$

$$\varphi p(e,1) = \varphi\left(\frac{3e+1}{2}\right) = \begin{cases} 3^{\frac{7}{2}} - 1 & \text{if } e = 2 \\ 3^{\frac{3e+1}{2}} & \text{if } 3 \leq e \leq n. \end{cases}$$

$$\varphi p(fe, f1) = 3^{\frac{3}{2}} - 1 \leq (\frac{1}{\sqrt{3}})\varphi p(e,1).$$

If $l \in \{2, 3, \cdots, n\}$, $e \in \{2, 3, \cdots, n\}$ and $e \neq l$, then:

$$\varphi p(f_1 e, f_1 l) = \varphi p(1,1) = \varphi(0) = 0.$$

$$\varphi p(e,l) = \varphi\left(\frac{3e+l}{2}\right) = \begin{cases} 3^{\frac{3e+l}{2}} - 1 & \text{if } 3e + l \leq 2n \\ 3^{\frac{3e+l}{2}} & \text{if } 3e + l > 2n. \end{cases}$$

Similarly, we can show that:

$$(1 - \frac{1}{\sqrt{3}})p\left(f^t e, e\right) \leq p(l,e) \Longrightarrow \varphi p(fl, fe) \leq k\varphi p(l,e).$$

Hence, f_1 satisfies (ω, φ)-Suzuki contraction. Therefore, f_1 has a unique fixed point. □

Next, we introduce the definition of a generalized (ω, φ)-Suzuki contraction.

Definition 11. *Equip (B,q) with an $m\omega$-distance mapping p. We call $f_1 : B \to B$ a generalized (ω, φ)-Suzuki contraction if there exists an ultra distance function φ and a constant $k \in [0,1)$ such that for all $e, l \in B$, $j \in \mathbb{N}$, we have:*

$$(1-k)p\left(e, f_1^j e\right) \leq p(e,l) \Longrightarrow \varphi p(f_1 e, f_1 l) \leq k \max\{\varphi p(e, f_1 e), \varphi p(l, f_1 l)\},$$

and:

$$(1-k)p\left(f_1^j e, e\right) \leq p(l,e) \Longrightarrow \varphi p(f_1 l, f_1 e) \leq k \max\{\varphi p(e, f_1 e), \varphi p(l, f_1 l)\}.$$

We introduce and prove the second result:

Theorem 2. *Equip (X,q) with an $m\omega$-distance mapping p. Assume that p is bounded with respect to the almost perfect function φ. Assume that f_1 is a generalized (ω, φ)-Suzuki contraction mapping. Furthermore, suppose that:*

(i) f_1 is continuous,
or

(ii) if $u^* \in B$ and $u^* \neq f_1 u^*$, then:

$$\inf\left\{p\left(e, u^*\right) + p\left(f_1 e, u^*\right) : e \in B\right\} > 0. \tag{13}$$

Then, f_1 has a unique fixed point in B.

Proof. Start with $l_0 \in B$ to construct (l_n) in B inductively by putting $l_{t+1} = f_1 l_t$ for all $t \in \mathbb{N} \cup \{0\}$. Given $t, m \in \mathbb{N} \cup \{0\}$ with $t < m$, let $m = t + j$ with $j \in \mathbb{N}$. We note that:

$$\begin{aligned}(1-k)p(l_{t-1}, l_{m-1}) &= (1-k)p(l_{t-1}, f_1^j l_{t-1}) \\ &\leq p(l_{t-1}, l_{m-1}).\end{aligned}$$

Since f_1 is a generalized (ω, φ)-Suzuki contraction, we have:

$$\begin{aligned}\varphi p(l_t, l_m) &= \varphi p(f_1 l_{t-1}, f_1 l_{m-1}) \\ &\leq k \max\{\varphi p(l_{t-1}, f_1 l_{t-1}), \varphi p(l_{m-1}, f_1 l_{m-1})\} \\ &= k \max\{\varphi p(l_{t-1}, l_t)), \varphi p(l_{m-1}, l_m)\}.\end{aligned} \tag{14}$$

Now,

$$\begin{aligned}(1-k)p(l_{t-2}, l_{t-1}) &= (1-k)p(l_{t-2}, f_1 l_{t-2}) \\ &\leq p(l_{t-2}, l_{t-1}).\end{aligned}$$

Therefore, we get that:

$$\begin{aligned}\varphi p(l_{t-1}, l_t) &= \varphi p(f_1 l_{t-2}, f_1 l_{t-1}) \\ &\leq k \max\{\varphi p(l_{t-2}, f_1 l_{t-2}), \varphi p(l_{t-1}, f_1 l_{t-1})\} \\ &= k \max\{\varphi p(l_{t-2}, l_{t-1}), \varphi p(l_{t-1}, l_t)\}.\end{aligned} \tag{15}$$

Since $k < 1$, we get that:

$$\varphi p(l_{t-1}, l_t) \leq k \varphi p(l_{t-2}, l_{t-1}). \tag{16}$$

Repeating (16) t-times, we get that:

$$\varphi p(l_{t-1}, l_t) \leq k^{t-1} \varphi p(l_0, l_1). \tag{17}$$

Similarly, we get that that:

$$\varphi p(l_{m-1}, l_m) \leq k^{m-1} \varphi p(l_0, l_1). \tag{18}$$

Using Equations (14), (17), and (18), we get:

$$\varphi p(l_t, l_m) \leq k \max\{k^{t-1} \varphi p(l_0, l_1), k^{m-1} \varphi p(l_0, l_1)\}. \tag{19}$$

Since $t < m$, we get that:

$$\varphi p(l_t, l_m) \leq k^t \varphi p(l_0, l_1). \tag{20}$$

The boundedness property of p with respect to φ implies that:

$$\varphi p(l_t, l_m) \leq k^t A \text{ for some integer } A \geq 0. \tag{21}$$

By letting $t, m \to \infty$, we get that:

$$\lim_{t,m \to \infty} \varphi p(l_t, l_m) = 0. \tag{22}$$

Thus,

$$\lim_{t,m \to \infty} p(l_t, l_m) = 0. \tag{23}$$

Since $t < m$, Lemma 1 implies that (l_t) is right Cauchy. In a similar manner, we can show that (l_t) is left Cauchy. Hence, (l_t) is Cauchy. The completeness of q ensures that there exists $l^* \in B$ such that

(l_t) converges to l^*. If f_1 is continuous, then $(l_{t+1}) = (f_1 l_t)$ converges to $f_1 l^*$. The uniqueness of the limit implies that $f_1 l^* = l^*$. Given $\varepsilon > 0$. Since $\lim_{t,m\to\infty} p(l_t, l_m)) = 0$, there exists $n_0 \in \mathbb{N}$ such that $p(l_t, l_m) \le \frac{\varepsilon}{2}$ for all $t, m \ge n_0$. The lower semi continuity of p implies that:

$$p(l_t, l^*) \le \lim_{i\to\infty} \inf p(l_t, l_i) \le \frac{\varepsilon}{2} \text{ for all } t \ge n_0.$$

Assume that $l^* \ne f_1 l^*$, then by (13), we have:

$$\begin{aligned} &\inf\{p(e, l^*) + p(f_1 e, l^*) : e \in B\} \\ &\qquad \le \inf\{p(l_m, l^*) + p(f_1 l_t, l^*) : t \in \mathbb{N}\} \\ &\qquad = \inf\{p(l_t, l^*) + p(l_{t+1}, l^*) : n \in \mathbb{N}\} \le \varepsilon, \end{aligned}$$

a contradiction. Therefore, $l^* = f_1 l^*$. Assume $z^* \in B$ such that $f_1 z^* = z^*$. First, we prove that $p(z^*, z^*) = 0$. Since:

$$(1-k)p(z^*, f_1^j z^*) = (1-k)p(z^*, z^*) \le p(z^*, z^*),$$

then:

$$\varphi p(z^*, z^*) = \varphi p(f_1 z^*, f_1 z^*) \le k\varphi p(z^*, z^*).$$

Since $k < 1$ and φ is an almost perfect function, then $p(z^*, z^*) = 0$. Therefore,

$$(1-k)p(z^*, f_1{}^t z^*) = (1-k)p(z^*, z^*) \le p(z^*, l^*).$$

Therefore,

$$\begin{aligned} \varphi p(z^*, l^*) &= \varphi p(f_1 z^*, f_1 l^*) \\ &\le k \max\{\varphi p(z^*, f_1 z^*), \varphi p(l^*, f_1 l^*)\} \\ &= k \max\{\varphi(p(z^*, z^*)), \varphi(p(v^*, v^*))\} \\ &= 0. \end{aligned}$$

The definition of φ informs us that $p(z^*, l^*) = 0$. The definition of p implies that $q(z^*, l^*) = 0$. Hence: $z^* = l^*$. □

Corollary 3. *Equip (B, q) with an mω-distance mapping p. Assume p is bounded with respect to the almost perfect function φ. Suppose that for all $e, l \in B$, we have:*

$$\varphi p(f_1 e, f_1 l) \le k \max\{\varphi p(e, f_1 e), \varphi p(l, f_1 l)\}, \text{ where } k \in [0, 1). \tag{24}$$

Furthermore, assume that:

(i) f is continuous;

or

(ii) if $u^* \in B$ and $u^* \ne f_1 u^*$, then:

$$\inf\{p(e, u^*) + p(f_1 e, u^*) : e \in B\} > 0.$$

Then, f_1 has a unique fixed point in B.

Corollary 4. *Equip (B, q) with an mω-distance mapping p. Assume that there exists $A > 0$ such that $p(e, l) \le A$ for all $e, l \in B$. Furthermore, assume that for all $e, l \in B$, we have:*

$$p(f_1 e, f_1 l) \le \alpha(p(e, f_1 e) + p(l, f_1 l)), \text{ where } 0 \le \alpha < \frac{1}{2}.$$

Assume that:

(i) f_1 *is continuous;*
or

(ii) *if* $u^* \in B$ *and* $u^* \neq f_1 u^*$, *then:*

$$\inf\{p(e,u^*) + p(f_1 e, u^*) : e \in B\} > 0.$$

Then, f_1 *has a unique fixed point in B.*

Proof. Define the almost perfect function φ via $\varphi(e) = e$ in Corollary 3. Then:

$$\begin{aligned}
\varphi(p(f_1 e, f_1 l)) &= p(f_1 e, f_1 l) \\
&\leq \lambda(p(e, f_1 e) + p(l, f_1 l)) \\
&\leq 2\lambda \max\{p(e, f_1 e), p(l, f_1)\} \\
&= 2\lambda \max\{\varphi(p(e, f_1 e)), \varphi(p(l, f_1 l)))\}.
\end{aligned}$$

□

3. Application

In this section, we utilize Corollaries 1 and 4 to give some applications of our work.

Theorem 3. *For any positive integer n, the equation:*

$$nx^n - x^{n-1} + 4nx - 2 = 0$$

has a unique solution in [0, 1].

Proof. Let $B = [0,1]$. Define $q : B \times B \to \mathbb{R}^+$ by $q(x,y) = |x-y|$. Then, (B,q) is a complete quasi metric space. Furthermore, define $p : B \times B \to [0,\infty)$ by $p(x,y) = |x-y|$. Then, p is an $m\omega$-distance mapping. Now, equip (B,q) with p.
Define $f_1 : B \to B$ by:

$$f_1(x) = \frac{x^{n-1}+2}{n(x^{n-1}+4)}.$$

Furthermore, define $\varphi : [0,\infty) \to [0,\infty)$ by:

$$\varphi(a) = \begin{cases} a^2 & \text{if } a \in [0,1]; \\ a^2 + \frac{1}{2} & \text{if } a > 1. \end{cases}$$

Note that φ is an almost perfect function and p is bounded with respect to φ. For $x, y \in B$, we have:

$$\begin{aligned}
\varphi p(f_1 x, f_1 y) &= \frac{1}{n^2}\left|\frac{x^{n-1}+2}{x^{n-1}+4} - \frac{y^{n-1}+2}{y^{n-1}+4}\right|^2 \\
&= \frac{1}{n^2}\left|\frac{2x^{n-1} - 2y^{n-1}}{(x^{n-1}+4)(y^{n-1}+4)}\right|^2 \\
&\leq \frac{4(n-1)^2}{n^2}\left(\frac{1}{(x^2+4)^2(y^2+4)^2}\right)\left|x-y\right|^2 \\
&\leq \frac{(n-1)^2}{64n^2}\left|x-y\right|^2 \\
&= \frac{(n-1)^2}{64n^2}\varphi p(x,y).
\end{aligned}$$

By taking $k = \frac{(n-1)^2}{64n^2}$ and noting that f_1 is continuous, we conclude that f_1 satisfies all conditions of Corollary 1. Thus, f_1 has a unique fixed point. Note that the unique fixed point of f_1 is the unique solution of:

$$nx^n - x^{n-1} + 4nx - 2 = 0.$$

□

Example 2. *The equation:*

$$1000x^{1000} - x^{999} + 4000x - 2 = 0$$

has a unique solution in [0, 1].

Proof. It follows from Theorem 3 by taking $n = 1000$. □

Let Y be the set of non-decreasing functions $\tau : \mathbb{R}^+ \to \mathbb{R}^+$ such that τ is Lebesgue integrable for all compact sets in $\mathbb{R}^+$ and:

$$\int_0^{\mu} \tau(v)dv > 0 \text{ where } \mu > 0.$$

Theorem 4. *Equip (B, q) with an $m\omega$-distance mapping p. Assume that there exists $A > 0$ such that $p(e, l) \leq A$ for all $e, l \in B$. Furthermore, suppose the following condition:*

(i) *f_1 is continuous.*

(ii) *There exists $\tau \in Y$ and $\alpha \in [0, 1/2)$ such that for all $e, l \in B$, we have:*

$$\int_0^{p(f_1e, f_1l)} \tau(v)dv \leq \alpha\left(\int_0^{p(e, f_1e)} \tau(v)dv + \int_0^{p(l, f_1l)} \tau(v)dv\right).$$

Then, f_1 has a unique fixed point in B.

Proof. Let $\varphi = \int_0^t \tau(v)dv$. Then, φ is an almost perfect function. Corollary 4 ensures that f_1 has a unique fixed point in B. □

4. Conclusions

The notions of (ω, φ)-Suzuki contraction and generalized (ω, φ)-Suzuki contraction are introduced. According to these nations many fixed point results are investigated. Some applications are introduced on the obtained results.

Author Contributions: Supervision: A.T. and W.S. Writing, original draft: T.Q., A.T., and W.S. Writing, review and editing: W.S.

Acknowledgments: The authors would like to thank the anonymous reviewers and Editor for their valuable remarks on our paper. Furthermore, the third author thanks Prince Sultan University for funding this work through research group NAMAM, Group Number RG-DES-2017-01-17.

References

1. Banach, S. Sur les Opération dans les ensembles abstraits et leur application aux equations integral. *Fundam. Math.* **1922**, *3*, 133–181. [CrossRef]
2. Suzuki, T. A generalized Banach contraction principle that characterizes metric completeness. *Proc. Am. Math. Soc.* **2008**, *136*, 1861–1869. [CrossRef]
3. Kannan, R. Some results on fixed point. *Bull. Calcutta Math. Soi.* **1968**, *60*, 71–76.

4. Wilson, W.A. On quasi-metric spaces. *Am. J. Math.* **1931**, *53*, 675–684. [CrossRef]
5. Aydi, H.; Jellali, M.; Karapınar, E. On fixed point results for α-implicit contractions in quasi metric spaces and consequences. *Nonlinear Anal. Model. Control* **2016**, *21*, 40–56. [CrossRef]
6. Jleli, M.; Samet, B. Remarks on G-metric spaces and fixed point theorems. *Fixed Point Theory Appl.* **2012**, *2012*, 210. [CrossRef]
7. Bilgili, N.; Karapınar, E.; Samet, B. Eneralized α-ψ contractive mapping in quasi metric spaces and related fixed point theorems. *J. Inequal. Appl.* **2014**, *2014*, 36. [CrossRef]
8. Shatanawi, W.; Noorani, M.S.; Alsamir, H.; Bataihah, A. Fixed and common fixed point theorems in partially ordered quasi metric spaces. *Am. J. Math.* **2012**, *16*, 516–528 [CrossRef]
9. Shatanawi, W.; Pitea, A. Some coupled fixed point theorems in quasi-partial metric spaces. *Fixed Point Theory Appl.* **2013**, *2013*, 153. [CrossRef]
10. Alegre, C.; Marin, J. Modified ω-distance on quasi metric spaces and fixed point theorems on complete quasi metric spaces. *Topol. Appl.* **2016**, *203*, 120–129. [CrossRef]
11. Nuseir, I.; Shatanawi, W.; Abu-Irwaq, I.; Batiahah, A. Nonlinear contraction and fixed point theorems with modified ω-distance mappings in complete quasi metric spaces. *J. Nonlinear Sci. Appl.* **2017**, *10*, 5342–5350. [CrossRef]
12. Abodayeh, K.; Batiahah, A.; Shatanawi, W. Generalized Ω-distance mappings and some fixed point theorems. *Politehn. Univ. Bucharest Sci. Bull. Ser. A Appl. Math. Phys.* **2017**, *79*, 223–232.
13. Mongkolkeha, C.; Cho, Y.J. Some coincidence point theorems in ordered metric spaces via ω-distance. *Carapthan J.* **2018**, *34*, 207–214.
14. Khan, M.S.; Swaleh, M.; Sessa, S. Some fixed point and common fixed point results through Ω-distance under nonlinear contraction. *Gazi Univ. J. Sci.* **2017**, *30*, 293–302.
15. Abodayeh, K.; Shatanawi, W.; Bataihah, A.; Ansari, A.H. Fixed point theorems by altering distances between the points. *Bull. Aust. Math. Soc.* **1984**, *30*, 1–9.

12

New Definitions about $A^{\mathcal{I}}$-Statistical Convergence with Respect to a Sequence of Modulus Functions and Lacunary Sequences

Ömer Kişi [1,*], Hafize Gümüş [2] and Ekrem Savas [3]

1 Department of Mathematics, Faculty of Science, Bartin University, Bartin 74100, Turkey
2 Department of Mathematics, Faculty of Eregli Education, Necmettin Erbakan University, Eregli, Konya 42060, Turkey; hgumus@konya.edu.tr
3 Department of Mathematics, Istanbul Ticaret University, Üsküdar, Istanbul 34840, Turkey; ekremsavas@yahoo.com
* Correspondence: okisi@bartin.edu.tr;

Abstract: In this paper, using an infinite matrix of complex numbers, a modulus function and a lacunary sequence, we generalize the concept of $\mathcal{I}$-statistical convergence, which is a recently introduced summability method. The names of our new methods are $A^{\mathcal{I}}$-lacunary statistical convergence and strongly $A^{\mathcal{I}}$-lacunary convergence with respect to a sequence of modulus functions. These spaces are denoted by $S_{\theta}^{A}(\mathcal{I},F)$ and $N_{\theta}^{A}(\mathcal{I},F)$, respectively. We give some inclusion relations between $S^{A}(\mathcal{I},F)$, $S_{\theta}^{A}(\mathcal{I},F)$ and $N_{\theta}^{A}(\mathcal{I},F)$. We also investigate Cesáro summability for $A^{\mathcal{I}}$ and we obtain some basic results between $A^{\mathcal{I}}$-Cesáro summability, strongly $A^{\mathcal{I}}$-Cesáro summability and the spaces mentioned above.

Keywords: lacunary sequence; statistical convergence; ideal convergence; modulus function; $\mathcal{I}$-statistical convergence

---

1. Introduction

As is known, convergence is one of the most important notions in mathematics. Statistical convergence extends the notion. After giving the definition of statistical convergence, we can easily show that any convergent sequence is statistically convergent, but not conversely. Let E be a subset of $\mathbb{N}$, and the set of all natural numbers $d(E) := \lim_{n\to\infty} \frac{1}{n} \sum_{j=1}^{n} \chi_E(j)$ is said to be a natural density of E whenever the limit exists. Here, χ_E is the characteristic function of E.

In 1935, statistical convergence was given by Zygmund in the first edition of his monograph [1]. It was formally introduced by Fast [2], Fridy [3], Salat [4], Steinhaus [5] and later was reintroduced by Schoenberg [6]. It has become an active research area in recent years. This concept has applications in different fields of mathematics such as number theory [7], measure theory [8], trigonometric series [1], summability theory [9], etc.

Following this very important definition, the concept of lacunary statistical convergence was defined by Fridy and Orhan [10]. In addition, Fridy and Orhan gave the relationships between the lacunary statistical convergence and the Cesàro summability. Freedman and Sember [9] established the connection between the strongly Cesàro summable sequences space $|\sigma_1|$ and the strongly lacunary summable sequence space N_{θ}.

$\mathcal{I}$-convergence has emerged as a generalized form of many types of convergences. This means that, if we choose different ideals, we will have different convergences. Koystro et al. [11] introduced this concept in a metric space. Also, Das et al. [12], Koystro et al. [13], Savaş and Das [14] studied ideal convergence. We will explain this situation with two examples later. Before defining $\mathcal{I}$-convergence, the definitions of ideal and filter will be needed.

An ideal is a family of sets $\mathcal{I} \subseteq 2^{\mathbb{N}}$ such that (i) $\emptyset \in \mathcal{I}$, (ii) $A, B \in \mathcal{I}$ implies $A \cup B \in \mathcal{I}$, (iii), and, for each $A \in \mathcal{I}$, each $B \subseteq A$ implies $B \in \mathcal{I}$. An ideal is called non-trivial if $\mathbb{N} \notin \mathcal{I}$ and a non-trivial ideal is called admissible if $\{n\} \in \mathcal{I}$ for each $n \in \mathbb{N}$.

A filter is a family of sets $\mathcal{F} \subseteq 2^{\mathbb{N}}$ such that (i) $\emptyset \notin \mathcal{F}$, (ii) $A, B \in \mathcal{F}$ implies $A \cap B \in \mathcal{F}$, (iii) For each $A \in \mathcal{F}$, each $A \subseteq B$ implies $B \in \mathcal{F}$.

If $\mathcal{I}$ is an ideal in $\mathbb{N}$, then the collection

$$F(\mathcal{I}) = \{A \subset \mathbb{N} : \mathbb{N} \backslash A \in \mathcal{I}\}$$

forms a filter in $\mathbb{N}$ that is called the filter associated with $\mathcal{I}$.

The notion of a modulus function was introduced by Nakano [15]. We recall that a modulus f is a function from $[0, \infty)$ to $[0, \infty)$ such that (i) $f(x) = 0$ if and only if $x = 0$; (ii) $f(x+y) = f(x) + f(y)$ for $x, y \geq 0$; (iii) f is increasing; and (iv) f is continuous from the right at 0. It follows that f must be continuous on $[0, \infty)$. Connor [16], Bilgin [17], Maddox [18], Kolk [19], Pehlivan and Fisher [20] and Ruckle [21] have used a modulus function to construct sequence spaces. Now, let S be the space of sequences of modulus functions $F = (f_k)$ such that $\lim_{x \to 0^+} \sup_k f_k(x) = 0$. Throughout this paper, the set of all modulus functions determined by F is denoted by $F = (f_k) \in S$ for every $k \in \mathbb{N}$.

In this paper, we aim to unify these approaches and use ideals to introduce the notion of $A^{\mathcal{I}}$-lacunary statistically convergence with respect to a sequence of modulus functions.

2. Definitions and Notations

First, we recall some of the basic concepts that will be used in this paper.

Let $A = (a_{ki})$ be an infinite matrix of complex numbers. We write $Ax = (A_k(x))$, if $A_k(x) = \sum_{i=1}^{\infty} a_{ki} x_k$ converges for each k.

Definition 1. *A number sequence $x = (x_k)$ is said to be statistically convergent to the number L if for every $\varepsilon > 0$,*

$$\lim_{n \to \infty} \frac{1}{n} |\{k \leq n : |x_k - L| \geq \varepsilon\}| = 0.$$

In this case, we write $st - \lim x_k = L$. As we said before, statistical convergence is a natural generalization of ordinary convergence i.e., if $\lim x_k = L$, then $st - \lim x_k = L$ (Fast, [2]).

By a lacunary sequence, we mean an increasing integer sequence $\theta = \{k_r\}$ such that $k_0 = 0$ and $h_r = k_r - k_{r-1} \to \infty$ as $r \to \infty$. Throughout this paper, the intervals determined by θ will be denoted by $I_r = (k_{r-1}, k_r]$.

Definition 2. *A sequence $x = (x_k)$ is said to be lacunary statistically convergent to the number L if, for every $\varepsilon > 0$,*

$$\lim_{r \to \infty} \frac{1}{h_r} |\{k \in I_r : |x_k - L| \geq \varepsilon\}| = 0.$$

In this case, we write $S_\theta - \lim x_k = L$ or $x_k \to L(S_\theta)$ (Fridy and Orhan, [10]).

Definition 3. *The sequence space N_θ is defined by*

$$N_\theta = \left\{ (x_k) : \lim_{r \to \infty} \frac{1}{h_r} \sum_{k \in I_r} |x_k - L| = 0 \right\}$$

(Fridy and Orhan, [10]).

Definition 4. *Let $\mathcal{I} \subset 2^{\mathbb{N}}$ be a proper admissible ideal in $\mathbb{N}$. The sequence (x_n) of elements of $\mathbb{R}$ is said to be $\mathcal{I}$-convergent to $L \in \mathbb{R}$ if, for each $\varepsilon > 0$, the set*

$$A(\varepsilon) = \{n \in \mathbb{N} : |x_n - L| \geq \varepsilon\} \in \mathcal{I}$$

(Kostyrko et al. [11]).

Example 1. *Define the set of all finite subsets of $\mathbb{N}$ by $\mathcal{I}_f$. Then, $\mathcal{I}_f$ is a non-trivial admissible ideal and $\mathcal{I}_f$-convergence coincides with the usual convergence.*

Example 2. *Define the set $\mathcal{I}_d$ by $\mathcal{I}_d = \{A \subset \mathbb{N} : d(A) = 0\}$. Then, $\mathcal{I}_d$ is an admissible ideal and $\mathcal{I}_d$-convergence gives the statistical convergence.*

Following the line of Savas et al. [22], some authors obtained more general results about statistical convergence by using A matrix and they called this new method $A^{\mathcal{I}}$-statistical convergence (see, e.g., [17,23]).

Definition 5. *Let $A = (a_{ki})$ be an infinite matrix of complex numbers and (f_k) be a sequence of modulus functions in S. A sequence $x = (x_k)$ is said to be $A^{\mathcal{I}}$-statistically convergent to $L \in X$ with respect to a sequence of modulus functions, for each $\varepsilon > 0$, for every $x \in X$ and $\delta > 0$,*

$$\left\{n \in \mathbb{N} : \frac{1}{n}|\{k \leq n : f_k(|A_k(x) - L|) \geq \varepsilon\}| \geq \delta\right\} \in \mathcal{I}.$$

In this case, we write $x_k \to L\left(S^A(\mathcal{I}, F)\right)$ (Yamancı et al. [23]).

3. Inclusions between $S^A(\mathcal{I}, F)$, $S_\theta^A(\mathcal{I}, F)$ and $N_\theta^A(\mathcal{I}, F)$ Spaces

We now consider our main results. We begin with the following definitions.

Definition 6. *Let $A = (a_{ki})$ be an infinite matrix of complex numbers, $\theta = \{k_r\}$ be a lacunary sequence and $F = (f_k)$ be a sequence of modulus functions in S. A sequence $x = (x_k)$ is said to be $A^{\mathcal{I}}$-lacunary statistically convergent to $L \in X$ with respect to a sequence of modulus functions, for each $\varepsilon > 0$, for each $x \in X$ and $\delta > 0$,*

$$\left\{r \in \mathbb{N} : \frac{1}{h_r}|\{k \in I_r : f_k(|A_k(x) - L|) \geq \varepsilon\}| \geq \delta\right\} \in \mathcal{I}.$$

Definition 7. *Let $A = (a_{ki})$ be an infinite matrix of complex numbers, $\theta = \{k_r\}$ be a lacunary sequence and $F = (f_k)$ be a sequence of modulus functions in S. A sequence $x = (x_k)$ is said to be strongly $A^{\mathcal{I}}$-lacunary convergent to $L \in X$ with respect to a sequence of modulus functions, if, for each $\varepsilon > 0$, for each $x \in X$,*

$$\left\{r \in \mathbb{N} : \frac{1}{h_r}\sum_{k \in I_r} f_k(|A_k(x) - L|) \geq \varepsilon\right\} \in \mathcal{I}.$$

We shall denote by $S_\theta^A(\mathcal{I}, F)$, $N_\theta^A(\mathcal{I}, F)$ the collections of all $A^{\mathcal{I}}$-lacunary statistically convergent and strongly $A^{\mathcal{I}}$-lacunary convergent sequences, respectively.

Theorem 1. *Let $A = (a_{ki})$ be an infinite matrix of complex numbers and (f_k) be a sequence of modulus functions in S. $\left(S_\theta^A(\mathcal{I}, F)\right) \cap m(X)$ is a closed subset of $m(X)$ if X is a Banach space where $m(X)$ is the space of all bounded sequences of X.*

Proof. Suppose that $(x^n) \subset \left(S_\theta^A(\mathcal{I}, F)\right) \cap m(X)$ is a convergent sequence and it converges to $x \in m(X)$. We need to show that $x \in \left(S_\theta^A(\mathcal{I}, F)\right) \cap m(X)$. Assume that $x^n \to L_n\left(S_\theta^A(\mathcal{I}, F)\right), \forall n \in \mathbb{N}$.

Take a sequence $\{\varepsilon_r\}_{n\in\mathbb{N}}$ of strictly decreasing positive numbers converging to zero. We can find an $r \in \mathbb{N}$ such that $\|x - x^j\|_\infty < \frac{\varepsilon_r}{4}$ for all $j \geq r$. Choose $0 < \delta < \frac{1}{5}$.

Now,

$$A = \left\{ r \in \mathbb{N} : \frac{1}{h_r} \left| \left\{ k \in I_r : f_k \left(|A_k(x^n) - L_n| \right) \geq \frac{\varepsilon_r}{4} \right\} \right| < \delta \right\} \in \mathcal{F}(\mathcal{I})$$

and

$$B = \left\{ r \in \mathbb{N} : \frac{1}{h_r} \left| \left\{ k \in I_r : f_k \left(\left|A_k \left(x^{n+1}\right) - L_{n+1}\right| \right) \geq \frac{\varepsilon_r}{4} \right\} \right| < \delta \right\} \in \mathcal{F}(\mathcal{I}).$$

Since $A \cap B \in \mathcal{F}(\mathcal{I})$ and $\emptyset \notin \mathcal{F}(\mathcal{I})$, we can choose $r \in A \cap B$. Then,

$$\frac{1}{h_r} \left| \left\{ k \in I_r : f_k \left(|A_k(x^n) - L_n| \right) \geq \frac{\varepsilon_r}{4} \vee f_k \left(\left|A_k \left(x^{n+1}\right) - L_{n+1}\right| \right) \geq \frac{\varepsilon_r}{4} \right\} \right| \leq 2\delta < 1.$$

Since $h_r \to \infty$ and $A \cap B \in \mathcal{F}(\mathcal{I})$ is infinite, we can actually choose the above r so that $h_r > 5$. Hence, there must exist a $k \in I_r$ for which we have simultaneously, $|x_k^n - L_n| < \frac{\varepsilon_r}{4}$ and $\left|x_k^{n+1} - L_{n+1}\right| < \frac{\varepsilon_r}{4}$.

Then, it follows that

$$\begin{aligned} |L_n - L_{n+1}| &\leq |L_n - x_k^n| + \left|x_k^n - x_k^{n+1}\right| + \left|x_k^{n+1} - L_{n+1}\right| \\ &\leq |x_k^n - L_n| + \left|x_k^{n+1} - L_{n+1}\right| + \|x - x^n\|_\infty + \left\|x - x^{n+1}\right\|_\infty \\ &\leq \tfrac{\varepsilon_r}{4} + \tfrac{\varepsilon_r}{4} + \tfrac{\varepsilon_r}{4} + \tfrac{\varepsilon_r}{4} = \varepsilon_r. \end{aligned}$$

This implies that $\{L_n\}_{n\in\mathbb{N}}$ is a Cauchy sequence in X. Since X is a Banach space, we can write $L_n \to L \in X$ as $n \to \infty$. We shall prove that $x_k \to L\left(S_\theta^A(\mathcal{I}, F)\right)$. Choose $\varepsilon > 0$ and $r \in \mathbb{N}$ such that $\varepsilon_r < \frac{\varepsilon}{4}$, $\|x - x_n\|_\infty < \frac{\varepsilon}{4}$. Now, since

$$\begin{aligned} &\tfrac{1}{h_r} \left|\{k \in I_r : f_k(|A_k(x) - L|) \geq \varepsilon\}\right| \\ &\leq \tfrac{1}{h_r} \left|\{k \in I_r : f_k(|A_k(x - x_n)|) + f_k(|A_k(x^n) - L_n|) + f_k(|L_n - L|) \geq \varepsilon\}\right| \\ &\leq \tfrac{1}{h_r} \left|\left\{k \in I_r : f_k(|A_k(x^n) - L_n|) \geq \tfrac{\varepsilon}{2}\right\}\right|. \end{aligned}$$

It follows that

$$\begin{aligned} &\left\{ r \in \mathbb{N} : \tfrac{1}{h_r} \left|\{k \in I_r : f_k(|A_k(x) - L|) \geq \varepsilon\}\right| \geq \delta \right\} \\ &\subset \left\{ r \in \mathbb{N} : \tfrac{1}{h_r} \left|\left\{k \in I_r : f_k(|A_k(x) - L|) \geq \tfrac{\varepsilon}{2}\right\}\right| \geq \delta \right\} \end{aligned}$$

for given $\delta > 0$. This shows that $x_k \to L\left(S_\theta^A(\mathcal{I}, F)\right)$ and this completes the proof of the theorem. □

Theorem 2. *Let $A = (a_{ki})$ be an infinite matrix of complex numbers, $\theta = \{k_r\}$ be a lacunary sequence and (f_k) be a sequence of modulus functions in S. Then, we have*

(i) If $x_k \to L\left(N_\theta^A(\mathcal{I}, F)\right)$, then $x_k \to L\left(S_\theta^A(\mathcal{I}, F)\right)$ and $N_\theta^A(\mathcal{I}, F) \subset S_\theta^A(\mathcal{I}, F)$ is proper for every ideal $\mathcal{I}$;
(ii) If $x \in m(X)$, the space of all bounded sequences of X and $x_k \to L\left(S_\theta^A(\mathcal{I}, F)\right)$, then $x_k \to L\left(N_\theta^A(\mathcal{I}, F)\right)$;
(iii) $S_\theta^A(\mathcal{I}, F) \cap m(X) = N_\theta^A(\mathcal{I}, F) \cap m(X)$.

Proof. (i) Let $\varepsilon > 0$ and $x_k \to L\left(N_\theta^A(\mathcal{I}, F)\right)$. Then, we can write

$$\begin{aligned}\sum_{k\in I_r} f_k\left(|A_k(x)-L|\right) &\geq \sum_{\substack{k\in I_r \\ f_k(|A_k(x)-L|)\geq\varepsilon}} f_k\left(|A_k(x)-L|\right)\\ &\geq \varepsilon\left|\left\{k\in I_r : f_k\left(|A_k(x)-L|\right)\geq\varepsilon\right\}\right|.\end{aligned}$$

Thus, for given $\delta > 0$,

$$\frac{1}{h_r}\left|\left\{k\in I_r : f_k\left(|A_k(x)-L|\right)\geq\varepsilon\right\}\right|\geq\delta \Longrightarrow \frac{1}{h_r}\sum_{k\in I_r} f_k\left(|A_k(x)-L|\right)\geq\varepsilon\delta,$$

i.e.,

$$\left\{r\in\mathbb{N} : \frac{1}{h_r}\left|\left\{k\in I_r : f_k\left(|A_k(x)-L|\right)\geq\varepsilon\right\}\right|\geq\delta\right\}\subseteq\left\{r\in\mathbb{N} : \frac{1}{h_r}\sum_{k\in I_r} f_k\left(|A_k(x)-L|\right)\geq\varepsilon\delta\right\}.$$

Since $x_k \to L\left(N_\theta^A(\mathcal{I}, F)\right)$, the set on the right-hand side belongs to $\mathcal{I}$ and so it follows that $x_k \to L\left(S_\theta^A(\mathcal{I}, F)\right)$.

To show that $\left(S_\theta^A(\mathcal{I}, F)\right) \subsetneq \left(N_\theta^A(\mathcal{I}, F)\right)$, take a fixed $K \in \mathcal{I}$. Define $x = (x_k)$ by

$$(x_k) = \begin{cases} ku, & \text{for } k_{r-1} < k \leq k_{r-1} + \left[\sqrt{h_r}\right], r = 1,2,3..., r \notin K, \\ ku, & \text{for } k_{r-1} < k \leq k_{r-1} + \left[\sqrt{h_r}\right], r = 1,2,3..., r \in K, \\ \theta, & \text{otherwise}, \end{cases}$$

where $u \in X$ is a fixed element with $\|u\| = 1$ and θ is the null element of X. Then, $x \notin m(X)$ and for every $0 < \varepsilon < 1$ since

$$\frac{1}{h_r}\left|\left\{k\in I_r : f_k\left(|A_k(x)-0|\right)\geq\varepsilon\right\}\right| = \frac{\left[\sqrt{h_r}\right]}{\sqrt{h_r}} \to 0.$$

As $r \to \infty$ and $r \notin K$, for every $\delta > 0$,

$$\left\{r\in\mathbb{N} : \frac{1}{h_r}\left|\left\{k\in I_r : f_k\left(|A_k(x)-0|\right)\geq\varepsilon\right\}\right|\geq\delta\right\}\subset M\cup\{1,2,...,m\}$$

for some $m \in \mathbb{N}$. Since $\mathcal{I}$ is admissible, it follows that $x_k \to \theta\left(S_\theta^A(\mathcal{I}, F)\right)$. Obviously,

$$\frac{1}{h_r}\sum_{k\in I_r} f_k\left(|A_k(x)-\theta|\right)\to\infty,$$

i.e., $x_k \nrightarrow \theta\left(N_\theta^A(\mathcal{I}, F)\right)$. Note that, if $K \in \mathcal{I}$ is finite, then $x_k \nrightarrow \theta\left(S_\theta^A\right)$. This example shows that $A^{\mathcal{I}}$-lacunary statistical convergence is more general than lacunary statistical convergence.

(ii) Suppose that $x \in l_\infty$ and $x_k \to L\left(S_\theta^A(\mathcal{I}, F)\right)$. Then, we can assume that

$$f_k\left(|A_k(x)-L|\right)\leq M$$

for each $x \in X$ and all k.

Given $\varepsilon > 0$, we get

$$\begin{aligned}\frac{1}{h_r}\sum_{k\in I_r} f_k\left(|A_k(x)-L|\right) &= \frac{1}{h_r}\sum_{\substack{k\in I_r\\ f_k(|A_kx-L|)\geq\varepsilon}} f_k\left(|A_k(x)-L|\right)\\ &\quad+\frac{1}{h_r}\sum_{\substack{k\in I_r\\ f_k(|A_kx-L|)<\varepsilon}} f_k\left(|A_k(x)-L|\right)\\ &\leq \frac{M}{h_r}\left|\left\{k\in I_r : f_k\left(|A_k(x)-L|\right)\geq\varepsilon\right\}\right|+\varepsilon.\end{aligned}$$

Note that

$$A(\varepsilon)=\left\{r\in\mathbb{N} : \frac{1}{h_r}\left|\left\{k\in I_r : f_k\left(|A_k(x)-L|\right)\geq\varepsilon\right\}\right|\geq\frac{\varepsilon}{M}\right\}\in\mathcal{I}.$$

If $n\in(A(\varepsilon))^c$, then

$$\frac{1}{h_r}\sum_{k\in I_r} f_k|A_k(x)-L|<2\varepsilon.$$

Hence,

$$\left\{r\in\mathbb{N} : \frac{1}{h_r}\sum_{k\in I_r} f_k|A_k(x)-L|\geq 2\varepsilon\right\}\subset A(\varepsilon)$$

and thus belongs to $\mathcal{I}$. This shows that $x_k\to L\left(N_\theta^A(\mathcal{I},F)\right)$.

(iii) This is an immediate consequence of (i) and (ii). □

Theorem 3. *Let $A=(a_{ki})$ be an infinite matrix of complex numbers and (f_k) be a sequence of modulus functions in S. If $\theta=\{k_r\}$ is a lacunary sequence with $\liminf_r q_r>1$, then*

$$x_k\to L\left(S^A(\mathcal{I},F)\right)\Rightarrow x_k\to L\left(S_\theta^A(\mathcal{I},F)\right).$$

Proof. Suppose first that $\liminf_r q_r>1$, then there exists $\delta>0$ such that $q_r\geq 1+\delta$ for sufficiently large r, which implies that

$$\frac{h_r}{k_r}\geq\frac{\delta}{1+\delta}.$$

If $x_k\to L\left(S_\theta^A(\mathcal{I},F)\right)$, then for every $\varepsilon>0$, for each $x\in X$ and for sufficiently large r, we have

$$\begin{aligned}\frac{1}{k_r}\left|\left\{k\leq k_r : f_k\left(|A_k(x)-L|\right)\geq\varepsilon\right\}\right| &\geq \frac{1}{k_r}\left|\left\{k\in I_r : f_k\left(|A_k(x)-L|\right)\geq\varepsilon\right\}\right|\\ &\geq \frac{\delta}{1+\delta}\frac{1}{h_r}\left|\left\{k\in I_r : f_k\left(|A_k(x)-L|\right)\geq\varepsilon\right\}\right|.\end{aligned}$$

Then, for any $\delta>0$, we get

$$\begin{aligned}&\left\{r\in\mathbb{N} : \tfrac{1}{h_r}\left|\left\{k\in I_r : f_k\left(|A_k(x)-L|\right)\geq\varepsilon\right\}\right|\geq\delta\right\}\\ &\subseteq\left\{r\in\mathbb{N} : \tfrac{1}{k_r}\left|\left\{k\leq k_r : f_k\left(|A_k(x)-L|\right)\geq\varepsilon\right\}\right|\geq\tfrac{\delta\alpha}{(\alpha+1)}\right\}\in\mathcal{I}.\end{aligned}$$

This completes the proof. □

For the next result, we assume that the lacunary sequence θ satisfies the condition that, for any set $C\in\mathcal{F}(\mathcal{I})$, $\cup\{n : k_{r-1}<n\leq k_r,\ r\in C\}\in\mathcal{F}(\mathcal{I})$.

Theorem 4. *Let $A = (a_{ki})$ be an infinite matrix of complex numbers and (f_k) be a sequence of modulus functions in S. If $\theta = \{k_r\}$ is a lacunary sequence with $\limsup_r q_r < \infty$, then*

$$x_k \to L\left(S_\theta^A(\mathcal{I}, F)\right) \text{ implies } x_k \to L\left(S^A(\mathcal{I}, F)\right).$$

Proof. If $\limsup_r q_r < \infty$, then, without any loss of generality, we can assume that there exists a $0 < M < \infty$ such that $q_r < M$ for all $r \geq 1$. Suppose that $x_k \to L\left(S_\theta^A(\mathcal{I}, F)\right)$, and for $\varepsilon, \delta, \delta_1 > 0$ define the sets

$$C = \left\{r \in \mathbb{N} : \frac{1}{h_r} |\{k \in I_r : f_k(|A_k(x) - L|) \geq \varepsilon\}| < \delta\right\}$$

and

$$T = \left\{n \in \mathbb{N} : \frac{1}{n} |\{k \leq n : f_k(|A_k(x) - L|) \geq \varepsilon\}| < \delta_1\right\}.$$

It is obvious from our assumption that $C \in \mathcal{F}(\mathcal{I})$, the filter associated with the ideal $\mathcal{I}$. Further observe that

$$K_j = \frac{1}{h_j} |\{k \in I_j : f_k(|A_k(x) - L|) \geq \varepsilon\}| < \delta$$

for all $j \in C$. Let $n \in \mathbb{N}$ be such that $k_{r-1} < n \leq k_r$ for some $r \in C$. Now,

$$\begin{aligned}
\frac{1}{n} |\{k \leq n : f_k |A_k(x) - L| \geq \varepsilon\}| &\leq \frac{1}{k_{r-1}} |\{k \leq k_r : f_k(|A_k(x) - L|) \geq \varepsilon\}| \\
&= \frac{1}{k_{r-1}} \{|\{k \in I_1 : f_k(|A_k(x) - L|) \geq \varepsilon\}|\} \\
&\quad + \frac{1}{k_{r-1}} \{|\{k \in I_2 : f_k(|A_k x - L|) \geq \varepsilon\}|\} \\
&\quad + \ldots + \frac{1}{k_{r-1}} \{|\{k \in I_r : f_k |A_k(x) - L| \geq \varepsilon\}|\} \\
&= \frac{k_1}{k_{r-1}} \frac{1}{h_1} |\{k \in I_1 : f_k |A_k(x) - L| \geq \varepsilon\}| \\
&\quad + \frac{k_2 - k_1}{k_{r-1}} \frac{1}{h_2} |\{k \in I_2 : f_k(|A_k(x) - L|) \geq \varepsilon\}| \\
&\quad + \ldots + \frac{k_r - k_{r-1}}{k_{r-1}} \frac{1}{h_r} |\{k \in I_r : f_k(|A_k(x) - L|) \geq \varepsilon\}| \\
&= \frac{k_1}{k_{r-1}} K_1 + \frac{k_2 - k_1}{k_{r-1}} K_2 + \ldots + \frac{k_r - k_{r-1}}{k_{r-1}} K_r \\
&\leq \left\{\sup_{j \in C} K_j\right\} \frac{k_r}{k_{r-1}} < M\delta.
\end{aligned}$$

Choosing $\delta_1 = \dfrac{\delta}{M}$ and in view of the fact that $\cup\{n : k_{r-1} < n \leq k_r, \ r \in C\} \subset T$ where $C \in \mathcal{F}(\mathcal{I})$, it follows from our assumption on θ that the set T also belongs to $\mathcal{F}(\mathcal{I})$ and this completes the proof of the theorem. □

Combining Theorems 3 and 4, we get the following theorem.

Theorem 5. *Let $A = (a_{ki})$ be an infinite matrix of complex numbers and (f_k) be a sequence of modulus functions in S. If $\theta = \{k_r\}$ is a lacunary sequence with $1 < \liminf_r q_r \leq \limsup_r q_r < \infty$, then*

$$x_k \to L\left(S_\theta^A(\mathcal{I},F)\right) = x_k \to L\left(S_\theta^A(\mathcal{I},F)\right).$$

4. Cesàro Summability for $A^{\mathcal{I}}$

Definition 8. *Let $A = (a_{ki})$ be an infinite matrix of complex numbers and (f_k) be a sequence of modulus functions in S. A sequence $x = (x_k)$ is said to be $A^{\mathcal{I}}$-Cesàro summable to L if, for each $\varepsilon > 0$ and for each $x \in X$,*

$$\left\{n \in \mathbb{N} : \left|\frac{1}{n}\sum_{k=1}^{n} f_k(A_k(x) - L)\right| \geq \varepsilon\right\} \in \mathcal{I}.$$

In this case, we write $x_k \to L\left((\sigma_1)_\theta^A(\mathcal{I},F)\right)$.

Definition 9. *Let $A = (a_{ki})$ be an infinite matrix of complex numbers and (f_k) be a sequence of modulus functions in S. A sequence $x = (x_k)$ is said to be strongly $A^{\mathcal{I}}$-Cesàro summable to L if, for each $\varepsilon > 0$ and for each $x \in X$,*

$$\left\{n \in \mathbb{N} : \frac{1}{n}\sum_{k=1}^{n} f_k(|A_k(x) - L|) \geq \varepsilon\right\} \in \mathcal{I}.$$

In this case, we write $x_k \to L\left(|\sigma_1|_\theta^A(\mathcal{I},F)\right)$.

Theorem 6. *Let θ be a lacunary sequence. If $\liminf_r q_r > 1$, then*

$$x_k \to L\left(|\sigma_1|_\theta^A(\mathcal{I},F)\right) \Rightarrow x_k \to L\left(N_\theta^A(\mathcal{I},F)\right).$$

Proof. If $\liminf_r q_r > 1$, then there exists $\delta > 0$ such that $q_r \geq 1+\delta$ for all $r \geq 1$. Since $h_r = k_r - k_{r-1}$, we have $\frac{k_r}{h_r} \leq \frac{1+\delta}{\delta}$ and $\frac{k_{r-1}}{h_r} \leq \frac{1}{\delta}$. Let $\varepsilon > 0$ and define the set

$$S = \left\{k_r \in \mathbb{N} : \frac{1}{k_r}\sum_{k=1}^{k_r} f_k(|A_k(x) - L|) < \varepsilon\right\}.$$

We can easily say that $S \in \mathcal{F}(\mathcal{I})$, which is a filter of the ideal $\mathcal{I}$,

$$\begin{aligned}
\frac{1}{h_r}\sum_{k\in I_r} f_k(|A_k(x) - L|) &= \frac{1}{h_r}\sum_{k=1}^{k_r} f_k(|A_k(x) - L|) - \frac{1}{h_r}\sum_{k=1}^{k_{r-1}} f_k(|A_k(x) - L|) \\
&= \frac{k_r}{h_r}\frac{1}{k_r}\sum_{k=1}^{k_r} f_k(|A_k(x) - L|) - \frac{k_{r-1}}{h_r}\frac{1}{k_{r-1}}\sum_{k=1}^{k_{r-1}} f_k(|A_k(x) - L|) \\
&\leq \left(\frac{1+\delta}{\delta}\right)\varepsilon - \frac{1}{\delta}\varepsilon'
\end{aligned}$$

for each $k_r \in S$. Choose $\eta = \left(\frac{1+\delta}{\delta}\right)\varepsilon - \frac{1}{\delta}\varepsilon'$. Therefore,

$$\left\{r \in \mathbb{N} : \frac{1}{h_r}\sum_{k\in I_r} f_k(|A_k(x) - L|) < \eta\right\} \in \mathcal{F}(\mathcal{I}),$$

and it completes the proof. □

Theorem 7. *Let $A = (a_{ki})$ be an infinite matrix of complex numbers and (f_k) be a sequence of modulus functions in S. If $(x_k) \in m(X)$ and $x_k \to L\left(S_\theta^A(\mathcal{I}, F)\right)$, then $x_k \to L\left((\sigma_1)_\theta^A(\mathcal{I}, F)\right)$.*

Proof. Suppose that $(x_k) \in m(X)$ and $x_k \to L\left(S_\theta^A(\mathcal{I}, F)\right)$. Then, we can assume that

$$f_k(|A_k x - L|) \leq M$$

for all $k \in \mathbb{N}$. In addition, for each $\varepsilon > 0$, we can write

$$\begin{aligned}
\left|\frac{1}{n}\sum_{k=1}^{n} f_k(A_k(x) - L)\right| &\leq \frac{1}{n}\sum_{k=1}^{n} f_k(|A_k(x) - L|) \\
&\leq \frac{1}{n}\sum_{\substack{k=1 \\ f_k(|A_k(x)-L|)\geq\frac{\varepsilon}{2}}}^{n} f_k(|A_k(x) - L|) \\
&\quad + \frac{1}{n}\sum_{\substack{k=1 \\ f_k(|A_k(x)-L|)<\frac{\varepsilon}{2}}}^{n} f_k(|A_k(x) - L|) \\
&\leq M\frac{1}{n}\left|\{k \leq n : f_k(|A_k(x) - L|) \geq \varepsilon\}\right| + \varepsilon.
\end{aligned}$$

Consequently, if $\delta > \varepsilon > 0$, δ and ε are independent, and, putting $\delta_1 = \delta - \varepsilon > 0$, we have

$$\left\{n \in \mathbb{N} : \left|\frac{1}{n}\sum_{k,l=1}^{n} f_k(A_k(x) - L)\right| \geq \delta\right\}$$

$$\subseteq \left\{n \in \mathbb{N} : \tfrac{1}{n}\left|\{k \leq n : f_k(|A_k(x) - L|) \geq \varepsilon\}\right| \geq \tfrac{\delta_1}{M}\right\} \in \mathcal{I}.$$

This shows that $x_k \to L\left((\sigma_1)_\theta^A(\mathcal{I}, F)\right)$. □

Theorem 8. *Let θ be a lacunary sequence. If $\limsup_r q_r < \infty$, then*

$$x_k \to L\left(N_\theta^A(\mathcal{I}, F)\right) \Rightarrow x_k \to L\left(|\sigma_1|_\theta^A(\mathcal{I}, F)\right).$$

Proof. If $\limsup_r q_r < \infty$, then there exists $M > 0$ such that $q_r < M$ for all $r \geq 1$. Let $x_k \to L\left(N_\theta^A(\mathcal{I}, F)\right)$ and define the sets T and R such that

$$T = \left\{r \in \mathbb{N} : \frac{1}{h_r}\sum_{k\in I_r} f_k(|A_k(x) - L|) < \varepsilon_1\right\}$$

and

$$R = \left\{n \in \mathbb{N} : \frac{1}{n}\sum_{k=1}^{n} f_k(|A_k(x) - L|) < \varepsilon_2\right\}.$$

Let

$$A_j = \frac{1}{h_j}\sum_{k\in I_j} f_k(|A_k(x) - L|) < \varepsilon_1$$

for all $j \in T$. It is obvious that $T \in \mathcal{F}(\mathcal{I})$. Choose n as being any integer with $k_{r-1} < n < k_r$, where $r \in T$,

$$\begin{aligned}
\frac{1}{n}\sum_{k=1}^{n} f_k\left(\left|A_k(x)-L\right|\right) &\leq \frac{1}{k_{r-1}}\sum_{k=1}^{k_r} f_k\left(\left|A_k(x)-L\right|\right) \\
&= \frac{1}{k_{r-1}}\left(\sum_{k\in I_1} f_k\left(\left|A_k(x)-L\right|\right) + \sum_{k\in I_2} f_k\left(\left|A_k(x)-L\right|\right)\right. \\
&\quad \left. +\ldots+\sum_{k\in I_r} f_k\left(\left|A_k(x)-L\right|\right)\right) \\
&= \frac{k_1}{k_{r-1}}\left(\frac{1}{h_1}\sum_{k\in I_1} f_k\left(\left|A_k(x)-L\right|\right)\right) + \frac{k_2-k_1}{k_{r-1}}\left(\frac{1}{h_2}\sum_{k\in I_2} f_k\left(\left|A_k(x)-L\right|\right)\right) \\
&\quad +\ldots+\frac{k_r-k_{r-1}}{k_{r-1}}\left(\frac{1}{h_r}\sum_{k\in I_r} f_k\left(\left|A_k(x)-L\right|\right)\right) \\
&= \frac{k_1}{k_{r-1}}A_1 + \frac{k_2-k_1}{k_{r-1}}A_2+\ldots+\frac{k_r-k_{r-1}}{k_{r-1}}A_r \\
&\leq \left(\sup_{j\in T} A_j\right)\frac{k_1}{k_{r-1}} \\
&< \varepsilon_1 M.
\end{aligned}$$

Choose $\varepsilon_2 = \frac{\varepsilon_1}{M}$ and in view of the fact that $\cup\{n : k_{r-1} < n < k_r, r \in T\} \subset R$, where $T \in \mathcal{F}(\mathcal{I})$, it follows from our assumption on θ that the set R also belongs to $\mathcal{F}(\mathcal{I})$ and this completes the proof of the theorem. □

Theorem 9. *If $x_k \to L\left(|\sigma_1|_\theta^A(\mathcal{I},F)\right)$, then $x_k \to L\left(S^A(\mathcal{I},F)\right)$.*

Proof. Let $x_k \to L\left(|\sigma_1|_\theta^A(\mathcal{I},F)\right)$ and $\varepsilon > 0$ is given. Then,

$$\begin{aligned}
\sum_{k=1}^{n} f_k\left(\left|A_k(x)-L\right|\right) &\geq \sum_{\substack{k=1 \\ f_k(|A_k x-L|)\geq\varepsilon}}^{n} f_k\left(\left|A_k(x)-L\right|\right) \\
&\geq \varepsilon\left|\left\{k\leq n : f_k\left(\left|A_k(x)-L\right|\right)\geq\varepsilon\right\}\right|
\end{aligned}$$

and so

$$\frac{1}{\varepsilon n}\sum_{k=1}^{n} f_k\left(\left|A_k(x)-L\right|\right) \geq \frac{1}{n}\left|\left\{k\leq n : f_k\left(\left|A_k(x)-L\right|\right)\geq\varepsilon\right\}\right|.$$

Thus, for a given $\delta > 0$,

$$\left\{n\in\mathbb{N} : \frac{1}{n}\left|\left\{k\leq n : f_k\left(\left|A_k(x)-L\right|\right)\geq\varepsilon\right\}\right|\geq\delta\right\}$$

$$\subseteq\left\{n\in\mathbb{N} : \frac{1}{n}\sum_{k=1}^{n} f_k\left(\left|A_k(x)-L\right|\right)\geq\varepsilon\delta\right\}\in\mathcal{I}.$$

Therefore, $x_k \to L\left(S^A(\mathcal{I},F)\right)$. □

Theorem 10. *Let* $(x_k) \in m(X)$. *If* $x_k \to L\left(S^A(\mathcal{I},F)\right)$. *Then,* $x_k \to L\left(|\sigma_1|_\theta^A(\mathcal{I},F)\right)$.

Proof. Suppose that (x_k) is bounded and $x_k \to L\left(S^A(\mathcal{I},F)\right)$. Then, there is an M such that $f_k(|A_k(x)-L|) \le M$ for all k. Given $\varepsilon > 0$, we have

$$\begin{aligned}
\frac{1}{n}\sum_{k=1}^{n} f_k(|A_k(x)-L|) &= \frac{1}{n}\sum_{\substack{k=1 \\ f_k(|A_k(x)-L|)\ge\varepsilon}}^{n} f_k(|A_k(x)-L|) \\
&+ \frac{1}{n}\sum_{\substack{k=1 \\ f_k(|A_k(x)-L|)<\varepsilon}}^{n} f_k(|A_k(x)-L|) \\
&\le \frac{1}{n}M\left|\{k\le n : f_k(|A_k(x)-L|)\ge\varepsilon\}\right| \\
&\quad +\frac{1}{n}\varepsilon\left|\{k\le n : f_k(|A_k(x)-L|)<\varepsilon\}\right| \\
&\le \frac{M}{n}\left|\{k\le n : f_k(|A_k(x)-L|)\ge\varepsilon\}\right| + \varepsilon.
\end{aligned}$$

Then, for any $\delta > 0$,

$$\begin{aligned}
&\left\{n\in\mathbb{N} : \frac{1}{n}\sum_{k=1}^{n} f_k(|A_k(x)-L|)\ge\delta\right\} \\
&\qquad\subseteq \left\{n\in\mathbb{N} : \frac{1}{n}\left|\{k\le n : f_k(|A_k(x)-L|)\ge\varepsilon\}\right| \ge \frac{\delta}{M}\right\} \in \mathcal{I}.
\end{aligned}$$

Therefore, $x_k \to L\left(|\sigma_1|_\theta^A(\mathcal{I},F)\right)$. □

5. Conclusions

$\mathcal{I}$-statistical convergence gained a different perspective after identification of the $A^{\mathcal{I}}$-statistical convergence with an infinite matrix of complex numbers. Some authors have studied this new method with different sequences. Our results in this paper were developed with lacunary sequences. By also using a modulus function, we obtain more interesting and general results. These definitions can be adapted to many different concepts such as random variables in order to have different results.

Acknowledgments: We would like to thanks the referees for their valuable comments.

Author Contributions: Both authors completed the paper together. Both authors read and approved the final manuscript.

References

1. Zygmund, A. *Trigonometric Series*; Cambridge University Press: Cambridge, UK, 1979.
2. Fast, H. Sur la convergence statistique. *Colloq. Math.* **1951**, *2*, 241–244.
3. Fridy, J.A. On statistical convergence. *Analysis* **1985**, *5*, 301–313.
4. Salat, T. On statistically convergent sequences of real numbers. *Math. Slovaca* **1980**, *30*, 139–150.
5. Steinhaus, H. Sur la convergence ordinaire et la convergence asymptotique. *Colloq. Math.* **1951**, *2*, 73–74.
6. Schoenberg, I.J. The integrability of certain functions and related summability methods. *Am. Math. Mon.* **1959**, *66*, 361–375.
7. Erdös, P.; Tenenbaum, G. Sur les densities de certaines suites d'entiers. *Proc. Lond. Math. Soc.* **1989**, *3*, 417–438.

8. Miller, H.I. A measure theoretical subsequence characterization of statistical convergence. *Trans. Am. Math. Soc.* **1995**, *347*, 1811–1819.
9. Freedman, A.R.; Sember, J.J. Densities and summability. *Pac. J. Math.* **1981**, *95*, 10–11.
10. Fridy, J.A.; Orhan, C. lacunary statistical convergence. *Pac. J. Math.* **1993**, *160*, 43–51.
11. Kostyrko, P.; Macaj, M.; Salat, T. $\mathcal{I}$-convergence. *Real Anal. Exchang.* **2000**, *26*, 669–686.
12. Das, P.; Ghosal, S. Some further results on $\mathcal{I}$-Cauchy sequences and condition (AP). *Comput. Math. Appl.* **2010**, *29*, 2597–2600.
13. Kostyrko, P.; Macaj, M.; Salat, T.; Sleziak, M. $\mathcal{I}$-convergence and extremal $\mathcal{I}$-limit points. *Math. Slovaca* **2005**, *55*, 443–464.
14. Savas, E.; Das, P. A generalized statistical convergence via ideals. *Appl. Math. Lett.* **2011**, *24*, 826–830.
15. Nakano, H. Concave modulars. *J. Math. Soc. Jpn.* **1953**, *5*, 29–49.
16. Connor, J.S. On strong matrix summability with respect to a modulus and statistical convergence. *Can. Math. Bull.* **1989**, *32*, 194–198.
17. Bilgin, T. Lacunary strong A-convergence with respect to a modulus. *Mathematica* **2001**, *4*, 39–46.
18. Maddox, I.J. Sequence spaces defined by a modulus. *Math. Proc. Camb. Philos. Soc.* **1986**, *100*, 161–166.
19. Kolk, E. On strong boundedness and summability with respect to a sequence moduli. *Tartu Ülikooli Toimetised* **1993**, *960*, 41–50.
20. Pehlivan, S.; Fisher, B. On some sequence spaces. *Indian J. Pure Appl. Math.* **1994**, *25*, 1067–1071.
21. Ruckle, W.H. FK spaces in which the sequence of coordinate vectors is bounded. *Can. J. Math.* **1973**, *25*, 973–978.
22. Das, P.; Savas, E.; Ghosal, S. On generalizations of certain summability methods using ideals. *Appl. Math. Lett.* **2011**, *24*, 1509–1514.
23. Yamancı, U.; Gürdal, M.; Saltan, S. $A^{\mathcal{I}}$-statistical convergence with respect to a sequence of modulus functions. *Contemp. Anal. Appl. Math.* **2014**, *2*, 136–145.

13

Some New Results Involving the Generalized Bose–Einstein and Fermi–Dirac Functions

Rekha Srivastava [1], Humera Naaz [2], Sabeena Kazi [2] and Asifa Tassaddiq [2,*]

1 Department of Mathematics and Statistics, University of Victoria, Victoria, BC V8W 3R4, Canada; rekhas@math.uvic.ca
2 College of Computer and Information Sciences, Majmaah University, Al Majmaah 11952, Saudiarabia; h.naaz@mu.edu.sa (H.N.); s.badesaheb@mu.edu.sa (S.K.)
* Correspondence: a.tassaddiq@mu.edu.sa

Abstract: In this paper, we obtain a new series representation for the generalized Bose–Einstein and Fermi–Dirac functions by using fractional Weyl transform. To achieve this purpose, we obtain an analytic continuation for these functions by generalizing the domain of Riemann zeta functions from $(0 < \Re(s) < 1)$ to $(0 < \Re(s) < \mu)$. This leads to fresh insights for a new generalization of the Riemann zeta function. The results are validated by obtaining the classical series representation of the polylogarithm and Hurwitz–Lerch zeta functions as special cases. Fractional derivatives and the relationship of the generalized Bose–Einstein and Fermi–Dirac functions with Apostol–Euler–Nörlund polynomials are established to prove new identities.

Keywords: Fermi–Dirac function; Bose–Einstein function; Weyl transform; series representation

1. Introduction

The importance of the Fermi–Dirac and Bose–Einstein functions emerges from their fundamental presence in quantum physics and related sciences. Unlike the classical mechanics of particles, where the Maxwell distribution is used to study the velocity of classical gas molecules, the quantum gas is analyzed by using the Fermi–Dirac and Bose–Einstein functions. The distinct particles obey Fermi–Dirac statistics, while the indistinct particles follow Bose–Einstein statistics. All particles have a spin in relation to the usual theory. Fermions have half-integer spin and bosons have integer spin. The Fermi–Dirac and Bose–Einstein distribution functions are used to analyze them in the language of mathematics and physics. Indistinguishable particles that are not categorized through either of the aforementioned types are called anyons. The extensions of the Bose–Einstein and Fermi–Dirac functions interpolate between the two. Therefore, Chaudhry et al. [1] proposed that the extensions of the Bose–Einstein and Fermi–Dirac functions may help to describe anyons. In this paper, we generalize the results of Chaudhry and Qadir [2] by proving a general representation theorem to establish a new series representation of the generalized Bose–Einstein and Fermi–Dirac functions. However, we also discuss the fractional derivative, and the relationship of the generalized Bose–Einstein and Fermi–Dirac functions with Apostol–Euler–Nörlund polynomials. Before we provide our research results, it is necessary to enlist all the basic definitions and preliminaries that are required to present and understand this work.

2. Materials and Methods

2.1. Generalized Bose–Einstein and Fermi–Dirac Functions

During the course of our investigation, we consider the subsequent usual notations:

$$\mathbb{N} := \{1, 2, 3 \ldots.\},\ \mathbb{N} \cup \{0\} = \mathbb{N}_0; \mathbb{Z}^- = \{-1, -2, -3 \ldots\}.$$

In addition, $\mathbb{Z}$ is the set of integers, $\mathbb{R}$ denotes the set of real numbers, $\mathbb{R}^+$ denotes the set of positive numbers, and $\mathbb{C}$ is the set of complex numbers, $s = \sigma + i\tau$. Gamma function $\Gamma(s)$ as a generalization of factorials is also used here as a basic special function. For a detailed study of gamma and related functions, we refer the interested reader to [3,4].

More recently, Bayad and Chikhi [5] introduced and studied the generalized Fermi–Dirac functions given by ([5], (p. 12), Equation (45))

$$\begin{gathered}\Theta_\nu\ (s, \mu; x)\ := \frac{\Gamma(\mu)}{\Gamma(s)} \int_0^\infty \frac{t^{s-1}e^{-\nu(x+t)}}{(e^{x+t}+1)^\mu}\, dt \\ (\Re(x) \geq 0,\ \Re(\nu) > -\Re(\mu) \wedge \Re(s) > \Re(\mu) > 0 \text{ when } e^{-x} \neq -1)\end{gathered} \tag{1}$$

and their series representation is given by ([5], (p. 12), Equation (46))

$$\Theta_\nu\ (s, \mu; x) := \sum_{n=1}^{\infty} \frac{(-1)^n \Gamma(\mu+n) e^{-(\nu+\mu+n)x}}{n!(\nu+\mu+n)^s}. \tag{2}$$

For $\mu = 1$, in Equation (1) the extended Fermi–Dirac functions ([6], (p. 113), Equation (3.14)) are given here by

$$\begin{gathered}\Theta_\nu(s; x) := \Theta_\nu\ (s, 1; x)\ = \frac{1}{\Gamma(s)} \int_0^\infty \frac{t^{s-1}e^{-\nu(t+x)}}{e^{t+x}+1}\, dt \\ (\Re(x) \geq 0,\ \Re(\nu) > -1),\end{gathered} \tag{3}$$

and for $\nu = 0$ and $\mu = 1$ in Equation (1), the original Fermi–Dirac function is given by ([6], (p. 109), Equation (1.12))

$$\mathcal{F}_{s-1}(x)\ := \Theta_0\ (s, 1; x) = \frac{1}{\Gamma(s)} \int_0^\infty \frac{t^{s-1}}{e^{t+x}+1} dt\ (\Re(x) \geq 0; \Re(s) > 0). \tag{4}$$

Similarly, the generalized Bose–Einstein functions $\Psi_\nu(s,\ \alpha\,; x)$, which are defined by ([5], (p. 13), Equation (51)), are as follows

$$\begin{gathered}\Psi_\nu(s, \mu; x)\ := \frac{\Gamma(\mu)}{\Gamma(s)} \int_0^\infty \frac{t^{s-1}e^{-\nu t}}{(e^{t+x}-1)^\mu}\, dt \\ (\Re(x) \geq 0, \Re(\nu) > -\Re(\mu) \wedge\ \Re(s) > \Re(\mu) > 0 \text{ when } e^{-x} = 1\ \wedge\ \Re(s) > 0\ ),\end{gathered} \tag{5}$$

and their series representation is given by ([5], (p. 13), Equation (52))

$$\Psi_\nu\ (s, \mu; x) := \sum_{n=1}^{\infty} \frac{\Gamma(\mu+n) e^{-(\nu+\mu+n)x}}{n!(\nu+\mu+n)^s}. \tag{6}$$

For $\mu = 1$, the extended Bose–Einstein functions ([6], (p. 115), Equation (4.4)) are given here by

$$\Psi_\nu(s; x)\ := \Psi_\nu(s, 1; x) = \frac{1}{\Gamma(s)} \int_0^\infty \frac{t^{s-1}e^{-\nu t}}{e^{t+x}-1}\, dt\ (\Re(x) \geq 0,\ \Re(\nu) > -1), \tag{7}$$

and the original Bose–Einstein function is given by ([6], (p. 109), Equation (1.13)).

$$\mathcal{B}_{s-1}(x)\ := \Psi_0\ (s, 1; x) = \frac{1}{\Gamma(s)} \int_0^\infty \frac{t^{s-1}}{e^{t+x}-1}\, dt\ (\Re(x) \geq 0; \Re(s) > 1). \tag{8}$$

For further study of the Fermi–Dirac and Bose–Einstein functions, we refer the interested reader to [7–9]. The reduction and duality theorems for these functions are given by ([5], (p. 12–13))

$$\Theta_{\nu-M}(s;\ \mu+M;x) = \sum_{m=0}^{M} R_1(M\,,\ m,-\nu)\Theta_\nu(s-m,\mu;x), \tag{9}$$

$$\Theta_\nu(s-M;\mu;x) = \sum_{m=0}^{M}\ (-1)^{M-m}R(M\,,\ m,-\nu)\Theta_{\nu-m}(s\,,\ \mu+m;x), \tag{10}$$

$$\Psi_{\nu-M}(s;\ \mu+M;x) = \sum_{m=0}^{M} R_1(M\,,\ m,-\nu)\Psi_\nu(s-m,\ \mu;x), \tag{11}$$

$$\Psi_\nu(s-M;\ \mu;x) = \sum_{m=0}^{M}(-1)^{M-m}R(M\,,\ m,-\nu)\Psi_{\nu-m}(s\,,\ \mu+m;x), \tag{12}$$

respectively, where $R_1(M\,,\ m,-\nu)$ and $R(M\,,\ m,-\nu)$ are the polynomials having explicit representations in terms of Stirling numbers. For examples and details see Carlitz [10,11]. More recently, Tassaddiq [12,13] considered the λ-generalized extended Fermi–Dirac functions and λ-generalized extended Bose–Einstein functions as a transformed form of Srivastava's λ-generalized Hurwitz–Lerch zeta functions ([14], (p. 1487), Equation (1.14)). In this research, we generalize the results of Chaudhry and Qadir [2]. To achieve this goal, it is important to briefly highlight their relationship with the zeta functions. It should be noted that for $x = 0$, the Bose–Einstein and Fermi–Dirac functions are related to the Riemann zeta functions respectively.

$$\zeta(s) := \mathcal{B}_{s-1}(0); \Re(s) > 1 \tag{13}$$

$$\zeta(s)\left(1-2^{1-s}\right) := \mathcal{F}_{s-1}(0); \Re(s) > 0. \tag{14}$$

The polylogarithm function is an important function in the study of theory of polymers that was introduced and investigated by Truesdell [15]

$$\mathrm{Li}_s(z) := \sum_{n=1}^{\infty} \frac{z^n}{n^s}\ (s \in \mathbb{C},\ |z| < 1; \Re(s) > 1,\ |z| = 1). \tag{15}$$

It generalizes the Riemann zeta function, as we have

$$\mathrm{Li}_s(1) = \phi(1,s) = \zeta(s)\ (\Re(s) > 1\ ), \tag{16}$$

and it can also be represented as an integral

$$\mathrm{Li}_s(z) = \frac{z}{\Gamma(s)} \int_0^\infty \frac{t^{s-1}}{e^t - z}\,dt\ (s \in \mathbb{C} \text{ when } |z| < 1; \Re(s) > 1 \text{ and when } |z| = 1). \tag{17}$$

In our present analysis, we are especially interested in Lindelöf's representation of these functions given by ([15], (p. 149), Equation (13)),

$$\begin{gathered}\mathrm{Li}_s(z) = \Gamma(1-s)(\log z)^{s-1} + \textstyle\sum_{n=0}^{\infty} \zeta(s-n)\frac{(\log z)^n}{n!} \\ \left(|\log z| < 2\pi, s \neq 1,2,3,\ldots, \nu \neq 0,-1,-2,\ldots,\right).\end{gathered} \tag{18}$$

The Hurwitz–Lerch zeta function ([16], (p. 27)), as a generalization of the polylogarithm, is given by

$$\Phi(z,s,a)=\sum_{n=0}^{\infty}\frac{z^n}{(n+a)^s}\ (a\in\mathbb{C}\backslash\mathbb{Z}^-;s\in\mathbb{C}\text{ when }|z|<1;\ \Re(s)>1\text{ when }|z|=1). \tag{19}$$

It has a meromorphic extension to the whole complex s-plane, while it has a simple singularity at $s = 1$ of residue 1. It is also represented by ([16], (p. 27), Equation (1.6))

$$\Phi(z,s,a)=\frac{1}{\Gamma(s)}\int_0^{\infty}\frac{t^{s-1}e^{-at}}{1-ze^{-t}}\,dt\ (|z|<1\Rightarrow\Re(s)>0;\Re(a)>0;z=1\Rightarrow\Re(s)>1\,). \tag{20}$$

Apart from other applications, the Hurwitz–Lerch zeta function is the most general function in the original zeta family. For example, different values of the involved parameters in (19–20) yield the following relationships with the polylogarithm, Hurwitz, and Riemann zeta functions, respectively:

$$\mathrm{Li}_s(z):=\sum_{n=1}^{\infty}\frac{z^n}{n^s}=z\Phi(s,z,1), \tag{21}$$

$$\zeta(s,a):=\sum_{n=0}^{\infty}\frac{1}{(n+a)^s}=\Phi(s,1,a), \tag{22}$$

$$\zeta(s):=\sum_{n=1}^{\infty}\frac{1}{n^s}=\Phi(s,1,1)=\zeta(s,1). \tag{23}$$

For our purposes, it is important to note that the Hurwitz–Lerch zeta function has a series representation ([16], (pp. 28–29))

$$\begin{gathered}\Phi(z,s,\nu)=\frac{\Gamma(1-s)}{z^{\nu}}\left(\log\tfrac{1}{z}\right)^{s-1}+z^{-\nu}\sum_{n=0}^{\infty}\zeta(s-n,\nu)\frac{(\log z)^n}{n!}\\ \left(\left|\log z\right|<2\pi\,,s\neq1,2,3,\ldots,\nu\neq0,-1,-2,\ldots,\right)\end{gathered} \tag{24}$$

that generalizes Lindelöf's representation (18).

Further to all of the above discussion, Chaudhry et al. [17] defined a new generalization of the Riemann zeta function in the critical strip by

$$\Xi_a(s;x):=\frac{1}{\Gamma(s)}\int_x^{\infty}(t-x)^{s-1}\left(\frac{1}{e^t-1}-\frac{1}{t}\right)e^{-at}\,dt\ (0<\ \Re(s)<1:x\geq0;a\geq0). \tag{25}$$

The Riemann hypothesis is a well-known unsolved problem in analytic number theory [18]. It states that "all the non-trivial zeros of the zeta function exist on the line $s = 1/2$". These zeros seem to be complex conjugates and are hence symmetrical on this line. The Riemann zeta function in the critical strip is defined and studied in [18]

$$\zeta(s):=\frac{1}{\Gamma(s)}\int_0^{\infty}t^{s-1}\left(\frac{1}{e^t-1}-\frac{1}{t}\right)dt\ (0<\ \Re(s)<1), \tag{26}$$

which can be obtained as a special case of Equation (25) by putting $x = a = 0$.

2.2. A Class $\Re_{\infty}(A,P,\delta)$ of Functions and the Representation Theorem

More recently Chaudhry and Qadir [19] discussed some important classes of functions. The statements of this section are taken from [19–21].

We first give a brief introduction to the function spaces $H(\xi;\eta)$ and $H(\infty;\eta)$. The elements of $H(\xi;\eta)$ are particular functions $f\in C^{\infty}(0,\infty)$ that satisfy the following conditions

1. $\int_0^T f(t)dt$ is well defined for $T \in [0, \infty)$;
2. $f(t) = O(t^{-\eta})$ $(t \to 0^+)$;
3. $f(t) = O(t^{-\xi})$ $(t \to \infty)$.

Furthermore, if $f(t) = O(t^{-\xi})$ $(t \to \infty; \xi \in \mathbb{R}_0^+)$, then $f(t) \in H(\infty;\eta)$. We can note that $H(\infty;\eta) \subset H(\xi;\eta)$ $(\forall \xi \in \mathbb{R}_0^+)$.

Clearly, we have

$$f(t) = e^{-bt} \in H(\infty, 0) \ (b > 0). \tag{27}$$

The Mellin transform of $f \in H(\xi;\eta)$ is defined by

$$f_M(s) = M[f(t);s] := \int_0^\infty f(t)t^{s-1}dt \ (s = \sigma + i\tau, \ \eta < \Re(s) < \xi). \tag{28}$$

The fractional Weyl transform of $f \in H(\xi;0)$ is defined by

$$\begin{gathered}\Omega(s;x) := W^{-s}[f(t)](x) := \tfrac{1}{\Gamma(s)}M[f(t+x);s] \\ = \tfrac{1}{\Gamma(s)}\int_0^\infty f(t+x)t^{s-1}dt = \tfrac{1}{\Gamma(s)}\int_x^\infty f(t)(t-x)^{s-1}dt; \ (s = \sigma + i\tau, \ 0 < \Re(s) < \xi \ , \ x \geq 0).\end{gathered} \tag{29}$$

Considering $\Re(s) \leq 0$, we define the Weyl transform of $\omega \in H(\xi;0)$ as follows,

$$\Omega(s;x) := W^{-s}[f(t)](x) := (-1)^n \frac{d^n}{dx^n}(\Omega(n+s;x)), \ (0 \leq n + \Re(s) < \xi), \tag{30}$$

and

$$\Omega(0;x) := \omega(x). \tag{31}$$

We can rewrite Equation (30) alternately as

$$\begin{gathered}\Omega(-s;x) := W^s[\omega(t)](x) = (-1)^n \tfrac{d^n}{dx^n}\left(W^{-(n-s)}[\omega(t)](x)\right) \\ =: (-1)^n \tfrac{d^n}{dx^n}(\Omega(n-s;x)) \ (0 \leq n - \Re(s) < \xi, \Re(s) > 0).\end{gathered} \tag{32}$$

For these formulae, $n \geq \Re(s)$ where n is the positive and smallest such integer. For $s = n$ in Equation (32), we get

$$\Omega(-n;x) := W^n[\omega(t)](x) := (-1)^n \frac{d^n}{dx^n}(\Omega(0;x)) = (-1)^n \frac{d^n}{dx^n}(\omega(x)). \tag{33}$$

Note that $\{W^s\}$ $(s \in \mathbb{C})$ satisfies

$$W^{-(\mu+s)}[\omega(t)](x) = W^{-\mu}[\Omega(s;t)](x) = \Omega(s+\mu;x) \tag{34}$$

the multiplicative group property. For further detailed study of Weyl and related integral transforms, we refer the interested reader to [22–24].

The space of analytic functions [20,21] as discussed by Hardy is reviewed here as follows: Let $0 < \delta < 1$ and $H(\delta) := \{s = \sigma + i\tau : \Re(s) \geq -\delta\}$ be the half space. Further, for an analytic function $\phi(s)$, $s \epsilon H(\delta)$, suppose that $0 < A < \pi$ and

$$\Re = \Re(A, P, \delta) := \left\{\phi(s) : |\phi(s)| \leq Ce^{P\sigma + A|\tau|}\right\} \tag{35}$$

is called the Hardy space of analytic functions that restricts the parameter A to lie in $(0, \pi)$. Consider a function $\phi \in \mathfrak{R}$ and define

$$\Phi(x) := \frac{1}{2\pi i} \int_{c-i\infty}^{c+i\infty} \frac{\pi}{\sin \pi s} \phi(-s) x^{-s} ds \; (0 < c < \delta), \tag{36}$$

such that the kernel is majorized by

$$e^{-(\pi - A)|\tau|} e^{-Pc} x^{-c}, \; (x > 0). \tag{37}$$

These are uniformly convergent in an interval of $0 < x \le x_0 \le X < \infty$. Therefore, the function $\Phi(x)$ is regular, and represented by the integral (36), for all positive x. We combine these classes to define a new class of functions for our purposes. Assume that $\omega(0) := \Omega(0;0)$ is well defined and $\omega(x) := \Omega(0;x)$ $(x \ge 0)$. Then, $\omega \in \mathfrak{R}_\infty(A, P, \delta)$ iff $\omega \in H(\delta;0)$ and

$$\left| \frac{\Omega(s;0)}{\Gamma(1-s)} \right| \le Ce^{\sigma p + A|\tau|} \; (0 \le \mathfrak{R}(s) < \delta). \tag{38}$$

Theorem 1. *Let* $\varphi \in H(\delta;0)$ *and* $\Phi(s;x)$ $(x > 0)$ *be its Weyl transform; then, the series representation is*

$$\Phi(s;x) = \sum_{n=0}^{\infty} \frac{\Phi(s-n;0)(-x)^n}{n!} \; (0 \le \mathfrak{R}(s) < \delta, 0 < x < \infty). \tag{39}$$

Proof. Since $\varphi \in \mathfrak{R}_\infty(A, P, \delta)$, the inverse Mellin transform is

$$\begin{aligned} \Phi(0;x) &:= \tfrac{1}{2\pi i} \int_{c-i\infty}^{c+i\infty} \varphi_M(s) x^{-s} ds \\ &= \tfrac{1}{2\pi i} \int_{c-i\infty}^{c+i\infty} \Gamma(s)\Phi(s;0) x^{-s} ds = \tfrac{1}{2\pi i} \int_{c-i\infty}^{c+i\infty} \tfrac{\pi\Phi(s;0)}{\sin(\pi s)\Gamma(1-s)} x^{-s} ds \\ &\quad (0 \le c < \delta \,,\; x > 0), \end{aligned} \tag{40}$$

well defined because the integrand is majorized by a constant multiple of $e^{-(\pi - A)|\tau|} e^{-Pc} x^{-c}$.

Familiar Cauchy's theorem for complex numbers is used to invert the Mellin transform in Equation (40). The integrand has singularities of order 1 at $s = -n$ $(n = 0, 1, 2, 3, \ldots)$ with residues $\frac{\Phi(-n;0)(-x)^n}{n!}$. Therefore,

$$\Phi(0;x) = \sum_{n=0}^{\infty} \frac{(-1)^n \Phi(-n;0) x^n}{n!} \; (0 < x < e^{-P}). \tag{41}$$

Because $\varphi \in H(\delta;0)$, the series (41) extends uniquely for the Weyl transform as

$$(s;x) = \sum_{n=0}^{\infty} \frac{\Phi(s-n;0)(-x)^n}{n!} \; (0 < x < e^{-P}, 0 \le \mathfrak{R}(s) < \delta). \tag{42}$$

□

Theorem 2. *Let* $\varphi \in \mathfrak{R}_\infty(A, P, \delta)$ *and*

$$\psi(t) = \lambda t^{-\mu} + \varphi(t) \; (\mu > 0). \tag{43}$$

Then, the Weyl transform $\Psi(s;x)$ *has a closed form representation*

$$\Psi(s;x) = \lambda \frac{\Gamma(\mu - s)}{\mu} x^{s-\mu} + \sum_{n=0}^{\infty} \frac{\Phi(s-n;0)(-x)^n}{n!} (0 \le \mathfrak{R}(s) < \min(\delta, \mu); 0 < x < e^{-P}). \tag{44}$$

Proof. An application of the linearity property of Weyl's transform to Equation (43) gives

$$\Psi(s;x) = W^{-s}[\psi(t)](x) = \lambda W^{-s}[t^{-\mu}](x) + \Phi(s;x) \quad (0 \le \Re(s) < \min(\delta,\mu); 0 < x < e^{-p}). \tag{45}$$

However, (see [22], (p. 249)) we have

$$W^{-s}[t^{-\mu}](x) = \frac{\Gamma(\mu - s)}{\mu} x^{s-\mu}; \ (0 < \Re(s) < \mu; 0 < x < \infty). \tag{46}$$

From Equations (38), (42), and (46) we arrive at Equation (44). □

Example 1. *Define*

$$\varphi(t) := \frac{1}{e^t - 1} - \frac{1}{t} \ (t > 0). \tag{47}$$

Note that $\varphi \in \Re_\infty(\frac{\pi}{2}, \ln(1/2\pi), \delta)$ *and*

$$\Phi(s,0) = \zeta(s) \ (0 < \Re(s) < 1). \tag{48}$$

Hence, we have an expansion

$$\Phi(0,x) = \frac{1}{e^x - 1} - \frac{1}{x} = \zeta(0) + \sum_{n=1}^{\infty} \frac{(-1)^n \zeta(-n) x^n}{n!} \ (0 < x < 2\pi), \tag{49}$$

which is the standard result. Using

$$\zeta(-n) = -\frac{B_n}{n+1} \ (n = 0, 1, 2, 3, \ldots), \tag{50}$$

we can rewrite $\Phi(s;x)$ *in terms of Bernoulli numbers.*

3. Results

Application of the General Representation Theorem to the Generalized Bose–Einstein and Fermi–Dirac and Related Functions

In this section, we first evaluate the fractional Weyl transform for the function involved in the integrand of generalized Bose–Einstein functions and then analytically continued this function in the interval $(0 < \Re(s) < \mu)$, namely the generalized critical strip.

Remark 1. *To apply the general representation theorem, we first discussed analytic continuation of the Bose–Einstein function in the critical strip. The integral representation (5) of the generalized Bose–Einstein function* $\Psi_\nu\ (s,\mu;0)$ *can be continued to the domain,* $0 < \Re(s) < \mu$*, where a particular case of this domain* $0 < \Re(s) < 1$ *is known as the critical strip for the zeta function. For* $\Re(s) > \mu$*, we may write in the usual sense as we write for the zeta function* ([18], (p. 37))

$$\Gamma(s)\Psi_\nu\ (s,\mu;0) = \int_0^1 \left(\frac{e^{-\nu t}}{(e^t - 1)^\mu} - \frac{1}{t^\mu} \right) t^{s-1} dt + \frac{1}{s-\mu} + \int_1^\infty \frac{e^{-\nu t}}{(e^t - 1)^\mu} t^{s-1} dt, \tag{51}$$

which is true by analytic continuation for $\Re(s) > 0$. *For these values* $0 < \Re(s) < \mu$, *we get*

$$\frac{1}{s-\mu} = \int_1^{\infty} \frac{t^{s-1}}{t^{\mu}} dt\,, \tag{52}$$

such that we can write

$$\Gamma(s)\Psi_{\nu}\,(s,\mu;0) = \int_0^{\infty}\left(\frac{e^{-\nu t}}{(e^t-1)^{\mu}} - \frac{1}{t^{\mu}}\right)t^{s-1}dt\ (0 < \Re(s) < \mu; \Re(\nu) > 0). \tag{53}$$

Putting $\nu = 0$ *in Equation (53), we get the representation*

$$\Gamma(s)\Psi_0\,(s,\mu;0) = \int_0^{\infty}\left(\frac{1}{(e^t-1)^{\mu}} - \frac{1}{t^{\mu}}\right)t^{s-1}dt\ (0 < \Re(s) < \mu). \tag{54}$$

Putting $\nu = 0$; $\mu = 1$ *in Equation (53), the classical representation (26) for the Riemann zeta function is recovered.*

Remark 2. *The series representation (24) for the Hurwitz–Lerch function is proved in* ([16], (p. 28)) *by using the following steps.*

1. *Using the contour integral to state the involved function*
2. *Using the Cauchy residue theorem from complex analysis*
3. *Using the following identity known as Hurwitz formula* [16]

$$\zeta(s,\nu) = 2(2\pi)^{s-1}\Gamma(1-s)\sum_{n=1}^{\infty}\frac{\sin(2\pi n\nu + \frac{\pi s}{2})}{n^{1-s}}\ (\Re(s) < 0\,,\ 0 < \nu \leq 1). \tag{55}$$

In this section, we have obtained a new series representation for the generalized Bose–Einstein and Fermi–Dirac functions. We have shown that the above stated results (18) and (24) for the polylogarithm and Hurwitz–Lerch functions are special cases by using the fractional Weyl transform.

Theorem 3. *Show that the generalized Fermi–Dirac functions have a series representation*

$$\Theta_{\nu}(s,\mu;x) := \Gamma(\mu)\sum_{M=0}^{\infty}\frac{\Theta_{\nu}\,(s-M,\mu;0)x^M}{M!} = \Gamma(\mu)\sum_{M=0}^{\infty}\frac{\sum_{m=0}^{M}\ (-1)^{M-m}R(M\,,\ m,-\nu)\Theta_{\nu-m}(s\,,\ \mu+m;0)}{M!}x^M \tag{56}$$
$$(0 \leq \Re(s) < \mu;\ \nu \neq 0,-1,-2,\ldots).$$

Proof. The generalized Fermi–Dirac function (1) can be written as

$$\begin{aligned}\Theta_{\nu}(s,\mu;x) &= \frac{\Gamma(\mu)}{\Gamma(s)}\int_x^{\infty} e^{-\nu t}\frac{(t-x)^{s-1}}{(e^t+1)^{\mu}}dt \\ &= \frac{\Gamma(\mu)}{\Gamma(s)}\int_x^{\infty}\left[\frac{e^{-\nu t}}{(e^t+1)^{\mu}}\right](t-x)^{s-1}dt \\ &= \Gamma(\mu)W^{-s}\left[\frac{e^{-\nu t}}{(e^t+1)^{\mu}}\right](x) = \Gamma(\mu)\sum_{M=0}^{\infty}\frac{\Theta_{\nu}\,(s-M,\mu;0)}{M!}x^M\,,\end{aligned} \tag{57}$$

which leads to the required result by using Equations (10) and (39). □

Corollary 1. *The Fermi–Dirac function has a representation* ([2], Equation (4.2)):

$$F_{s-1}(x) := \sum_{M=0}^{\infty}\frac{(1-2^{M-s+1})\zeta(s-M)x^M}{M!}\,. \tag{58}$$

Proof. This result follows by putting $\mu = 1; \nu = 0$ in Equation (56) and using Equation (14). □

Theorem 4. *Show that the generalized Bose–Einstein functions have a series representation:*

$$\Psi_\nu(s,\mu;x) = \frac{\Gamma(\mu)\Gamma(\mu-s)}{\mu}x^{s-\mu} + \Gamma(\mu)\sum_{M=0}^{\infty}\frac{(-1)^M\Psi_\nu(s-M,\mu;0)}{M!}x^M$$
$$(0 \le \Re(s) < \mu;\ \nu \ne 0,-1,-2,\ldots). \tag{59}$$

Proof. First, we note that the integral representation (5) can be rewritten as

$$\Psi_\nu(s,\mu;x) := \frac{\Gamma(\mu)}{\Gamma(s)}\int_0^\infty \frac{t^{s-1}e^{-\nu t}}{(e^{t+x}-1)^\mu}dt = \frac{\Gamma(\mu)}{\Gamma(s)}\int_x^\infty e^{-\nu t}\frac{(t-x)^{s-1}}{(e^t-1)^\mu}dt, \tag{60}$$

which can be rearranged as follows

$$= \frac{\Gamma(\mu)}{\Gamma(s)}\int_x^\infty\left[\frac{e^{-\nu t}}{(e^t-1)^\mu} - \frac{1}{t^\mu} + \frac{1}{t^\mu}\right](t-x)^{s-1}dt\,. \tag{61}$$

Next, by making use of the definition of Weyl transform, we get

$$\Psi_\nu(s,\mu;x) = \Gamma(\mu)W^{-s}\left[\frac{e^{-\nu t}}{(e^t-1)^\mu} - \frac{1}{t^\mu}\right](x) + \Gamma(\mu)W^{-s}\left[\frac{1}{t^\mu}\right](x). \tag{62}$$

However, an application of the Weyl transform (46) along with an application of the general representation theorem (39) on the left hand side of the above Equation (62) leads to the required series representation. □

Corollary 2. *The Bose–Einstein function has a representation* ([2], Equation (4.7)):

$$\mathcal{B}_{s-1}(x) = \Gamma(1-s)x^{s-1} + \sum_{M=0}^{\infty}\frac{(-1)^n\zeta(s-M)x^M}{M!}. \tag{63}$$

Proof. The result follows by putting $\mu = 1; \nu = 0$ in Equation (59) and using Equation (13). □

Remark 3. *Putting* $\mu = 1; x = \log\frac{1}{z} \Rightarrow z = e^{-x}; -x = \log z$, *replacing* ν *by* $\nu - 1$ *in (59), and using the relation, ([6], Equation (4.5))* $\Psi_\nu(s;\ x) = e^{-(\nu+1)x}\Phi(e^{-x},\ s,\ \nu\ +\ 1)$, *we obtain*

$$z^\nu\Phi(z,s,\nu) = \Gamma(1-s)(\log\tfrac{1}{z})^{s-1} + \sum_{M=0}^{\infty}\zeta(s-M,\nu)\frac{(\log z)^M}{M!}\left(|\log z| < 2\pi, s \ne 1,2,3,\ldots,\nu \ne 0,-1,-2,\ldots\right), \tag{64}$$

which is exactly Equation (24). Further, by putting $\nu = 0$, *we deduce Lindelöf's representation (18) for the polylogarithm function.*

Remark 4. *The use of fractional derivatives and fractional integrals has become vital to solve many physical problems that were unsolvable otherwise, see for example* [25,26]. *For our interest, the Riemann–Liouville fractional derivative is defined by* ([22], (p. 70)) *and* [23].

$$\mathfrak{D}_z^\mu\{f(z)\} := \begin{cases} \frac{1}{\Gamma(-\mu)}\int_0^z (z-t)^{\mu-1} f(t)dt\ \Re(\mu) > 0 \\ \frac{d^m}{dz^m}\left\{\mathfrak{D}_z^{\mu-m}\{f(z)\}\right\}\ (m-1 \leqq \Re(\mu) < m(m \in \mathbb{N})). \end{cases} \tag{65}$$

It is important to notice from integral representations (1) and (5) that the functions $\Theta_\nu(s,\mu;x)$ and $\Psi_\nu(s,\mu;x)$ are in effect a Riemann–Liouville fractional derivative of the Fermi–Dirac and Bose–Einstein functions respectively given by

$$\Theta_\nu(s,\mu;x) = \frac{1}{\Gamma(\mu)}\mathfrak{D}_x^{\mu-1}\left\{e^{-x(\mu-1)}\Theta_\nu(s;x)\right\}; \Re(\mu) > 0, \tag{66}$$

$$\Theta_\nu(s,\mu;x) = \frac{1}{\Gamma(\mu)}\mathfrak{D}_x^{\mu-1}\left\{e^{-x(\mu-1)}\Theta_\nu(s;x)\right\}; \Re(\mu) > 0. \tag{67}$$

Remark 5. *The Apostol–Euler–Nörlund polynomials* $E_n^{(\mu)}(x;\lambda)$ [27,28] *are defined by the generating function*

$$\left(\frac{2}{\lambda e^t+1}\right)^{\mu} e^{vt} = \sum_{n=0}^{\infty} E_n^{(\mu)}(v;\lambda)\frac{t^n}{n!}; |t| < \left|\log(-\lambda)\right|; \lambda \neq -1 \tag{68}$$

and Bernoulli–Nörlund polynomials [27,28] are defined by

$$\left(\frac{t}{e^t-1}\right)^{\mu} e^{vt} = \sum_{n=0}^{\infty} B_n^{(\mu)}(v)\frac{t^n}{n!}; |t| < |2\pi|. \tag{69}$$

It is important to further mention that the relation of the generalized Fermi–Dirac and Bose–Einstein functions with Apostol—-Euler–Nörlund [27,28] polynomials can be established in view of integral representations (1) and (5), respectively, as follows.

Consider ([5], Equation (47))

$$\Theta_v(s,\mu;x) = e^{-x(v+\mu)}\zeta(s,\mu;\nu+\mu,-e^{-x}). \tag{70}$$

Now, replace ν by $\nu-\mu$ and $s=-m$ in Equation (70); we get

$$\Theta_{v-\mu}(-m,\mu;x) = e^{-xv}\zeta(-m,\mu;\nu,-e^{-x}). \tag{71}$$

Next, by using $\lambda = e^{-x}; \alpha = \mu$ in ([5], Equation (27)), we get

$$\Theta_{v-\mu}(-m,\mu;x) = e^{-xv}\Gamma(\mu)2^{-\mu}E_m^{(\mu)}(\nu;e^{-x}). \tag{72}$$

Similarly, by considering (5], Equation (53)) and replacing ν by $\nu-\mu$, $s=-m$, we get

$$\Psi_{v-\mu}(-m,\mu;x) = e^{-xv}\zeta(-m,\mu\,;\nu,e^{-x}). \tag{73}$$

Next, by using $\lambda = e^{-x}$ in ([5], Equation (27)) and using the result in the above Equation (73), we get

$$\Psi_{v-\mu}(-m,\mu;x) = e^{-xv}\Gamma(\mu)2^{-\mu}E_m^{(\mu)}(\nu;-e^{-x}). \tag{74}$$

For $x = 0$, in Equations (72) and (74), we get

$$\Theta_{v-\mu}(-m,\mu;0) = \Gamma(\mu)2^{-\mu}E_m^{(\mu)}(\nu;1)\Psi_{v-\mu}(-m,\mu;0) = \Gamma(\mu)2^{-\mu}E_m^{(\mu)}(\nu;-1),$$

which can be used in Equations (56) and (59) to obtain the representation in terms of special cases of Apostol–Euler–Nörlund polynomials $E_m^{(\mu)}(v;\mp 1)$. Considering the further restrictions $v=\mu=1$, we can get these relations in terms of commonly used Bernoulli and Euler numbers.

Remark 6. *One can note that Equation (59) may be stated alternately in terms of Stirling numbers by using Equation (13) as follows*

$$\Psi_{\nu}(s,\mu;x) = \frac{\Gamma(\mu)\Gamma(\mu-s)}{\mu}x^{s-\mu} + \Gamma(\mu)\sum_{M=0}^{\infty}\frac{(-1)^{M}\sum_{m=0}^{M}(-1)^{M-m}R(M,m,-\nu)\Psi_{\nu-m}(s,\mu+m;0)}{M!}x^{M} \tag{75}$$
$$(0 \leq \Re(s) < \mu;\ \nu \neq 0,-1,-2,\ldots).$$

4. Concluding Remarks

One important aspect in relation to the analysis of special functions is to study their representations. These special functions can be studied in different regions by using their series, asymptotic, and integral representations. This fact is also important when writing simpler mathematical proofs of known results. Here, we have provided a new series representation of the generalized Bose–Einstein and Fermi–Dirac functions by using a general representation theorem. To accomplish this work, we discussed an analytic continuation for these functions by generalizing the Riemann zeta function from $(0 < \Re(s) < 1)$ to $(0 < \Re(s) < \mu)$. This gives new insights for a possible generalization of the Rieman zeta function

$$\zeta_{\mu}^{*}(s) := \frac{1}{\Gamma(s)}\int_{0}^{\infty} t^{s-1}\left(\frac{1}{(e^{t}-1)^{\mu}} - \frac{1}{t^{\mu}}\right)dt\ (0 < \Re(s) < \mu)$$

and will be discussed in more detail in our future research. Our results were validated by obtaining known series representations for the polylogarithm and the Hurwitz–Lerch zeta functions as special cases. A comparison of the known proof of their series representation was given with this new proof. It is hoped that the general representation theorem can also be applied to analyze other special functions.

Author Contributions: R.S. did project administration and supervision of this research. All the authors (R.S., H.N., S.K., and A.T.) participated equally in the methodology and conceptualization of this research. H.N. and A.T. wrote, reviewed, and edited the manuscript. S.K. checked the validation of the results. All the authors finalized the manuscript after its internal evaluation and contributed substantially to the work reported.

Acknowledgments: The authors are thankful to the anonymous reviewers for their useful comments. They significantly improved the quality of this manuscript.

References

1. Chaudhry, M.A.; Iqbal, A.; Qadir, A. A representation for the anyon integral function. *arXiv* **2005**, arXiv:math-ph/0504081.
2. Chaudhry, M.A.; Qadir, A. Operator representation of Fermi-Dirac and Bose-Einstein integral functions with applications. *Int. J. Math. Math. Sci.* **2007**, 80515. [CrossRef]
3. Tassaddiq, A. A new representation of the k-gamma functions. *Mathematics* **2019**, *7*, 133. [CrossRef]
4. Tassaddiq, A.; Qadir, A. Fourier transform and distributional representation of the generalized gamma function with some applications. *Appl Math. Comput.* **2011**, *218*, 1084–1088. [CrossRef]
5. Bayad, A.; Chikhi, J. Reduction and duality of the generalized Hurwitz-Lerch zetas. *Fixed Point Theory Appl.* **2013**, *82*, 1–14. [CrossRef]
6. Srivastava, H.M.; Chaudhry, M.A.; Qadir, A.; Tassaddiq, A. Some extensions of the Bose-Einstein and Fermi-Dirac functions with applications to zeta and related functions. *Russ. J. Math. Phys.* **2011**, *18*, 107–121. [CrossRef]
7. Tassaddiq, A.; Qadir, A. Fourier transform representation of the extended Bose-Einstein and Fermi-Dirac functions with applications to the family of the zeta and related functions. *Integral Transforms Spec. Funct.* **2011**, *22*, 453–466. [CrossRef]
8. Tassaddiq, A. Some Representations of the Extended Fermi-Dirac and Bose-Einstein Functions with Applications. Ph.D. Dissertation, National University of Sciences and Technology, Islamabad, Pakistan, 2012.

9. Tassaddiq, A. A New Representation of the Extended Fermi-Dirac and Bose-Einstein Functions. *Int. J. Appl. Math.* **2017**, *5*, 435–446.
10. Carlitz, L. Weighted Stirling numbers of the first and second kind—I. *Fibonacci Q.* **1980**, *18*, 147–162.
11. Carlitz, L. Weighted Stirling numbers of the first and second kind—II. *Fibonacci Q.* **1980**, *18*, 242–257.
12. Tassaddiq, A. A new representation of the Srivastava λ-generalized Hurwitz-Lerch zeta functions. *Symmetry* **2018**, *10*, 733. [CrossRef]
13. Tassaddiq, A. Some difference equations for Srivastava's λ-generalized Hurwitz–Lerch zeta functions with applications. *Symmetry* **2019**, *11*, 311. [CrossRef]
14. Srivastava, H.M. A new family of the λ-generalized Hurwitz-Lerch Zeta functions with applications. *Appl. Math. Inf. Sci.* **2014**, *8*, 1485–1500. [CrossRef]
15. Truesdell, C. On a function which occurs in the theory of the structure of polymers. *Ann. Math.* **1945**, *46*, 144–157. [CrossRef]
16. Erdelyi, A.; Magnus, W.; Oberhettinger, F.; Tricomi, F.G. *Higher Transcendental Functions*; McGraw-Hill Book Company: New York, NY, USA; Toronto, ON, Canada; London, UK, 1953; Volume I.
17. Chaudhry, M.A.; Qadir, A.; Tassaddiq, A. A new generalization of the Riemann zeta function. *Adv. Differ. Equ.* **2011**, *2011*, 20. [CrossRef]
18. Titchmarsh, E.C. *The Theory of the Riemann Zeta Function*; Oxford University Press: Oxford, UK, 1951.
19. Chaudhry, M.A.; Qadir, A. Extension of Hardy's class for Ramanujan's interpolation formula and master theorem with applications. *J Inequal Appl.* **2012**, *2012*, 52. [CrossRef]
20. Hardy, G.H.; Littlewood, J.E. Contributions to the Theory of the Riemann Zeta function and the Theory of the Distribution of Primes. *Acta Math.* **1918**, *41*, 119–196. [CrossRef]
21. Hardy, G.H. *Ramanujan: Twelve Lectures on Subjects Suggested by His Life and Work*; Chelsea Publishing Company: New York, NY, USA, 1959.
22. Kilbas, A.A.; Srivastava, H.M.; Trujillo, J.J. *Theory and Applications of Fractional Differential Equations, North-Holland Mathematical Studies*; Elsevier (North- Holland) Science Publishers: Amsterdam, The Netherlands; London, UK; New York, NY, USA, 2006; Volume 204.
23. Samko, S.G.; Kilbas, A.A.; Marichev, O.I. *Fractional Integrals and Derivatives: Theory and Applications*; Translated from the Russian: Integrals and Derivatives of Fractional Order and Some of Their Applications ("Nauka i Tekhnika", Minsk, 1987); Gordon and Breach Science Publishers: Reading, UK; Tokyo, Japan; Paris, French; Berlin, Germany; Langhorne, PA, USA, 1993.
24. Zayed, A.I. *Handbook of Functions and Generalized Function Transforms*; CRC Press: Boca Raton, FL, USA, 1996.
25. Gomez-Aguilar, J.F.; Abro, K.A.; Kolebaje, O.; Yildirim, A. Chaos in a calcium oscillation model via Atangana-Baleanu operator with strong memory. *Eur. Phys. J. Plus.* **2019**, *134*, 140. [CrossRef]
26. Tassaddiq, A. MHD flow of a fractional second grade fluid over an inclined heated plate. *Chaos Solitons Fractals* **2019**, *123*, 341–346. [CrossRef]
27. He, Y.; Araci, S.; Srivastava, H.M.; Abdel-Aty, M. Higher-order convolutions for apostol-bernoulli, apostol-euler and apostol-genocchi polynomials. *Mathematics* **2018**, *6*, 329. [CrossRef]
28. Liu, G.-D.; Srivastava, H.M. Explicit formulas for the Nörlund polynomials $B_n^{(x)}$ and $b_n^{(x)}$. *Comput. Math. Appl.* **2006**, *51*, 1377–1384.

14

Periodically Forced Nonlinear Oscillatory Acoustic Vacuum

Makrina Agaoglou [1,2], Michal Fečkan [1,3,*], Michal Pospíšil [1,3], Vassilis M. Rothos [2] and Alexander F. Vakakis [4]

[1] Mathematical Institute of Slovak Academy of Sciences, Štefánikova 49, 814 73 Bratislava, Slovakia; makrina_agao@hotmail.com (M.A.); michal.pospisil@fmph.uniba.sk (M.P.)
[2] Department of Mechanical Engineering, Faculty of Engineering, Aristotle University of Thessaloniki, 54124 Thessaloniki, Greece; rothos@auth.gr
[3] Department of Mathematical Analysis and Numerical Mathematics, Comenius University in Bratislava, Mlynská dolina, 842 48 Bratislava, Slovakia
[4] Department of Mechanical Science and Engineering, University of Illinois at Urbana-Champaign, Urbana, IL 61801, USA; avakakis@illinois.edu
* Correspondence: michal.feckan@fmph.uniba.sk

Abstract: In this work, we study the in-plane oscillations of a finite lattice of particles coupled by linear springs under distributed harmonic excitation. Melnikov-type analysis is applied for the persistence of periodic oscillations of a reduced system.

Keywords: nonlinear oscillatory acoustic vacuum; periodic oscillations; Melnikov function; symmetry

1. Introduction

We analytically study the persistence of periodic oscillations for certain three-dimensional systems of ordinary differential equations (ODEs) with periodic perturbations and a slowly-varying variable. The considered ODEs are derived from a model of a finite lattice of particles coupled by linear springs under distributed harmonic excitation, which is described in detail in Section 2. This model presents a low-energy nonlinear acoustic vacuum. We refer the reader for more motivations, further details and applications to [1,2]. Following the computations of [1], we consider just two modes in Section 3, and we postpone higher modes investigation to our future paper, since another approach will be used. Melnikov analysis is demonstrated in Section 4 for finding conditions for the existence of periodic solutions for the perturbed ODEs corresponding to two modes. More precisely, following [1], we derive a three-dimensional periodically-perturbed system of ODEs with a slowly-varying variable. Then, we analyze an unperturbed autonomous system of ODEs to compute its family of periodic solutions by revising the results of [1] in more detail. Since we are interested in the persistence of periodic solutions for the perturbed ODEs, we compute the corresponding Melnikov functions by [3]. Due to the difficulty of finding simple roots of these Melnikov functions explicitly, we outline an asymptotic approach for the location of some of them. Note that the simple roots of Melnikov functions predict the persistence and location of periodic solutions for perturbed ODEs. This is a novelty and a contribution of our paper. Section 5 outlines possible future research along with summarizing our achievements in this paper.

2. The Model

We consider a lattice consisting of N identical particles coupled by identical linear springs (they are un-stretched when the lattice is in the horizontal position) and executing in-plane oscillations (see Figure 1). Fixed boundary conditions and dissipative terms are imposed, and the transverse harmonic forces are also applied. The equations of motion can be expressed as follows,

$$\begin{aligned} m\frac{d^2u_i}{dt^2} + \left(T_i - \xi\frac{d\epsilon_i}{dt}\right)\cos\phi_i - \left(T_{i+1} - \xi\frac{d\epsilon_{i+1}}{dt}\right)\cos\phi_{i+1} &= 0 \\ m\frac{d^2v_i}{dt^2} + \left(T_i - \xi\frac{d\epsilon_i}{dt}\right)\sin\phi_i - \left(T_{i+1} - \xi\frac{d\epsilon_{i+1}}{dt}\right)\sin\phi_{i+1} - F_i &= 0 \end{aligned} \tag{1}$$

with u_i, v_i being the longitudinal and transversal displacements of the i-th particle, respectively, ϕ_i the angle between the i-th spring and the horizontal direction, ξ the damping coefficient, $\epsilon_i = l_i' - l_i$ the deformation of the i-th spring, F_i the exciting transverse force and m the mass of each particle of the lattice. The tensile forces are proportional to the deformations of the springs, and considering the geometry of the deformed state of the lattice (see Figure 1), one may write:

$$\begin{aligned} T_i &= k(l_i' - l_i), \\ \epsilon_i = l_i' - l_i &= [(v_i - v_{i-1})^2 + (l_i + u_i - u_{i-1})^2]^{1/2} - l_i \end{aligned}$$

with l_i being the equilibrium length of the i-th spring (each spring has the same length) and k the linear stiffness coefficient of each coupling spring. Introducing $\delta_i = \epsilon_i/l_i$, $s_i = u_i/l_i$, $w_i = v_i/l_i$, $c = \xi/(km)^{1/2}$, where s_i and w_i are the normalized axial and transverse displacements, and the "slow" time scale $\tau = (\frac{k}{m})^{1/2}t$, Equation (1) can be rewritten in normalized form:

$$\begin{aligned} \frac{d^2s_i}{d\tau^2} &= \delta_i\cos\phi_i - \delta_{i-1}\cos\phi_{i-1} + c\delta_i'\cos\phi_i - c\delta_{i-1}'\cos\phi_{i-1} \\ \frac{d^2w_i}{d\tau^2} &= \delta_i\sin\phi_i - \delta_{i-1}\sin\phi_{i-1} + c\delta_i'\sin\phi_i - c\delta_{i-1}'\sin\phi_{i-1} + f_i, \end{aligned} \tag{2}$$

where:

$$\begin{aligned} \delta_i &= [(w_{i+1} - w_i)^2 + (1 + s_{i+1} - s_i)]^{1/2} - 1, \\ \delta_i' &= \frac{(w_{i+1}-w_i)(w_{i+1}'-w_i')+(1+s_{i+1}-s_i)(s_{i+1}'-s_i')}{[(w_{i+1}-w_i)^2+(1+s_{i+1}-s_i)^2]^{1/2}} \end{aligned}$$

and:

$$\begin{aligned} \cos\phi_i &= \frac{1+s_{i+1}-s_i}{[(w_{i+1}-w_i)^2+(1+s_{i+1}-s_i)^2]^{1/2}}, \\ \sin\phi_i &= \frac{w_{i+1}-w_i}{[(w_{i+1}-w_i)^2+(1+s_{i+1}-s_i)^2]^{1/2}}, \\ f_i &= \frac{F_i}{k}. \end{aligned}$$

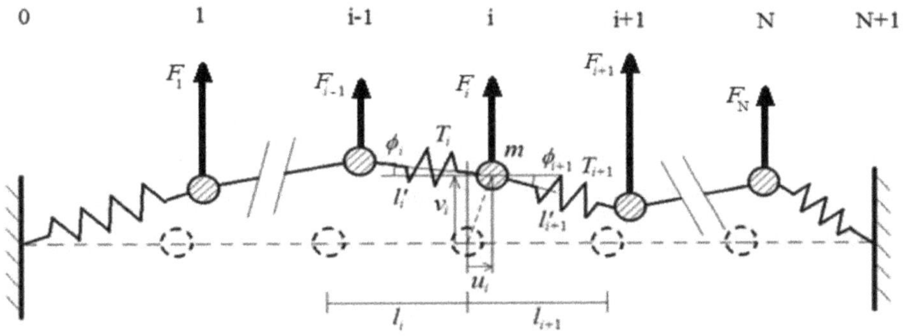

Figure 1. Forced and damped lattice oscillating in the plane (see [2]).

The normalized system (2) is referred to as the "exact lattice" in the following sections.

According to the previous research [1], when we introduce this system (2) without extra transverse force and damping terms, an interesting feature is that in the low energy limit and under the assumption that the axial displacements are assumed to be an order of magnitude smaller compared to the transverse ones, it was shown that, correct to the leading order of approximation, the transverse oscillations decouple from the axial ones and are governed by the following reduced system of equations for predominantly transverse oscillations of the particles:

$$\begin{gathered}\frac{d^2 w_i}{d\tau^2} + 2^{-1}(N+1)^{-1} \sum_{q=0}^{N} (w_{q+1} - w_q)^2 (2w_i - w_{i+1} - w_{i-1}) = 0, \\ i = 1, \cdots, N, \quad w_0(0) = w_{N+1}(0) = 0.\end{gathered} \tag{3}$$

Then, the nearly linear axial oscillations are driven by the transverse responses (see [1,2] for more details). Therefore, we focus our analysis like in [1] just on Equation (3), which presents *a low-energy nonlinear acoustic vacuum*, because in the absence of linear terms, it possesses zero speed of sound in the context of classical linear acoustics. What is more, it is notable that the existence of the strongly nonlocal multiplier $2^{-1}(N+1)^{-1}\sum_{q=0}^{N}(w_{q+1} - w_q)^2$ indicates that the response of each particle is dependent (and hence, it is coupled) on the responses of all other particles. Equation (3) admits N exact nonlinear standing waves, or nonlinear normal modes (NNMs), in the form:

$$w_i(\tau) = A_p(\tau) \sin \frac{\pi p i}{N+1}, \quad i = 1, \cdots, N$$

for the p-th NNM, $1 \leq p \leq N$, where $A_p(\tau)$ denotes the p-th modal amplitude. These, by construction, are mutually orthogonal, and there are no other NNMs in this system, nor any NNM bifurcations [1]. Substituting this NNM ansatz into Equation (3) yields a set of N uncoupled nonlinear equations governing the time-dependent amplitudes of the NNMs:

$$A_p''(\tau) + (1/4)\omega_p^4 A_p^3(\tau) = 0$$

with:

$$\omega_p = 2 \sin \frac{\pi p}{2(N+1)},$$

which is the p-th natural frequency of the corresponding linear system Equation (3) and the prime denoting differentiation with respect to τ. The exciting force, which is applied on each particle in the transverse direction, is expressed as:

$$f_i = F_p \cos \omega_p \tau \sin \frac{p i \pi}{n+1}$$

where $i = 1, \cdots, N$, for the p-th NNM, $1 \leq p \leq N$, and this exciting force includes NNMs in the form $\sin \frac{pi\pi}{n+1}$, $i = 1, \cdots, N$, for the p-th NNM, $1 \leq p \leq N$ and the p-th natural linear frequency ω_p.

The frequency of the p-th NNM is tunable with the force and energy, and it also paves the way for nonlinear resonances between NNMs widely separated in the nonlinear spectrum, given that their energies tune their frequencies to satisfy certain rational relationships.

Summarizing, we can write (3) as:

$$\frac{d^2 w}{d\tau^2} + 2^{-1}(N+1)^{-1}\langle Mw, w\rangle Mw = 0,$$

where $w = [w_1, \cdots, w_N] \in \mathbb{R}^N$, M is a symmetric matrix given by

$$M = \begin{pmatrix} 2 & -1 & 0 & \cdots & 0 \\ -1 & 2 & -1 & \cdots & 0 \\ \vdots & \vdots & \vdots & \vdots & \vdots \\ 0 & \cdots & -1 & 2 & -1 \\ 0 & \cdots & 0 & -1 & 2 \end{pmatrix}$$

and $\langle \cdot, \cdot \rangle$ is the standard scalar product on $\mathbb{R}^N$. The eigenvectors of M are $\underline{\phi}_p = [\sin \frac{p\pi}{N+1}, \cdots, \sin \frac{pN\pi}{N+1}]$ with the corresponding eigenvalues ω_p^2, $1 \le p \le N$. Moreover, it holds (see [4], p. 37):

$$\langle \underline{\phi}_p, \underline{\phi}_p \rangle = \sum_{i=1}^{N} \sin^2 \frac{pi\pi}{N+1} = \frac{N+1}{2},$$

$$\langle \underline{\phi}_p, \underline{\phi}_k \rangle = \sum_{i=1}^{N} \sin \frac{pi\pi}{N+1} \sin \frac{ki\pi}{N+1}$$

$$= \frac{1}{2} \sum_{i=1}^{N} \left(\cos \frac{(p-k)i\pi}{N+1} - \cos \frac{(p+k)i\pi}{N+1} \right) = 0, \quad p \neq k.$$

The forced (3) has the form:

$$\frac{d^2 w}{d\tau^2} + 2^{-1}(N+1)^{-1} \langle Mw, w \rangle Mw = \sum_{p=1}^{N} F_p \cos \omega_p \tau \underline{\phi}_p. \tag{4}$$

Therefore, considering the basis $\{\underline{\phi}_p\}_{p=1}^N$ of $\mathbb{R}^N$, we take $w(\tau) = \sum_{p=1}^N C_p(\tau) \underline{\phi}_p$ in (4) to get:

$$C_p''(\tau) + \frac{1}{4} \left(\sum_{i=1}^{N} C_i^2(\tau) \omega_i^2 \right) \omega_p^2 C_p(\tau) = F_p \cos \omega_p \tau, \quad 1 \le p \le N. \tag{5}$$

Next, applying the coordinate transformation to (5):

$$A_p(\tau) = \frac{\omega_p}{2} C_p(\tau), \quad 1 \le p \le N,$$

we get:

$$A_p''(\tau) + \left(\sum_{i=1}^{N} A_i^2(\tau) \right) \omega_p^2 A_p(\tau) = \frac{F_p \omega_p}{2} \cos \omega_p \tau, \quad 1 \le p \le N.$$

3. Two-Mode System

In this paper, we consider just two modes: k and p, so we study the system:

$$\begin{aligned} A_k''(\tau) + [A_k^2(\tau) + A_p^2(\tau)] \omega_k^2 A_k(\tau) + \epsilon \mu_1 \cos(\omega_k \tau) &= 0 \\ A_p''(\tau) + [A_k^2(\tau) + A_p^2(\tau)] \omega_p^2 A_p(\tau) + \epsilon \mu_2 \cos(\omega_p \tau) &= 0, \end{aligned} \tag{6}$$

for $\epsilon \neq 0$ small and parameters μ_1, μ_2. Using the transformation:

$$\begin{aligned} \psi_1(\tau) &= A_k'(\tau) + j\Omega A_k(\tau) \equiv \zeta_1(\tau) e^{j\Omega\tau} \\ \psi_2(\tau) &= A_p'(\tau) + j\Omega A_p(\tau) \equiv \zeta_2(\tau) e^{j\Omega\tau}, \end{aligned}$$

and performing an averaging approach like in [1], Equation (6) is modified to the form:

$$\begin{aligned}
\zeta_1' + \frac{j\Omega}{2}\zeta_1 - \frac{j\omega_k^2}{8\Omega^3}(\zeta_2^2\zeta_1^* + 2|\zeta_2|^2\zeta_1 + 3|\zeta_1|^2\zeta_1) + \epsilon\mu_1\cos(\omega_k\tau)e^{-j\Omega\tau} &= 0 \\
\zeta_2' + \frac{j\Omega}{2}\zeta_2 - \frac{j\omega_p^2}{8\Omega^3}(\zeta_1^2\zeta_2^* + 2|\zeta_1|^2\zeta_2 + 3|\zeta_2|^2\zeta_2) + \epsilon\mu_2\cos(\omega_p\tau)e^{-j\Omega\tau} &= 0.
\end{aligned}$$

Introducing $\zeta_i = a_i e^{j\beta_i}$ and $\Delta = \beta_2 - \beta_1$, we get:

$$\begin{aligned}
a_1' + \frac{\omega_k^2}{8\Omega^3}a_2^2 a_1\sin 2\Delta + \epsilon\mu_1\cos(\omega_k\tau)\cos(\Omega\tau+\beta_1) &= 0 \\
a_2' - \frac{\omega_p^2}{8\Omega^3}a_1^2 a_2\sin 2\Delta + \epsilon\mu_2\cos(\omega_p\tau)\cos(\Omega\tau+\beta_1+\Delta) &= 0 \\
\Delta' - \frac{\omega_p^2}{8\Omega^3}(3a_2^2 + a_1^2\cos 2\Delta + 2a_1^2) + \frac{\omega_k^2}{8\Omega^3}(3a_1^2 + a_2^2\cos 2\Delta + 2a_2^2) & \\
-\frac{\epsilon\mu_2}{a_2}\cos(\omega_p\tau)\sin(\Omega\tau+\beta_1+\Delta) + \frac{\epsilon\mu_1}{a_1}\cos(\omega_k\tau)\sin(\Omega\tau+\beta_1) &= 0
\end{aligned} \tag{7}$$

where we consider β_1 as a constant parameter. Now, by introducing the coordinate transformations $a_1 = (\frac{\rho}{\omega_p})\sin\theta$ and $a_2 = (\frac{\rho}{\omega_k})\cos\theta$ into Equation (7), we get:

$$\begin{aligned}
\rho' &= -\epsilon\mu_1\omega_p\sin\theta\cos(\omega_k\tau)\cos(\Omega\tau+\beta_1) - \epsilon\mu_2\omega_k\cos\theta\cos(\omega_p\tau)\cos(\Omega\tau+\Delta+\beta_1) \\
\theta' &+ \frac{\rho^2}{16\Omega^3}\sin 2\theta\sin 2\Delta + \epsilon\mu_1\omega_p\frac{\cos\theta}{\rho}\cos(\omega_k\tau)\cos(\Omega\tau+\beta_1) \\
&-\epsilon\mu_2\omega_k\frac{\sin\theta}{\rho}\cos(\omega_p\tau)\cos(\Omega\tau+\Delta+\beta_1) = 0 \\
\Delta' &- \frac{\rho^2}{8\Omega^3}\left[\frac{3\omega_p^2}{\omega_k^2}\cos^2\theta - \frac{3\omega_k^2}{\omega_p^2}\sin^2\theta - \cos 2\theta(2+\cos 2\Delta)\right] \\
&-\frac{\epsilon\mu_2\omega_k}{\rho\cos\theta}\cos(\omega_p\tau)\sin(\Omega\tau+\Delta+\beta_1) + \frac{\epsilon\mu_1\omega_p}{\rho\sin\theta}\cos(\omega_k\tau)\sin(\Omega\tau+\beta_1) = 0.
\end{aligned}$$

In the rest of the paper, we assume $\omega_p = \omega_k = P$ and $\Omega = kP$ for a natural number k, so we study the periodically-perturbed system:

$$\begin{aligned}
\rho' &= -\epsilon\mu_1\sin\theta\cos(P\tau)\cos(kP\tau+\beta_1) - \epsilon\mu_2\cos\theta\cos(P\tau)\cos(kP\tau+\Delta+\beta_1) \\
\theta' &+ \frac{\rho^2}{16k^3P^3}\sin 2\theta\sin 2\Delta + \epsilon\mu_1\frac{\cos\theta}{\rho}\cos(P\tau)\cos(kP\tau+\beta_1) \\
&-\epsilon\mu_2\frac{\sin\theta}{\rho}\cos(P\tau)\cos(kP\tau+\Delta+\beta_1) = 0 \\
\Delta' &- \frac{\rho^2}{4k^3P^3}\cos 2\theta\sin^2\Delta + \frac{\epsilon\mu_1}{\rho\sin\theta}\cos(P\tau)\sin(kP\tau+\beta_1) \\
&-\frac{\epsilon\mu_2}{\rho\cos\theta}\cos(P\tau)\sin(kP\tau+\Delta+\beta_1) = 0
\end{aligned} \tag{8}$$

where we scaled $P\mu_i \leftrightarrow \mu_i, i = 1,2$. We may suppose:

$$\mu_1^2 + \mu_2^2 = 1.$$

First, consider the unperturbed case where $\epsilon = 0$, so the system:

$$\begin{aligned} \rho' &= 0 \\ \theta' + \frac{\rho^2}{16k^3P^3}\sin 2\theta \sin 2\Delta &= 0 \\ \Delta' - \frac{\rho^2}{4k^3P^3}\cos 2\theta \sin^2\Delta &= 0. \end{aligned} \tag{9}$$

By introducing the temporal variable $\tau_2 = \frac{\rho^2}{8k^3P^3}\tau$ in Equation (9), we get:

$$\begin{aligned} \frac{d\theta}{d\tau_2} &= -\tfrac{1}{2}\sin 2\theta \sin 2\Delta \\ \frac{d\Delta}{d\tau_2} &= 2\cos 2\theta \sin^2\Delta \end{aligned} \tag{10}$$

which is fully integrable and gives us the first integral $I = \sin 2\theta \sin\Delta = K = const.$ of the degenerate slow flow. If we consider the following initial conditions: $\theta(0) = \theta_0$, where $0 < \theta_0 < \frac{\pi}{4}$ and $\Delta(0) = \pi/2$, then we get $K = \sin 2\theta_0 \in (0,1)$. To find exact solutions of (10), we first derive:

$$\begin{aligned} \frac{d\tau_2}{d\theta} &= -\frac{2}{\sin 2\theta \sin 2\Delta} = -\frac{1}{K\cos\Delta} = \frac{1}{K\sqrt{1-\sin^2\Delta}} \\ &= \frac{1}{K\sqrt{1-\frac{K^2}{\sin^2 2\theta}}} = \frac{1}{K}\frac{\sin 2\theta}{\sqrt{\sin^2 2\theta - K^2}}, \end{aligned}$$

for $\tau_2 > 0$ small, since (10) gives $0 < \theta < \frac{\pi}{2}$, $0 < \Delta < \pi$ (see Figure 2) and $\frac{d\Delta}{d\tau_2}(0) = 2\cos 2\theta_0 > 0$, so for $\tau_2 > 0$ small, we have $\Delta(\tau_2) > \frac{\pi}{2}$.

By using the formula in [4] ((2.599.4) p. 205), we obtain:

$$\tau_2 = \int_{\theta_0}^{\theta} \frac{1}{K}\frac{\sin 2\theta}{\sqrt{\sin^2 2\theta - K^2}}d\theta = \frac{1}{2K}\left(\frac{\pi}{2} - \sin^{-1}\left(\frac{\cos 2\theta}{\sqrt{1-K^2}}\right)\right),$$

which gives:

$$\theta(\tau_2) = \frac{1}{2}\cos^{-1}\left(\sqrt{1-K^2}\cos(2K\tau_2)\right), \tag{11}$$

recalling $0 < \theta < \frac{\pi}{2}$. Of course, Formula (11) holds for all τ_2, not just for small positive ones. Next, using (10) and (11), we have:

$$\frac{d\Delta}{d\tau_2} = 2\sqrt{1-K^2}\cos(2K\tau_2)\sin^2\Delta,$$

which can be easily solved to arrive at:

$$\Delta(\tau_2) = \pi - \cot^{-1}\left(\frac{\sqrt{1-K^2}}{K}\sin(2K\tau_2)\right).$$

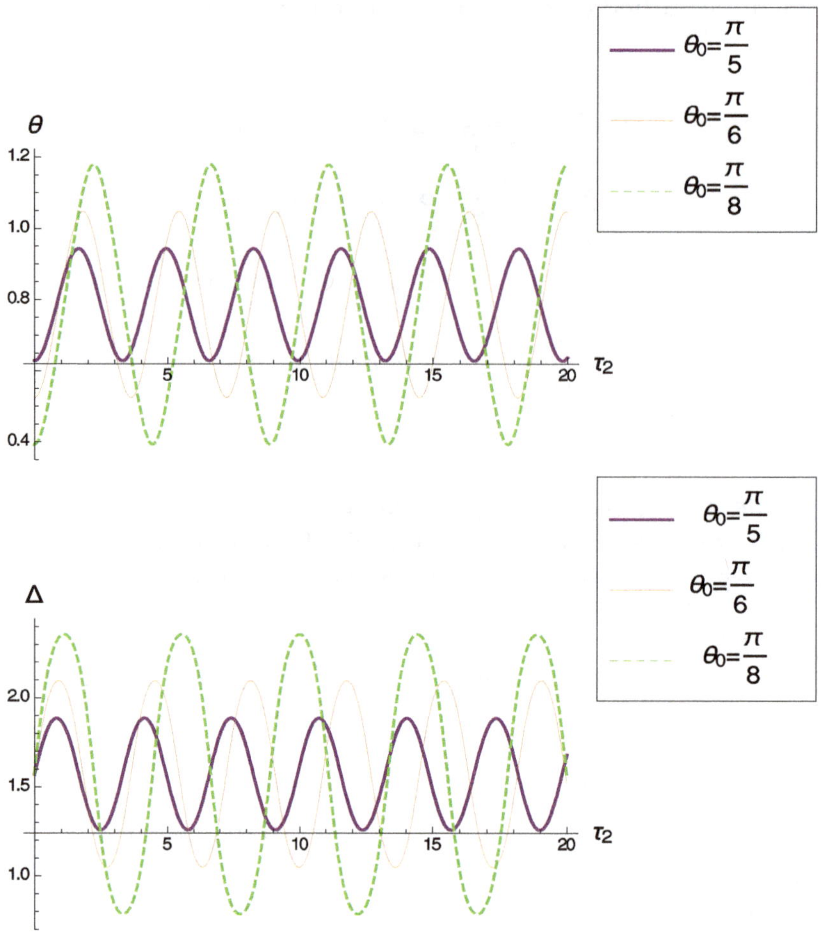

Figure 2. Top panel: Graph for θ for different initial values of θ_0. Bottom panel: Graph for Δ for different initial values of θ_0.

Hence, the exact solution of the system (10) is given as:

$$\theta(\tau_2) = \frac{1}{2}\cos^{-1}\left(\sqrt{1-K^2}\cos(2K\tau_2)\right)$$
$$\Delta(\tau_2) = \pi - \cot^{-1}\left(\frac{\sqrt{1-K^2}}{K}\sin(2K\tau_2)\right)$$

where the period is:

$$T(\theta_0) = \frac{2}{K}\int_{\theta_0}^{\frac{\pi}{2}-\theta_0} \frac{d\theta}{\left(1-\frac{\sin^2 2\theta_0}{\sin^2 2\theta}\right)^{1/2}} = \frac{\pi}{K}.$$

It is easy to verify the following symmetry property (see Figure 3):

$$\Delta\left(\tau_2+\frac{T}{2}\right) = \pi - \Delta(\tau_2), \quad \theta\left(\tau_2+\frac{T}{2}\right) = \frac{\pi}{2} - \theta(\tau_2).$$

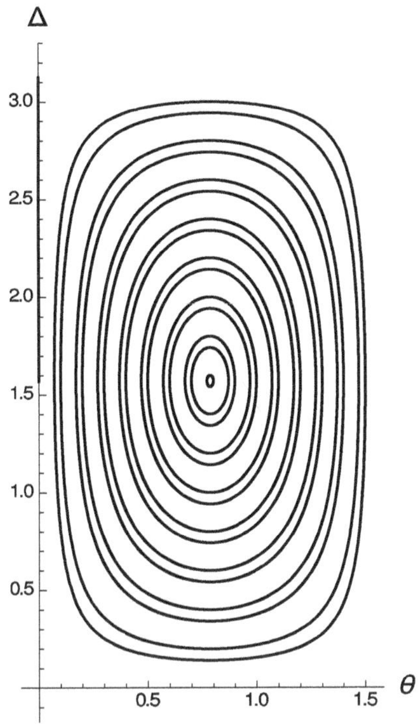

Figure 3. Orbits in the phase portrait of (10), where $(\theta, \Delta) \in [0, \frac{\pi}{2}] \times [0, \pi]$.

Summarizing, the exact solution of the unperturbed (9) is the following:

$$\rho(\tau) = const.$$
$$\theta(\tau) = \frac{1}{2}\cos^{-1}\left(\sqrt{1-K^2}\cos\left(\frac{K\rho^2}{4k^3P^3}\tau\right)\right)$$
$$\Delta(\tau) = \pi - \cot^{-1}\left(\frac{\sqrt{1-K^2}}{K}\sin\left(\frac{K\rho^2}{4k^3P^3}\tau\right)\right)$$

with the period:

$$T = \frac{8k^3P^3\pi}{K\rho^2}.$$

Consequently, for any:

$$\rho > 2kP^2\sqrt{k},$$

taking:

$$K(\rho) = \frac{4k^3P^4}{\rho^2} \in (0,1),$$

Equation (9) has the $T = 2\pi/P$-periodic solution:

$$\theta(\rho,\tau) = \frac{1}{2}\cos^{-1}\left(\frac{\sqrt{\rho^4 - 16k^6P^8}}{\rho^2}\cos(P\tau)\right)$$
$$\Delta(\rho,\tau) = \pi - \cot^{-1}\left(\frac{\sqrt{\rho^4 - 16k^6P^8}}{4k^3P^4}\sin(P\tau)\right). \tag{12}$$

4. Melnikov Analysis for Periodic Oscillations

Writing (8) as:

$$\begin{aligned}\theta' + \frac{\rho^2}{16k^3P^3}\sin 2\theta \sin 2\Delta &= \epsilon\rho^{-1}g_1(\rho,\theta,\Delta,\tau)\\ \Delta' - \frac{\rho^2}{4k^3P^3}\cos 2\theta \sin^2\Delta &= \epsilon\rho^{-1}g_2(\rho,\theta,\Delta,\tau)\\ \rho' &= \epsilon g_3(\rho,\theta,\Delta,\tau)\end{aligned}$$

and using [5], (3.5.11), p. 111, with $\alpha = 0$, and [6], Lemma 2.5, p. 283, we compute the Melnikov function:

$$M(\beta_1,\rho) = (M_1(\beta_1,\rho), M_2(\beta_1,\rho))$$

as:

$$\begin{aligned}M_1(\beta_1,\rho) &= \int_0^T \left(\frac{\partial I}{\partial\theta}\rho^{-1}g_1 + \frac{\partial I}{\partial\Delta}\rho^{-1}g_2 - \frac{\partial I}{\partial\theta}\frac{\partial\theta}{\partial\rho}g_3 - \frac{\partial I}{\partial\Delta}\frac{\partial\Delta}{\partial\rho}g_3\right)d\tau\\ &= \int_0^T \left(\frac{\partial I}{\partial\theta}\rho^{-1}g_1 + \frac{\partial I}{\partial\Delta}\rho^{-1}g_2\right)d\tau - \frac{dK}{d\rho}\int_0^T g_3 d\tau\end{aligned} \tag{13}$$

since differentiating by ρ the identity:

$$I(\theta(\rho,\tau),\Delta(\rho,\tau)) = K(\rho),$$

we get:

$$\frac{\partial I}{\partial\theta}\frac{\partial\theta}{\partial\rho} + \frac{\partial I}{\partial\Delta}\frac{\partial\Delta}{\partial\rho} = \frac{dK}{d\rho},$$

which is independent of τ, and:

$$M_2(\beta_1,\rho) = \int_0^T g_3 d\tau. \tag{14}$$

Formulas (13) and (14) are similar to [3], (2.7). We are looking for a simple zero of M, which is equivalent to considering:

$$\begin{gathered}\bar{M}(\beta_1,\rho) = (\bar{M}_1(\beta_1,\rho), \bar{M}_2(\beta_1,\rho))\\ \bar{M}_1(\beta_1,\rho) = \int_0^T \left(\frac{\partial I}{\partial\theta}g_1 + \frac{\partial I}{\partial\Delta}g_2\right)d\tau, \quad \bar{M}_2(\beta_1,\rho) = \int_0^T g_3 d\tau.\end{gathered}$$

Since:

$$g_i = \mu_1 g_{i1} + \mu_2 g_{i2}, \quad i = 1,2,3$$

for:

$$\begin{aligned}g_{11}(\rho,\theta,\Delta,\tau) &= -\cos\theta\cos(P\tau)\cos(kP\tau+\beta_1)\\ g_{12}(\rho,\theta,\Delta,\tau) &= \sin\theta\cos(P\tau)\cos(kP\tau+\Delta+\beta_1)\\ g_{21}(\rho,\theta,\Delta,\tau) &= -\frac{\cos(P\tau)}{\sin\theta}\sin(kP\tau+\beta_1)\\ g_{22}(\rho,\theta,\Delta,\tau) &= \frac{\cos(P\tau)}{\cos\theta}\sin(kP\tau+\Delta+\beta_1)\\ g_{31}(\rho,\theta,\Delta,\tau) &= -\sin\theta\cos(P\tau)\cos(kP\tau+\beta_1)\\ g_{32}(\rho,\theta,\Delta,\tau) &= -\cos\theta\cos(P\tau)\cos(kP\tau+\Delta+\beta_1)\end{aligned}$$

we get:

$$\bar{M}_i(\beta_1,\rho) = \mu_1\bar{M}_{i1}(\beta_1,\rho) + \mu_2\bar{M}_{i2}(\beta_1,\rho), \quad i = 1,2,$$

for
$$\bar{M}_{1j}(\beta_1,\rho)=\int_0^T\left(\frac{\partial I}{\partial\theta}g_{1j}+\frac{\partial I}{\partial\Delta}g_{2j}\right)d\tau,\quad \bar{M}_{2j}(\beta_1,\rho)=\int_0^T g_{3j}d\tau,\quad j=1,2.$$

To solve:
$$\mu_1\bar{M}_{i1}(\beta_1,\rho)+\mu_2\bar{M}_{i2}(\beta_1,\rho)=0,\quad i=1,2,$$
we first solve the scalar equation:
$$\tilde{M}(\beta_1,\rho)=\bar{M}_{11}(\beta_1,\rho)\bar{M}_{22}(\beta_1,\rho)-\bar{M}_{12}(\beta_1,\rho)\bar{M}_{21}(\beta_1,\rho)=0 \tag{15}$$
to get its root $\beta_{1,0}$ and ρ_0. Then, we look for $\mu_{1,0}$ and $\mu_{2,0}$ with $\mu_{1,0}^2+\mu_{2,0}^2=1$ such that:
$$\begin{gathered}\mu_{1,0}\bar{M}_{i1}(\beta_{1,0},\rho_0)+\mu_{2,0}\bar{M}_{i2}(\beta_{1,0},\rho_0)=0,\quad i=1,2\\ \det\begin{pmatrix}\mu_{1,0}\nabla\bar{M}_{11}(\beta_{1,0},\rho_0)^\top+\mu_{2,0}\nabla\bar{M}_{12}(\beta_{1,0},\rho_0)^\top\\ \mu_{1,0}\nabla\bar{M}_{21}(\beta_{1,0},\rho_0)^\top+\mu_{2,0}\nabla\bar{M}_{22}(\beta_{1,0},\rho_0)^\top\end{pmatrix}\neq 0.\end{gathered} \tag{16}$$

Summarizing, we have the following result.

Theorem 1. *If there are $\beta_{1,0}\in[0,2\pi)$, ρ_0 satisfying* (15)*, $\mu_{1,0}$ and $\mu_{2,0}$ with $\mu_{1,0}^2+\mu_{2,0}^2=1$ solving* (16)*, then for any μ_1 near $\mu_{1,0}$ and μ_2 near $\mu_{2,0}$ with $\mu_1^2+\mu_2^2=1$ and $\epsilon\neq 0$ small, there are $\beta_1(\epsilon)$ near $\beta_{1,0}$ and $\rho(\epsilon)$ near ρ_0 such that* (8) *with $\beta_1=\beta_1(\epsilon)$ and $\rho=\rho(\epsilon)$ has a $T=2\pi/P$-periodic solution near* (12) *with $\rho=\rho(\epsilon)$.*

Note:
$$\begin{aligned}\bar{M}_{11}(\beta_1,\rho)&=-\int_0^T\Big(2\cos(2\theta(\rho,\tau))\sin\Delta(\rho,\tau)\cos\theta(\rho,\tau)\cos(P\tau)\cos(kP\tau+\beta_1)\\&\quad+2\cos\theta(\rho,\tau)\cos\Delta(\rho,\tau)\cos(P\tau)\sin(kP\tau+\beta_1)\Big)d\tau\\ \bar{M}_{12}(\beta_1,\rho)&=\int_0^T\Big(2\cos(2\theta(\rho,\tau))\sin\Delta(\rho,\tau)\sin\theta(\rho,\tau)\cos(P\tau)\cos(kP\tau+\Delta(\rho,\tau)+\beta_1)\\&\quad+2\sin\theta(\rho,\tau)\cos\Delta(\rho,\tau)\cos(P\tau)\sin(kP\tau+\Delta(\rho,\tau)+\beta_1)\Big)d\tau\\ \bar{M}_{21}(\beta_1,\rho)&=-\int_0^T\sin\theta(\rho,\tau)\cos(P\tau)\cos(kP\tau+\beta_1)d\tau\\ \bar{M}_{22}(\beta_1,\rho)&=-\int_0^T\cos\theta(\rho,\tau)\cos(P\tau)\cos(kP\tau+\Delta(\rho,\tau)+\beta_1)d\tau.\end{aligned}$$

Next, taking $\rho\to\infty$ in (12), we obtain:
$$\begin{gathered}\theta(\infty,\tau)=\begin{cases}\dfrac{P\tau}{2} & \tau\in\left[0,\dfrac{\pi}{P}\right],\\ \pi-\dfrac{P\tau}{2} & \tau\in\left[\dfrac{\pi}{P},\dfrac{2\pi}{P}\right],\end{cases}\\ \Delta(\infty,\tau)=\begin{cases}\pi & \tau\in\left(0,\dfrac{\pi}{P}\right),\\ 0 & \tau\in\left(\dfrac{\pi}{P},\dfrac{2\pi}{P}\right),\\ \dfrac{\pi}{2} & \tau\in\left\{0,\dfrac{\pi}{P},\dfrac{2\pi}{P}\right\},\end{cases}\\ \frac{\partial\theta}{\partial\rho}(\infty,\tau)=0,\quad \frac{\partial\Delta}{\partial\rho}(\infty,\tau)=0.\end{gathered} \tag{17}$$

Hence:

$$\begin{aligned}
\bar{M}_{11}(\beta_1,\infty) &= \int_0^{\frac{2\pi}{P}} 2\cos\frac{P\tau}{2}\cos(P\tau)\sin(kP\tau+\beta_1)d\tau \\
&= \frac{16k(4k^2-5)\cos\beta_1}{(16k^4-40k^2+9)P} \\
\bar{M}_{12}(\beta_1,\infty) &= \int_0^{\frac{2\pi}{P}} 2\sin\frac{P\tau}{2}\cos(P\tau)\sin(kP\tau+\beta_1)\Big)d\tau \\
&= -\frac{8(4k^2+3)\sin\beta_1}{(16k^4-40k^2+9)P} \\
\bar{M}_{21}(\beta_1,\infty) &= -\int_0^{\frac{2\pi}{P}} \sin\frac{P\tau}{2}\cos(P\tau)\cos(kP\tau+\beta_1)d\tau \\
&= \frac{4(4k^2+3)\cos\beta_1}{(16k^4-40k^2+9)P} \\
\bar{M}_{22}(\beta_1,\infty) &= \int_0^{\frac{2\pi}{P}} \cos\frac{P\tau}{2}\cos(P\tau)\cos(kP\tau+\beta_1)d\tau \\
&= -\frac{8k(4k^2-5)\sin\beta_1}{(16k^4-40k^2+9)P}.
\end{aligned}$$

Then, (15) gives as $\rho \to \infty$,

$$\tilde{M}(\beta_1,\infty) = -\frac{16\sin 2\beta_1}{(4k^2-9)P^2}$$

with asymptotic solutions $\beta_{1,0}^{\infty}$ satisfying either $\sin\beta_{1,0}^{\infty}=0$ or $\cos\beta_{1,0}^{\infty}=0$. The asymptotic equation of (16) is as follows:

$$\begin{aligned}
0 &= \mu_{1,0}\bar{M}_{11}(\beta_{1,0}^{\infty},\infty)+\mu_{2,0}\bar{M}_{12}(\beta_{1,0}^{\infty},\infty) \\
&= \mu_{1,0}\frac{16k(4k^2-5)\cos\beta_{1,0}^{\infty}}{(16k^4-40k^2+9)P} - \mu_{2,0}\frac{8(4k^2+3)\sin\beta_{1,0}^{\infty}}{(16k^4-40k^2+9)P},
\end{aligned}$$

which has solutions: $\mu_{1,0}^{\infty}=0$ and $\mu_{2,0}^{\infty}=\pm 1$ when $\sin\beta_{1,0}^{\infty}=0$ or $\mu_{1,0}^{\infty}=\pm 1$ and $\mu_{2,0}^{\infty}=0$ when $\cos\beta_{1,0}^{\infty}=0$. However, $\frac{\partial \bar{M}_{ij}}{\partial\rho}(\beta_1,\infty)=0$ for $i,j\in\{1,2\}$, so Theorem 1 cannot be applied directly. Note we consider just the first asymptotic equation of (16), since the second one is a scalar multiple of the first one.

On the other hand, following the method of [5], p. 111, we get the following result.

Corollary 1. *For any $\rho>0$ sufficiently large and $\epsilon\neq 0$ sufficiently small, there is $\mu_1(\rho,\epsilon)$, $\mu_2(\rho,\epsilon)$ with $\mu_1^2(\rho,\epsilon)+\mu_2^2(\rho,\epsilon)=1$ and $\beta_1(\rho,\epsilon)$ such that (8) has a T-periodic solution near (17) for $\epsilon\neq 0$ small and either $\mu_1(\infty,0)=0$, $\mu_2(\infty,0)=\pm 1$, $\sin\beta_1(\infty,0)=0$ or $\mu_1(\infty,0)=\pm 1$, $\mu_2(\infty,0)=0$, $\cos\beta_1(\infty,0)=0$.*

Proof. The bifurcation equation has the form (see [5], p. 111):

$$\mu_1\bar{M}_{i1}(\beta_1,\rho)+\mu_2\bar{M}_{i2}(\beta_1,\rho)=O(\epsilon),\quad i=1,2,\quad \mu_1^2+\mu_2^2=1$$

i.e.,

$$\sin\Gamma\bar{M}_{i1}(\beta_1,\rho)+\cos\Gamma\bar{M}_{i2}(\beta_1,\rho)=O(\epsilon),\quad i=1,2 \tag{18}$$

for $\mu_1=\sin\Gamma$ and $\mu_2=\cos\Gamma$. For $\rho=\infty$ and $\epsilon=0$, (18) takes the form:

$$\begin{aligned}
\frac{16k(4k^2-5)}{(16k^4-40k^2+9)P}\sin\Gamma\cos\beta_1 - \frac{8(4k^2+3)}{(16k^4-40k^2+9)P}\cos\Gamma\sin\beta_1 = 0 \\
\frac{4(4k^2+3)\cos\beta_1}{(16k^4-40k^2+9)P}\sin\Gamma\cos\beta_1 - \frac{8k(4k^2-5)}{(16k^4-40k^2+9)P}\cos\Gamma\sin\beta_1 = 0.
\end{aligned} \tag{19}$$

The determinant of (19) is $-\frac{32}{(4k^2-9)P^2} \neq 0$, so (19) has the only solutions:

$$\sin\Gamma_\infty = \sin\beta_{1,0}^\infty = 0, \quad \cos\Gamma_\infty = \cos\beta_{1,0}^\infty = 0.$$

The determinants of Jacobians of (19) at these zeros are as follows:

$$\det\begin{pmatrix} \frac{16k(4k^2-5)}{(16k^4-40k^2+9)P}\cos\Gamma_\infty\cos\beta_{1,0}^\infty & -\frac{8(4k^2+3)}{(16k^4-40k^2+9)P}\cos\Gamma_\infty\cos\beta_{1,0}^\infty \\ \frac{4(4k^2+3)\cos\beta_{1,0}^\infty}{(16k^4-40k^2+9)P}\cos\Gamma_\infty\cos\beta_{1,0}^\infty & -\frac{8k(4k^2-5)}{(16k^4-40k^2+9)P}\cos\Gamma_\infty\cos\beta_{1,0}^\infty \end{pmatrix}$$
$$= -\frac{32\cos^2\Gamma_\infty\cos^2\beta_{1,0}^\infty}{(4k^2-9)P^2} \neq 0$$

and:

$$\det\begin{pmatrix} \frac{8(4k^2+3)}{(16k^4-40k^2+9)P}\sin\Gamma_\infty\sin\beta_{1,0}^\infty & -\frac{16k(4k^2-5)}{(16k^4-40k^2+9)P}\sin\Gamma_\infty\sin\beta_{1,0}^\infty \\ \frac{8k(4k^2-5)}{(16k^4-40k^2+9)P}\sin\Gamma_\infty\sin\beta_{1,0}^\infty & -\frac{4(4k^2+3)\cos\beta_{1,0}^\infty}{(16k^4-40k^2+9)P}\sin\Gamma_\infty\sin\beta_{1,0}^\infty \end{pmatrix}$$
$$= \frac{32\sin^2\Gamma_\infty\sin^2\beta_{1,0}^\infty}{(4k^2-9)P^2} \neq 0,$$

respectively. Hence, the zeroes Γ_∞ and $\beta_{1,0}^\infty$ are simple, so we can apply the implicit function theorem to get the result. The proof is complete. □

5. Discussion

Melnikov analysis is applied for the persistence of periodic oscillations for periodically-perturbed systems of ODEs with a slowly-varying variable. The ODEs are obtained from a model of a finite lattice of particles coupled by linear springs under distributed harmonic excitation, which presents a low-energy nonlinear acoustic vacuum (see [1]), but we consider just two modes. We extend the study of [1] to a problem with small exciting harmonic forces. We apply an analytical method based on derivation of Melnikov functions and then on the location of their simple roots. Melnikov functions are derived by using the approach of [3,5] developed for slowly-varying ODEs. Since the Melnikov functions are rather complicated, we follow an asymptotic way for solving the corresponding Melnikov equations. It would be nice to solve these Melnikov equations numerically for finding other simple roots, which is postponed to our next research. These roots determine and locate periodic solutions of the periodically-perturbed systems of ODEs derived from the two-mode low-energy nonlinear acoustic vacuum system. Moreover, our next investigation will be also to consider higher numbers of modes represented by a system of ODEs in (5). The method used will be different from that in this paper, since it will be based on the results of Section 3.3 of [5]. Note that higher modes of (2) were numerically studied in [2], which is another challenge for our study.

Author Contributions: The authors contributed equally to this work.

References

1. Manevich, L.I.; Vakakis, A.F. Nonlinear oscillatory acoustic vacuum. *SIAM J. Appl. Math.* **2014**, *74*, 1742–1762. [CrossRef]
2. Zhang, Z.; Manevitch, L.I.; Smirnov, V.; Bergman, L.; Vakakis, A.F. Extreme nonlinear energy exchanges in a geometrically nonlinear lattice oscillating in the plane. *J. Mech. Phys. Solids* **2018**, *110*, 1–20. [CrossRef]
3. Wiggins, S.; Holmes, P. Periodic orbits in slowly varying oscillators. *SIAM J. Math. Anal.* **1987**, *18*, 592–611. [CrossRef]
4. Gradshteyn, I.S.; Ryzhik, I.M. *Table of Integrals, Series, and Products*, 7th ed.; Academic Press: Cambridge, MA, USA, 2007.
5. Fečkan, M. *Topological Degree Approach to Bifurcation Problems*; Springer: Berlin, Germany, 2008.
6. Hale, J.K. *Ordinary Differential Equations*, 2nd ed.; Robert E. Krieger Publishing Company, Inc.: Malabar, FL, USA, 1980.

15

Differential Equations for Classical and Non-Classical Polynomial Sets

Paolo Emilio Ricci

Section of Mathematics, International Telematic University UniNettuno, Corso Vittorio Emanuele II, 39, 00186 Roma, Italy; paoloemilioricci@gmail.com
† Dedicated to Prof. Dr. Hari M. Srivastava.

Abstract: By using the monomiality principle and general results on Sheffer polynomial sets, the differential equation satisfied by several old and new polynomial sets is shown.

Keywords: differential equations; Sheffer polynomial sets; generating functions; monomiality principle

---

1. Introduction

In this survey article, a uniform method is presented for constructing the differential equations satisfied by several sets of classical and non classical polynomials. This has been done by starting from the basic elements of the relevant generating functions, using the monomiality principle by G. Dattoli [1] and a general result by Y. Ben Cheikh [2]. Of course, the polynomials considered in this paper are only examples for showing that the method works, but obviously this technique can be theoretically extended to every polynomial set.

This method has been recently applied in several articles (see [3–9]), which include works in collaboration with several authors. The most outstanding of them is Prof. Dr. Hari M. Srivastava, to whom this article is dedicated.

The derived differential equations are generally of infinite order, but they reduce to finite order when applied to polynomials.

It is worth noting that the differential equations for Sheffer polynomial sets have been studied even with different methods (see [10–13]), but here we use only elements directly connected with the theory of polynomials.

We start recalling, in Section 2, the definitions relevant to Sheffer polynomials, the G. Dattoli monomiality principle, and a general result by Y. Ben Cheikh.

The classical polynomial sets, considered in Section 3, are the Bernoulli, Euler, Genocchi and Mittag–Leffler polynomials. In Section 4, we show some new polynomial sets derived from non-classical generating functions.

2. Sheffer Polynomials

The Sheffer polynomials $\{s_n(x)\}$ are introduced [14] by means of the exponential generating function [15] of the type:

$$A(t)\exp(xH(t)) = \sum_{n=0}^{\infty} s_n(x)\,\frac{t^n}{n!}\,, \tag{1}$$

where

$$\begin{aligned} A(t) &= \sum_{n=0}^{\infty} a_n\,\frac{t^n}{n!}\,, \qquad (a_0 \neq 0)\,, \\ H(t) &= \sum_{n=0}^{\infty} h_n\,\frac{t^n}{n!}\,, \qquad (h_0 = 0)\,. \end{aligned} \tag{2}$$

According to a different characterization (see [16], p. 18), the same polynomial sequence can be defined by means of the pair $(g(t), f(t))$, where $g(t)$ is an invertible series and $f(t)$ is a delta series:

$$\begin{aligned} g(t) &= \sum_{n=0}^{\infty} g_n\,\frac{t^n}{n!}\,, \qquad (g_0 \neq 0)\,, \\ f(t) &= \sum_{n=0}^{\infty} f_n\,\frac{t^n}{n!}\,, \qquad (f_0 = 0,\ f_1 \neq 0)\,. \end{aligned} \tag{3}$$

Denoting by $f^{-1}(t)$ the compositional inverse of $f(t)$ (i.e., such that $f\left(f^{-1}(t)\right) = f^{-1}\left(f(t)\right) = t$), the exponential generating function of the sequence $\{s_n(x)\}$ is given by

$$\frac{1}{g[f^{-1}(t)]}\,\exp\left(xf^{-1}(t)\right) = \sum_{n=0}^{\infty} s_n(x)\,\frac{t^n}{n!}\,, \tag{4}$$

so that

$$A(t) = \frac{1}{g[f^{-1}(t)]}\,, \qquad H(t) = f^{-1}(t)\,. \tag{5}$$

When $g(t) \equiv 1$, the Sheffer sequence corresponding to the pair $(1, f(t))$ is called the associated Sheffer sequence $\{\sigma_n(x)\}$ for $f(t)$, and its exponential generating function is given by

$$\exp\left(xf^{-1}(t)\right) = \sum_{n=0}^{\infty} \sigma_n(x)\,\frac{t^n}{n!}\,. \tag{6}$$

A list of known Sheffer polynomial sequences and their associated ones can be found in [17].

Shift Operators and Differential Equation

We recall that a polynomial set $\{p_n(x)\}$ is called quasi-monomial if and only if there exist two operators $\hat{P}$ and $\hat{M}$ such that

$$\hat{P}\left(p_n(x)\right) = np_{n-1}(x)\,, \qquad \hat{M}\left(p_n(x)\right) = p_{n+1}(x)\,, \qquad (n = 1, 2, \ldots)\,. \tag{7}$$

$\hat{P}$ is called the *derivative* operator and $\hat{M}$ the *multiplication* operator, as they act in the same way as classical operators on monomials.

This definition traces back to a paper by J.F. Steffensen [18] recently improved by G. Dattoli and widely used in several applications [19,20].

Y. Ben Cheikh proved that every polynomial set is quasi-monomial under the action of suitable derivative and multiplication operators. In particular, in the same article, the following result is proved, as a particular case of Corollary 3.2 in [2]:

Theorem 1. *Let $(p_n(x))$ denote a Sheffer polynomial set, defined by the generating function*

$$A(t)\exp(xH(t)) = \sum_{n=0}^{\infty} p_n(x)\,\frac{t^n}{n!}\,, \tag{8}$$

where

$$A(t) = \sum_{n=0}^{\infty} \tilde{a}_n t^n\,, \qquad (\tilde{a}_0 \neq 0)\,, \tag{9}$$

and

$$H(t) = \sum_{n=0}^{\infty} \tilde{h}_n\, t^{n+1}\,, \qquad (\tilde{h}_0 \neq 0)\,. \tag{10}$$

Denoting, as before, by $f(t)$ the compositional inverse of $H(t)$, the Sheffer polynomial set $\{p_n(x)\}$ is quasi-monomial under the action of the operators

$$\hat{P} = f(D_x)\,, \qquad \hat{M} = \frac{A'[f(D_x)]}{A[f(D_x)]} + xH'[f(D_x)]\,, \tag{11}$$

where prime denotes the ordinary derivatives with respect to t.

Furthermore, according to the monomiality principle, the quasi-monomial polynomials $\{p_n(x)\}$ satisfy the differential equation

$$\hat{M}\hat{P}\, p_n(x) = n\, p_n(x)\,. \tag{12}$$

3. Differential Equations of Classical Polynomials

3.1. Bernoulli Polynomials

The Bernoulli polynomials are defined by the generating function

$$G(t,x) = \frac{t}{e^t - 1}\, e^{x\,t}\,, \tag{13}$$

so that

$$\begin{aligned}
A(t) &= \frac{t}{e^t - 1} = \sum_{k=0}^{\infty} b_k \frac{t^k}{k!}, \\
G(t,x) &= \sum_{k=0}^{\infty} B_k(x) \frac{t^k}{k!} = \sum_{k=0}^{\infty} \left[\sum_{h=0}^{k} \binom{k}{h} b_{k-h}\, x^h \right] \frac{t^k}{k!}, \\
B_k(x) &= \sum_{h=0}^{k} \binom{k}{h} b_{k-h}\, x^h,
\end{aligned} \tag{14}$$

where b_k are the Bernoulli numbers.

Differential Equation of the $B_k(x)$

Note that, recalling that $B_n(1) = (-1)^n\, b_n$, the following expansion holds:

$$\begin{aligned}
t\,\frac{A'(t)}{A(t)} &= \frac{e^t - te^t - 1}{e^t - 1} = 1 - \frac{te^t}{e^t - 1} = 1 - \sum_{n=0}^{\infty} B_n(1)\, \frac{t^n}{n!} = \\
&= \sum_{n=1}^{\infty} (-1)^{n+1}\, b_n\, \frac{t^n}{n!}.
\end{aligned} \tag{15}$$

The shift operators for the Bernoulli polynomials are given by

$$\begin{aligned}
\hat{P} &= D_x, \\
\hat{M} &= \frac{e^{D_x} - D_x e^{D_x} - 1}{D_x\,(e^{D_x} - 1)}.
\end{aligned} \tag{16}$$

Therefore, by using the factorization method, we find

Theorem 2. *The Bernoulli polynomials $\{B_n(x)\}$ satisfy the differential equation*

$$\left(\frac{e^{D_x} - D_x e^{D_x} - 1}{e^{D_x} - 1} + x\, D_x \right) B_n(x) = n\, B_n(x), \tag{17}$$

that is

$$\left(\sum_{k=1}^{\infty} \frac{(-1)^{k+1}\, b_k}{k!}\, D_x^k + x\, D_x \right) B_n(x) = n\, B_n(x), \tag{18}$$

or, in equivalent form:

$$\left(\sum_{k=1}^{n} \frac{(-1)^{k+1}\, b_k}{k!}\, D_x^k + x\, D_x \right) B_n(x) = n\, B_n(x). \tag{19}$$

Proof. It is sufficient to expand in series the operator (17). Equation (19) follows because, for any fixed n, the series expansion in Equation (18) reduces to a finite sum when applied to a polynomial of degree n. □

3.2. Euler Polynomials

The Euler polynomials are defined by the generating function

$$G(t,x) = \frac{2}{e^t+1}\, e^{x\,t}\,, \tag{20}$$

so that

$$\begin{aligned} A(t) &= \frac{2}{e^t+1} = \sum_{k=0}^{\infty} e_k\, \frac{t^k}{k!}\,, \\ G(t,x) &= \sum_{k=0}^{\infty} E_k(x)\, \frac{t^k}{k!} = \sum_{k=0}^{\infty} \left[\sum_{h=0}^{k} \binom{k}{h} e_{k-h}\, x^h \right] \frac{t^k}{k!}\,, \\ E_k(x) &= \sum_{h=0}^{k} \binom{k}{h} e_{k-h}\, x^h\,, \end{aligned} \tag{21}$$

where e_k are the Euler numbers.

Differential Equation of the $E_k(x)$

Note that the following expansion holds:

$$\begin{aligned} \frac{A'(t)}{A(t)} &= -\frac{e^t}{e^t+1} = -1 + \frac{1}{e^t+1} = -1 + \frac{1}{2}\sum_{n=0}^{\infty} e_n\, \frac{t^n}{n!} = \\ &= -\frac{1}{2} + \frac{1}{2}\sum_{n=1}^{\infty} e_n\, \frac{t^n}{n!} = \sum_{n=0}^{\infty} c_n\, \frac{t^n}{n!}\,, \end{aligned} \tag{22}$$

where $c_0 = -1/2$, and $c_n = e_n/2$.

The shift operators for the Euler polynomials are given by

$$\begin{aligned} \hat{P} &= D_x\,, \\ \hat{M} &= -\frac{e^{D_x}}{e^{D_x}+1}\,. \end{aligned} \tag{23}$$

Therefore, by using the factorization method, we find

Theorem 3. *The Euler polynomials $\{E_n(x)\}$ satisfy the differential equation*

$$\left(-\frac{e^{D_x}\, D_x}{e^{D_x}+1} + x\, D_x \right) E_n(x) = n\, E_n(x)\,. \tag{24}$$

that is

$$\left(\sum_{k=0}^{\infty} \frac{c_k}{k!}\, D_x^{k+1} + x\, D_x \right) E_n(x) = n\, E_n(x)\,, \tag{25}$$

or, in equivalent form:

$$\left(\sum_{k=0}^{n-1} \frac{c_k}{k!}\, D_x^{k+1} + x\, D_x \right) E_n(x) = n\, E_n(x)\,. \tag{26}$$

Proof. It is sufficient to expand in series the operator (24). Equation (26) follows because, for any fixed n, the series expansion in Equation (25) reduces to a finite sum when applied to a polynomial of degree n. □

3.3. Genocchi Polynomials

The Genocchi polynomials are defined by the generating function

$$G(t,x) = \frac{2\,t}{e^t+1}\, e^{x\,t}\,, \tag{27}$$

so that

$$\begin{aligned} A(t) &= \frac{2\,t}{e^t+1} = \sum_{k=0}^{\infty} g_k\, \frac{t^k}{k!}\,, \\ G(t,x) &= \sum_{k=0}^{\infty} G_k(x)\, \frac{t^k}{k!} = \sum_{k=0}^{\infty} \left[\sum_{h=0}^{k} \binom{k}{h} g_{k-h}\, x^h \right] \frac{t^k}{k!}\,, \\ G_k(x) &= \sum_{h=0}^{k} \binom{k}{h} g_{k-h}\, x^h\,, \end{aligned} \tag{28}$$

where g_k are the Genocchi numbers.

Differential Equation of the $G_k(x)$

Note that the following expansion holds:

$$\begin{aligned} t\,\frac{A'(t)}{A(t)} &= \frac{e^t - te^t + 1}{e^t+1} = 1 - \frac{te^t}{e^t+1} = 1 - \frac{1}{2}\,t + \frac{1}{2}\sum_{n=2}^{\infty} e_n\, \frac{t^{n+1}}{n!} = \\ &= \sum_{n=0}^{\infty} d_n\, \frac{t^n}{n!}\,, \end{aligned} \tag{29}$$

where $d_0 = 1$, $d_1 = -1/2$, and $d_k = e_k/2$, $(k \geq 2)$.

The shift operators for the Genocchi polynomials are given by

$$\begin{aligned} \hat{P} &= D_x\,, \\ \hat{M} &= \frac{e^{D_x} - D_x e^{D_x} + 1}{D_x\,(e^{D_x}+1)}\,, \end{aligned} \tag{30}$$

so that the Genocchi polynomials satisfy the differential equation

$$\left(\frac{e^{D_x} - D_x e^{D_x} + 1}{e^{D_x}+1} + x\,D_x \right) G_n(x) = n\,G_n(x)\,. \tag{31}$$

Therefore, by using the factorization method, we find

Theorem 4. *The Genocchi polynomials* $\{G_n(x)\}$ *satisfy the differential equation*

$$\left(\frac{e^{D_x} - D_x e^{D_x} + 1}{e^{D_x}+1} + x\,D_x \right) G_n(x) = n\,G_n(x)\,, \tag{32}$$

that is

$$\left(\sum_{k=0}^{\infty} \frac{d_k}{k!} D_x^k + x D_x\right) G_n(x) = n\, G_n(x)\,, \tag{33}$$

or, in equivalent form:

$$\left(\sum_{k=0}^{n} \frac{d_k}{k!} D_x^k + x D_x\right) G_n(x) = n\, G_n(x)\,. \tag{34}$$

Proof. It is sufficient to expand in series the operator (32). Equation (34) follows because, for any fixed n, the series expansion in Equation (33) reduces to a finite sum when applied to a polynomial of degree n. □

3.4. The Mittag–Leffler Polynomials

We recall that the Mittag–Leffler polynomials [21] are a special case of associated Sheffer polynomials, defined by the generating function

$$\begin{aligned} &A(t) = 1, \qquad H(t) = \log\frac{1+t}{1-t}, \\ &G(t,x) = \left(\frac{1+t}{1-t}\right)^x = \exp\left(x\,\log\frac{1+t}{1-t}\right) = \sum_{n=0}^{\infty} \mathcal{M}_n(x)\,\frac{t^n}{n!}\,. \end{aligned} \tag{35}$$

Therefore, we have

$$\frac{A'(t)}{A(t)} = 0, \qquad H'(t) = \frac{2}{1-t^2}, \qquad H^{-1}(t) = f(t) = \frac{e^t - 1}{e^t + 1}, \tag{36}$$

so that, for the Mittag-Leffer polynomials, we find the shift operators:

$$\begin{aligned} \hat{P} &= \frac{e^{D_x} - 1}{e^{D_x} + 1} = \tanh\left(\frac{D_x}{2}\right), \\ \hat{M} &= x\,\frac{(e^{D_x} + 1)^2}{2\,e^{D_x}} = x\,\left[1 + \cosh(D_x)\right]. \end{aligned} \tag{37}$$

3.5. Differential Equation of the $\mathcal{M}_n(x)$

In the present case, according to the identity:

$$[1 + \cosh x]\ \tanh\,(x/2) = \sinh x\,,$$

we can write

$$\hat{M}\,\hat{P} = x\,\frac{e^{2\,D_x} - 1}{2\,e^{D_x}} = x\,\sinh(D_x)\,, \tag{38}$$

so that we have the theorem

Theorem 5. *The Mittag–Leffler polynomials* $\{\mathcal{M}_n(x)\}$ *satisfy the differential equation*

$$x\,\sinh(D_x)\,\mathcal{M}_n(x) = n\,\mathcal{M}_n(x)\,, \tag{39}$$

that is

$$x\sum_{k=0}^{\infty}\frac{D_x^{2k+1}}{(2k+1)!}\,\mathcal{M}_n(x) = n\,\mathcal{M}_n(x)\,, \tag{40}$$

or

$$x\sum_{k=0}^{\left[\frac{n-1}{2}\right]}\frac{D_x^{2k+1}}{(2k+1)!}\,\mathcal{M}_n(x) = n\,\mathcal{M}_n(x)\,, \tag{41}$$

where $\left[\frac{n-1}{2}\right]$ *denotes the* integral part *of* $(n-1)/2$.

Proof. It is sufficient to expand in series the operator (39). Equation (41) follows because, for any fixed n, the series expansion in Equation (40) reduces to a finite sum when applied to a polynomial of degree n. □

4. Differential Equations of Non-Classical Polynomials

4.1. Euler-Type Polynomials

Here, we introduce a Sheffer polynomial set connected with the classical Euler polynomials. Assuming:

$$A(t) = \frac{1}{\cosh t}, \qquad H(t) = \sinh t\,, \tag{42}$$

we consider the Euler-type polynomials $\tilde{E}_n(x)$, defined by the generating function

$$G(t,x) = \frac{1}{\cosh t}\exp\left[x\,\sinh t\right] = \sum_{k=0}^{\infty}\tilde{E}_k(x)\,\frac{t^k}{k!}\,. \tag{43}$$

Note that the Euler numbers are recovered, since we have:

$$G(t,0) = \frac{2}{e^t+e^{-t}} = \sum_{k=0}^{\infty}\tilde{E}_k(0)\,\frac{t^k}{k!}\,, \tag{44}$$

so that $\tilde{E}_n(0) = E_n$.

In what follows, we use the expansions

$$\sinh t = \sum_{k=0}^{\infty}\frac{t^{2k+1}}{(2k+1)!} = \sum_{k=0}^{\infty}\left(\frac{1+(-1)^{k+1}}{2}\right)\frac{t^k}{k!}\,, \tag{45}$$

$$\cosh t = \sum_{k=0}^{\infty}\frac{t^{2k}}{(2k)!} = \sum_{k=0}^{\infty}\left(\frac{1+(-1)^{k}}{2}\right)\frac{t^k}{k!}\,. \tag{46}$$

Note that, in our case, we are dealing with a Sheffer polynomial set, so that, since we have $\psi(t) = e^t$, the operator σ defined by Equation (6) simply reduces to the derivative operator D_x. Furthermore, we have:

$$A(t) = \frac{1}{\cosh t}, \qquad \frac{A'(t)}{A(t)} = -\tanh t\,,$$

$$H(t) = \sinh t = \sum_{k=0}^{\infty} \frac{t^{2k+1}}{(2k+1)!}, \qquad \left(\tilde{h}_k = \left(\frac{1+(-1)^{k+1}}{2}\right)\frac{1}{k!}\right),$$

$$H'(t) = \cosh t\,, \qquad f(t) = H^{-1}(t) = \log(t + \sqrt{t^2+1})\,,$$

so that we have the theorem

Theorem 6. *The Euler-type polynomial set* $\{\tilde{E}_n(x)\}$ *is quasi-monomial under the action of the operators*

$$\hat{P} = \log(D_x + \sqrt{D_x^2+1})\,, \qquad \hat{M} = -\tanh(\operatorname{arcsinh} D_x) + x \operatorname{arcsinh} D_x \tag{47}$$

(by $\operatorname{arcsinh} t = \log(t+\sqrt{t^2+1})$*, we denote the inverse of the function* $\sinh t$*), i.e.,*

$$\hat{P} = \sum_{k=0}^{\infty} (-1)^k \frac{(2k)!}{4^k (k!)^2 (2k+1)} D_x^{2k+1}\,,$$

$$\hat{M} = -\frac{D_x}{\sqrt{1+D_x^2}} + x\sqrt{1+D_x^2} = (xD_x^2 - D_x + x)(1+D_x^2)^{-1/2}\,, \tag{48}$$

$$\hat{M} = (xD_x^2 - D_x + x) \sum_{k=0}^{\infty} \binom{-1/2}{k} D_x^{2k}\,.$$

There is no problem about the convergence of the above series, since they reduce to finite sums when applied to polynomials.

4.2. Differential Equation of the $\tilde{E}_n(x)$

In the present case, we have

Theorem 7. *The Euler-type polynomials* $\{\tilde{E}_n(x)\}$ *satisfy the differential equation*

$$\left\{\left[(xD_x^2 - D_x + x)\sum_{k=0}^{\infty}\binom{-1/2}{k} D_x^{2k}\right] \sum_{k=0}^{\infty}(-1)^k \frac{(2k)!}{4^k(k!)^2(2k+1)} D_x^{2k+1}\right\} \tilde{E}_n(x) = n\,\tilde{E}_n(x)\,, \tag{49}$$

i.e.,

$$(xD_x^2 - D_x + x) \sum_{k=0}^{\left[\frac{n-1}{2}\right]} \sum_{h=0}^{k} (-1)^h \binom{-1/2}{k-h} \frac{(2h)!}{4^h (h!)^2 (2h+1)} D_x^{2k+1} \tilde{E}_n(x) \\ = n\, \tilde{E}_n(x)\,. \tag{50}$$

Note that, for any fixed n, the Cauchy product of series expansions in Equation (49) reduces to a finite sum, with upper limit $\left[\frac{n-1}{2}\right]$, when it is applied to a polynomial of degree n, because the successive addends vanish.

Remark 1. *The first few Euler-type polynomials are as follows:*

$$\begin{aligned}
\tilde{E}_0(x) &= 1,\\
\tilde{E}_1(x) &= x,\\
\tilde{E}_2(x) &= x^2 - 1,\\
\tilde{E}_3(x) &= x^3 - 2x,\\
\tilde{E}_4(x) &= x^4 - 2x^2 + 5,\\
\tilde{E}_5(x) &= x^5 + 16x,\\
\tilde{E}_6(x) &= x^6 + 5x^4 + 31x^2 - 61,\\
\tilde{E}_7(x) &= x^7 + 14x^5 + 56x^3 - 272x,\\
\tilde{E}_8(x) &= x^8 + 28x^6 + 126x^4 - 692x^2 + 1385,\\
\tilde{E}_9(x) &= x^9 + 48x^7 + 336x^5 - 1280x^3 + 7936x,\\
\tilde{E}_{10}(x) &= x^{10} + 75x^8 + 882x^6 - 1490x^4 + 25,261x^2 - 50,521.
\end{aligned}$$

5. Adjointness for Sheffer Polynomial Sequences

According to the above considerations, Sheffer polynomials are characterized both by the ordered couples $(A(t), H(t))$, or by $(g(t), f(t))$.

Definition 1. *Adjoint Sheffer polynomials are defined by interchanging the ordered couple $(A(t), H(t))$ with $(g(t), f(t))$, when writing the generating function.*

Here and in the following the *tilde* "$\sim$" above the symbol of a polynomial set stands for the adjective "*adjoint*" (see e.g., [4]).

5.1. Adjoint Hahn Polynomials

Assuming:

$$A(t) = \sec t, \qquad H(t) = \tan t\,, \tag{51}$$

we consider the adjoint Hahn $\tilde{R}_n(x)$, defined by the generating function

$$G(t,x) = \sec t\, \exp(x \tan t) = \sum_{n=0}^{\infty} \tilde{R}_n(x) \frac{t^n}{n!}\,. \tag{52}$$

It is a Sheffer set.

We have:

$$\frac{\partial G}{\partial x} = \frac{1}{\cos^3 t} \exp(x \tan t) = \frac{1}{\cos t} G(t, x) \, .$$

Note that, in this case, we have:

$$A(t) = \sec t \, , \qquad H(t) = \tan t \, ,$$

$$H'(t) = \sec^2 t \, , \qquad f(t) = H^{-1}(t) = \arctan t \, ,$$

$$\frac{A'(t)}{A(t)} = \tan t \, ,$$

so that we have the theorem

Theorem 8. *The adjoint Hahn polynomial set* $\{\tilde{R}_n(x)\}$ *is quasi-monomial under the action of the operators*

$$\begin{aligned} \hat{P} &= \arctan D_x \, , \\ \hat{M} &= \tan(\arctan D_x) + x \sec^2(\arctan D_x) \, , \end{aligned} \tag{53}$$

i.e.,

$$\begin{aligned} \hat{P} &= \arctan D_x = \sum_{k=0}^{\infty} \frac{(-1)^k}{2k+1} D_x^{2k+1} \, , \\ \hat{M} &= D_x + x\,(1 + D_x^2) = x\,D_x^2 + D_x + x. \end{aligned} \tag{54}$$

5.2. Differential Equation of the $\tilde{R}_n(x)$

In the present case, we have

Theorem 9. *The Sheffer-type adjoint Hahn polynomials* $\{\tilde{R}_n(x)\}$ *satisfy the differential equation*

$$\left(x\,D_x^2 + D_x + x\right) \sum_{k=0}^{\left[\frac{n-1}{2}\right]} \frac{(-1)^k}{2k+1} D_x^{2k+1} \tilde{R}_n(x) = n\,\tilde{R}_n(x) \, . \tag{55}$$

Note that, for any fixed n, in Equation (55), a finite sum appears, with upper limit $\left[\frac{n-1}{2}\right]$, instead of a complete series expansion since, when this series is applied to a polynomial of degree n, the subsequent addends vanish.

Remark 2. *The first few values of the adjoint Hahn polynomials are as follows:*

$$\begin{aligned}
\tilde{R}_0(x) &= 1, \\
\tilde{R}_1(x) &= x, \\
\tilde{R}_2(x) &= x^2 + 1, \\
\tilde{R}_3(x) &= x^3 + 5x, \\
\tilde{R}_4(x) &= x^4 + 14x^2 + 5, \\
\tilde{R}_5(x) &= x^5 + 30x^3 + 61x, \\
\tilde{R}_6(x) &= x^6 + 55x^4 + 331x^2 + 61, \\
\tilde{R}_7(x) &= x^7 + 91x^5 + 1211x^3 + 1385x, \\
\tilde{R}_8(x) &= x^8 + 140x^6 + 3486x^4 + 12,284x^2 + 1385, \\
\tilde{R}_9(x) &= x^9 + 204x^7 + 8526x^5 + 68,060x^3 + 50,521x, \\
\tilde{R}_{10}(x) &= x^{10} + 285x^8 + 18,522x^6 + 281,210x^4 + 663,061x^2 + 50,521.
\end{aligned}$$

Remark 3. *Table of adjoint Hahn numbers*

$$\begin{array}{lll}
\tilde{R}_0(0) = 1 & \tilde{R}_1(0) = 0 & \tilde{R}_2(0) = 1, \\
\tilde{R}_3(0) = 0 & \tilde{R}_4(0) = 5 & \tilde{R}_{2k+1}(0) = 0\,,\ \forall k \geq 2, \\
\tilde{R}_6(0) = 61 & \tilde{R}_8(0) = 1385 & \tilde{R}_{10}(0) = 50,521.
\end{array}$$

Note that the sequence $\{1, 1, 5, 61, 1385, 50,521, \dots\}$ appears in the Encyclopedia of Integer Sequences [22] under #A000364—Euler (or secant numbers): $a(n)$ = number of downup permutations of $[2n]$.

Example 1. $a(2) = 5$ *counts 4231, 4132, 3241, 3142, 2143. - David Callan, Nov 21, 2011.*

5.3. Adjoint Bernoulli Polynomials of the Second Kind

Assuming

$$A(t) = \frac{t}{e^t - 1}, \qquad H(t) = e^t - 1\,, \tag{56}$$

we consider the adjoint Bernoulli polynomials of the second kind $\{\tilde{b}_k(x)\}$, defined by the generating function

$$G(t,x) = \frac{t}{e^t - 1} \exp\left[x(e^t - 1)\right] = \sum_{k=0}^{\infty} \tilde{b}_k(x)\, \frac{t^k}{k!}\,. \tag{57}$$

Note that, in our case, we are dealing with a Sheffer polynomial set, so that, since we have $\psi(t) = e^t$, the operator σ defined by Equation (6) simply reduces to the derivative operator D_x. Furthermore, we have

$$A(t) = \frac{t}{e^t - 1}, \qquad H(t) = e^t - 1 = \sum_{k=1}^{\infty} \frac{t^k}{k!}\,, \qquad (\tilde{h}_k = 1/(k+1)!)\,,$$

$$H'(t) = e^t\,, \qquad f(t) = H^{-1}(t) = \log(t+1)\,,$$

$$\frac{A'(t)}{A(t)} = \frac{e^t - te^t - 1}{t(e^t - 1)} = \frac{1}{t} - \frac{1}{e^t - 1} - 1 .$$

so that we have the theorem

Theorem 10. *The adjoint Bernoulli polynomials of the second kind $\{\tilde{b}_n(x)\}$ are quasi-monomials under the action of the operators*

$$\hat{P} = \log(D_x + 1) ,$$
$$\hat{M} = \frac{1}{\log(D_x + 1)} - \frac{1}{D_x} - 1 + x\,(D_x + 1) , \tag{58}$$

that is

$$\hat{P} = \log(D_x + 1) = \sum_{k=1}^{\infty} \frac{(-1)^{k+1}}{k} D_x^k ,$$
$$\hat{M} = \frac{1}{\log(D_x + 1)} + \left(x - \frac{1}{D_x}\right) (D_x + 1) . \tag{59}$$

5.4. Differential Equation of the $\tilde{b}_n(x)$

In the present case, we have

$$\hat{M}\hat{P} = 1 + \left(x - \frac{1}{D_x}\right) (D_x + 1) \log(D_x + 1), \tag{60}$$

so that we have the theorem

Theorem 11. *The adjoint Bernoulli polynomials of the second kind $\{\tilde{b}_n(x)\}$ satisfy the differential equation*

$$\left[1 + \left(x - \frac{1}{D_x}\right) (D_x + 1) \log(D_x + 1)\right] \tilde{b}_n(x) = n\,\tilde{b}_n(x) , \tag{61}$$

that is

$$\left[1 + (xD_x - 1)\,(D_x + 1) \sum_{k=0}^{n} \frac{(-1)^k}{k+1} D_x^k\right] \tilde{b}_n(x) = n\,\tilde{b}_n(x) , \tag{62}$$

because, for any fixed n, the series expansion in Equation (61) *reduces to a finite sum when it is applied to a polynomial of degree n.*

Note that, in this case, due to the presence of the operator D_x^{-1}, it is necessary to consider derivatives up to the order $n + 1$.

Remark 4. *The first few values of the adjoint Bernoulli polynomials of the second kind are as follows:*

$$\begin{aligned}
\tilde{b}_0(x) &= 1,\\
\tilde{b}_1(x) &= x - \tfrac{1}{2},\\
\tilde{b}_2(x) &= x^2 + \tfrac{1}{6},\\
\tilde{b}_3(x) &= x^3 + \tfrac{3}{2}x^2,\\
\tilde{b}_4(x) &= x^4 + 4x^3 + 2x^2 - \tfrac{1}{30},\\
\tilde{b}_5(x) &= x^5 + \tfrac{15}{2}x^4 + \tfrac{35}{3}x^3 + \tfrac{5}{2}x^2,\\
\tilde{b}_6(x) &= x^6 + 12x^5 + \tfrac{75}{2}x^4 + 30x^3 + 3x^2 + \tfrac{1}{42},\\
\tilde{b}_8(x) &= x^8 + 24x^7 + \tfrac{560}{3}x^6 + 560x^5 + 602x^4 + 168x^3 + 4x^2 - \tfrac{1}{30},\\
\tilde{b}_{10}(x) &= x^{10} + 40x^9 + \tfrac{1155}{2}x^8 + 3780x^7 + 11,585x^6 + 15,540x^5 + \tfrac{15,125}{2}x^4 + 850x^3 +\\
&\quad + 5x^2 + \tfrac{5}{66},\\
\tilde{b}_{12}(x) &= x^{12} + 60x^{11} + 1386x^{10} + 15,840x^9 + \tfrac{191,961}{2}x^8 + 307,692x^7 + 493,460x^6 +\\
&\quad + 349,800x^5 + 85,503x^4 + 4092x^3 + 6x^2 - \tfrac{691}{2730}.
\end{aligned}$$

Note that for $x = 0$ the generating function becomes

$$G(t,0) = \frac{t}{e^t - 1} = \sum_{n=0}^{\infty} \tilde{b}_n(0)\,\frac{t^n}{n!}\,, \tag{63}$$

so that $\tilde{b}_n(0) = B_n$, namely the nth classical Bernoulli number.

6. Conclusions

In this survey article, it has been shown that the common belonging of some polynomial sets to the Sheffer class allows to construct, in a uniform way, the differential equations they verify. This follows from the fact that it is possible to construct their shift operators, on the basis of general results due to G. Dattoli and Y. Ben Cheikh.

The equations derived in such a way are, in general, of infinite order, but they reduce to finite-order equations when they are applied to polynomials of the considered set. This means that the order of the equation increases with the degree of the polynomial, in a similar way to what happens for the order of the recurrence they verify.

Both classic and other polynomials—the so-called associated Sheffer polynomials—have been examined. In fact, it has been shown that, for the polynomials of the Sheffer class, the differential equation follows from the basic elements of their generating function, in a constructive way, using a simple and general method linked to the monomiality principle.

References

1. Dattoli, G. Hermite-Bessel and Laguerre-Bessel functions: A by-product of the monomiality principle. In *Advanced Special Functions and Applications, Proceedings of the Melfi School on Advanced Topics in Mathematics and Physics, Melfi, Italy, 9–12 May 1999*; Cocolicchio, D., Dattoli, G., Srivastava, H.M., Eds.; Aracne Editrice: Rome, Italy, 2000; pp. 147–164.
2. Cheikh, Y.B. Some results on quasi-monomiality. *Appl. Math. Comput.* **2003**, *141*, 63–76.
3. Bretti, G.; Natalini, P.; Ricci, P.E. New sets of Euler-type polynomials. *J. Anal. Number Theor.* **2018**, *6*, 51–54. [CrossRef]
4. Natalini, P.; Bretti, G.; Ricci, P.E. Adjoint Hermite and Bernoulli polynomials. *Bull. Allahabad Math. Soc.* **2018**, *33*, 251–264.
5. Natalini, P.; Ricci, P.E. New Bell-Sheffer polynomial sets. *Axioms* **2018**, *7*, 71. [CrossRef]

6. Natalini, P.; Ricci, P.E. Bell-Sheffer exponential polynomials of the second kind. *Georgian Math. J.* **2018** (to appear).
7. Ricci, P.E. Logarithmic-Sheffer polynomials of the second kind. *Tbilisi Math. J.* **2018**, *11*, 95–106. [CrossRef]
8. Ricci, P.E. Mittag–Leffler polynomial differential equation. *Jñānābha* **2018**, *48*, 1–6.
9. Srivastava, H.M.; Ricci, P.E.; Natalini, P. A Family of Complex Appell Polynomial Sets. *Real Acad. Sci. Exact. Fis Nat. Ser. A Math.* **2018**. [CrossRef]
10. Costabile, F.A.; Longo, E. An algebraic approach to Sheffer polynomial sequences. *Integral Transform. Spec. Funct.* **2014**, *25*, 295–311. [CrossRef]
11. Costabile, F.A.; Longo, E. An algebraic exposition of Umbral calculus with applications to general interpolation problem. A Survey. *Publ. Inst. Math. Nouv. Ser.* **2014**, *96*, 67–83. [CrossRef]
12. Youn, H.; Yang, Y. Differential equations and recursive formulas of Sheffer polynomial sequences. *ISRN Discret. Math.* **2011**, *2011*, 476462. [CrossRef]
13. Yang, S.L. Recurrence relation for the Sheffer sequences. *Linear Algebra Appl.* **2012**, *437*, 2986–2996. [CrossRef]
14. Sheffer, I.M. Some properties of polynomials sets of zero type. *Duke Math. J.* **1939**, *5*, 590–622. [CrossRef]
15. Srivastava, H.M.; Manocha, H.L. *A Treatise on Generating Functions*; Halsted Press (Ellis Horwood Limited, Chichester), John Wiley and Sons: New York, NY, USA; Chichester, UK; Brisbane, Australia; Toronto, ON, Canada, 1984.
16. Roman, S.M. *The Umbral Calculus*; Academic Press: New York, NY, USA, 1984.
17. Boas, R.P.; Buck, R.C. *Polynomial Expansions of Analytic Functions*; Springer: Berlin/Gottingen/Heidelberg, Germany; New York, NY, USA, 1958.
18. Steffensen, J.F. The poweroid, an extension of the mathematical notion of power. *Acta Math.* **1941**, *73*, 333–366. [CrossRef]
19. Dattoli, G.; Ricci, P.E.; Srivastava, H.M. (Eds.) Advanced Special Functions and Related Topics in Probability and in Differential Equations. In *Applied Mathematics and Computation, Proceedings of the Melfi School on Advanced Topics in Mathematics and Physics, Melfi, Italy, 24–29 June 2001*; Aracne Editrice: Rome, Italy, 2003; Volume 141, pp. 1–230.
20. Dattoli, G.; Germano, B.; Martinelli, M.R.; Ricci, P.E. Monomiality and partial differential equations. *Math. Comput. Model.* **2009**, *50*, 1332–1337. [CrossRef]
21. Weisstein, E.W. Mittag–Leffler Polynomial, From *MathWorld—A Wolfram Web Resource*. Available online: http://mathworld.wolfram.com/Mittag-LefflerPolynomial.html (accessed on 4 February 2019).
22. Sloane, N.J.A.; Plouffe, S. *The Encyclopedia of Integer Sequences*; Academic Press: San Diego, CA, USA, 1995.

16

Sub-Optimal Control in the Zika Virus Epidemic Model using Differential Evolution

Nonthamon Chaikham [1,2] and Wannika Sawangtong [1,2,*]

1 Department of Mathematics, Faculty of Science, Mahidol University, Bangkok 10400, Thailand; nonthamon.chi@student.mahidol.ac.th
2 Centre of Excellence in Mathematics, Commission on Higher Education, Ministry of Education, 328 Sri Ayuthaya Road, Bangkok 10400, Thailand
* Correspondence: wannika.saw@mahidol.ac.th;

Abstract: A dynamical model of Zika virus (ZIKV) epidemic with direct transmission, sexual transmission, and vertical transmission is developed. A sub-optimal control problem to counter against the disease is proposed including three controls: vector elimination, vector-to-human contact reduction, and sexual contact reduction. Each control variable is discretized into piece-wise constant intervals. The problem is solved by Differential Evolution (DE), which is one of the evolutionary algorithm developed for optimization. Two scenarios, namely four time horizons and eight time horizons, are compared and discussed. The simulations show that models with controls lead to decreasing the number of patients as well as epidemic period length. From the optimal solution, vector elimination is the prioritized strategy for disease control.

Keywords: Zika virus; sexual transmission; vertical transmission; sub-optimal control problem; differential evolution

---

1. Introduction

Despite being discovered in Africa and named after a forest in Uganda in 1947, there are only several reported cases of Zika virus (ZIKV) infection before 2007. The virus itself is a mosquito-borne virus from the genus flavivirus that is transmitted primarily through the bite of Aedes mosquitoes. At first, the ZIKV infection was not considered a threat due to its mild symptoms such as rashes, headache, malaise, and fever [1]. Besides, the mortality rate of ZIKV infection is considered negligible, unlike other Flavivirus infections such as dengue and yellow fever. However, the announcement [2] on the possible association between the ZIKV infection and the increasing number of newborn babies with microcephaly in 2015 raised the alarm and public awareness of this outbreak. There is some correlative evidence reported from Brazil, Colombia, and other Latin America countries [3,4].

Some previous studies suggest that ZIKV could pass from pregnant mothers to their baby through the intrauterine infection, and microcephaly also happened during this process. Beside vector-mediated transmission and intrauterine (or vertical) transmission, ZIKV could also transmit through sexual transmission [5]. Apart from being detected in blood and semen [6], researchers found that the virus persists in semen and urine after it disappeared from the serum [7]. Moreover, it could pass on even before the patient develops the symptoms [8]. To control the outbreak, it can be seen that only relying on reducing the number of mosquito population or preventing the contact between host and vectors might be insufficient. The control strategy should include reducing the sexual transmission and vertical transmission as well.

A dynamical model was a popular tool for describing and analyzing various diseases. It was used to describe the mechanism functions and causes abnormality in some non-infectious diseases such as diabetes [9,10]. In prior publications [11–13], models with systems of ordinary differential equations were utilized for describing the behavior of infectious virus such as Zika. Optimal control is one of the popular methods for finding the solution that will minimize the objective function value. According to some publications [14,15], optimal control can be used to find the best strategy associated with the implemented cost and the provided situations. The optimal solutions for some flavivirus control such as chikungunya [16] and dengue [17] have been studied.

In this paper, we develop the ZIKV virus infection model using some previous works [12,13,18]. The model consists of three populations: adult humans, newborn babies human, and mosquitoes. Three controls are selected from the suggestions of the World Health Organization(WHO) [19] and health authorities [20]: using larvicide and adulticide to reduce the mosquito population, using insect repellents to prevent mosquito biting, and abstaining from sexual activity or using protection to prevent pregnancy.

Regarding optimal control and sub-optimal control problem, there are direct and indirect methods for solving the problems. The indirect approaches, such as Pontryagin's Maximum Principle [21] and shooting method, are popular to find optimal solutions of some flavivirus control problems, for example, chikungunya [16] and dengue [17]. Previously, the authors also studied the optimal control problem of the Zika virus infection [18] by the indirect approach. However, the implementations for health authorities by these studies [16,17] are difficult to practically apply since the policies have to be varied continuously. Considering this drawback, we propose a sub-optimal control problem and find the solution that is both close to optimal and practical for real situation.

A sub-optimal control problem can be proposed through Model Predictive Control (MPC). MPC is the mathematical tool for controlling the optimization process in various fields such as industrial and medical technology [22]. It can be used from the simple to complex dynamic process, and the result control law is easy to implement. In our work, the control framework consists of a nonlinear system without constraints on spending cost or the final result of the system variables. This framework is related to the nonlinear unconstrained MPC schemes, which are those without terminal constraints and cost [23]. Furthermore, the study of Herty [24] estimated the suboptimality condition and suggested the guideline for the system containing a quadratic cost function.

In sub-optimal control problem, the previously continuous time horizon is discretized into N time horizons. Finding the solution of the controls in each time horizon helps archive the strategy which only applies the control at the constant level in each period. Caetano [25] and Yan [26] used the direct approach in finding the solutions to the sub-optimal disease control problem, and Yan [26] also used the Genetic Algorithm (GA).

Differential Evolution (DE) algorithm is an improved GA by Storn and Price [27]. DE algorithm is simple and fast, requires only a few control parameters and has high convergence attribute [28]. Thus, we find the solution of the sub-optimal control problem using the DE algorithm. The dynamical model is developed in Section 2 and the Differential Evolution is introduced in Section 3. The sub-optimal control problem is proposed through MPC and the numerical solutions are solved in Section 4. Lastly, the discussion and conclusion are presented in Section 5.

2. Dynamical Control Model for ZIKV Infection

To describe the behavior of ZIKV infection, the dynamical compartment model with control is developed from the models of Gao et al. [12] and Agusto et al. [13]. The developed model consists of two human populations, adults and babies, and one mosquito (vector) population. The total newborn baby population consists of six classes: susceptible $S_b(t)$, exposed $E_b(t)$, asymptomatically infected $A_b(t)$, symptomatically infected without microcephaly, referred to as infected $I_b(t)$, infected with microcephaly $I_{bm}(t)$, and recovered $R_b(t)$. The total adult population consists of six classes: susceptible $S_w(t)$, exposed $E_w(t)$, asymptomatically infected $A_w(t)$, symptomatically infected, referred to as infected $I_w(t)$, infected with microcephaly $I_{wm}(t)$, and recovered $R_w(t)$. The total vector population consists of three classes: susceptible $S_v(t)$, exposed $E_v(t)$, and infected $I_v(t)$.

Hence, the total newborn baby population is $N_b(t) = S_b(t) + E_b(t) + A_b(t) + I_b(t) + I_{bm}(t) + R_b(t)$, the total adult population is $N_w(t) = S_w(t) + E_w(t) + A_w(t) + I_w(t) + I_{wm}(t) + R_w(t)$, the total vector population is $N_v(t) = S_v(t) + E_v(t) + I_v(t)$, and the total human population is $N_h(t) = N_b(t) + N_w(t)$. The flow diagram of the compartment model is shown in Figure 1. Therefore, the ZIKV disease model with sexual transmission and vertical transmission is written by a system of differential equations under defined assumptions.

$$
\begin{aligned}
S_b'(t) =& \pi_b\left[(1-q_E)E_w(t) + (1-q_I)I_w(t) + (1-q_R)R_w(t)\right] \\
& - \lambda_b(I_v, N_b)S_b(t)(1-u_1(t)) - (\alpha+\mu_b)S_b(t) \\
E_b'(t) =& \pi_b q_E E_w(t) + p\lambda_b(I_v, N_b)S_b(t)(1-u_1(t)) \\
& - (\alpha+\sigma_b+\mu_b)E_b(t) \\
A_b'(t) =& (1-p)\lambda_b(I_v, N_b)S_b(t)(1-u_1(t)) - (\alpha+\gamma_b+\mu_b)A_b(t) \\
I_b'(t) =& \pi_b q_I I_w(t) + \sigma_b E_b(t) - (\alpha+\gamma_b+\mu_b)I_b(t) \\
I_{bm}'(t) =& (1-r)\pi_b q_R R_w(t) - (\alpha+\mu_b) \\
R_b'(t) =& r\pi_b q_R R_w(t) + \gamma_b(A_b(t)+I_b(t)) - (\alpha+\mu_b)R_b(t) \\
S_w'(t) =& \alpha S_b(t) - \lambda_w(I_v, N_w)S_w(t)(1-u_1(t)) \\
& - \lambda_s(E_w, I_w, N_w)S_w(t)(1-u_2(t)) - \mu_w S_w(t) \\
E_w'(t) =& pS_w(t)[\lambda_w(I_v, N_w)(1-u_1(t)) - \\
& \lambda_s(E_w, I_w, N_w)(1-u_2(t))] - (\sigma_w+\mu_w)E_w(t) \\
A_w'(t) =& (1-p)S_w(t)[\lambda_w(I_v, N_w)(1-u_1(t)) - \\
& \lambda_s(E_w, I_w, N_w)(1-u_2(t))] - (\gamma_w+\mu_w)A_w(t) \\
I_w'(t) =& \sigma_w E_w(t) - (\gamma_w+\mu_w)I_w(t) \\
I_{wm}'(t) =& \alpha I_{bm}(t) - \mu I_{wm}(t) \\
R_w'(t) =& \alpha R_b(t) + \gamma_w(A_w(t)+I_w(t)) - \mu_w R_w(t) \\
S_v'(t) =& \pi_v(1-u_3(t)) - \lambda_v(E_b, I_b, E_w, I_w, N_b, N_w)S_v(t)(1-u_1(t)) \\
& - (\mu_v + r_0 u_3(t))S_v(t) \\
E_v'(t) =& \lambda_v(E_b, I_b, E_w, I_w, N_b, N_w)S_v(t)(1-u_1(t)) - \\
& (\sigma_v+\mu_v+r_0 u_3(t))E_v(t) \\
I_v'(t) =& \sigma_v E_v(t) - (\mu_v + r_0 u_3(t))I_v(t)
\end{aligned}
\tag{1}
$$

where

$$
\begin{aligned}
&\lambda_b(I_v, N_b) = \frac{\eta\beta_b b I_v}{N_b}, \qquad \lambda_w(I_v, N_w) = \frac{\beta_w b I_v}{N_w}, \\
&\lambda_s(E_w, I_w, N_w) = \beta_s\left[\frac{I_w + \rho_s E_w}{N_w}\right], \\
&\lambda_v(E_b, I_b, E_w, I_w, N_b, N_w) = \beta_v b\left[\frac{I_w + \rho_w E_w + \eta(I_b + \rho_b Eb)}{N_w + N_b}\right]
\end{aligned}
$$

are the force of infection rates.

In the system, a susceptible adult can be infected either through the bite of an infected vector or sexual contact with an exposed or symptomatically infected adult. The infected class is more contagious due to the higher load of viremia [29], hence the infection modification parameters ρ_w, ρ_s are introduced for vector transmission and sexual transmission of an adult.

The newborn baby could be infected directly by mosquitoes or intrauterine via pregnant mother [30,31]. We assumed that the exposure rate of babies is different from adults, represented by the modification parameter $\eta > 0$. The infection modification ρ_b is introduced as well. In the vertical transmission in babies, it occurred in some cases that newborn baby is infected during pregnancy [31,32].

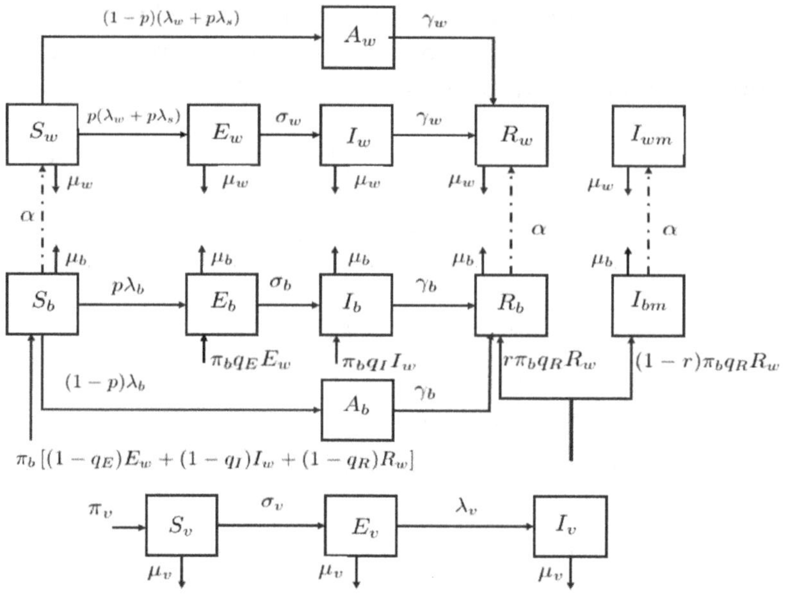

Figure 1. Flow diagram of the Zika virus model.

We assumed that babies are born with infection due to vertical transmission by the parameter $0 \leq q_E, q_I, q_R \leq 1$. Let the newborn birth rate be π_b, the fraction of healthy newborn babies are $\pi_b((1-q_E)E_w + (1-q_I)I_w + (1-q_R)R_w)$, and the infected babies are the remaining $\pi_b q_E E_w$, $\pi_b q_I Iw$, and $\pi_b q_R R_w$ fractions. A recovered mother might give birth to the baby with microcephaly [30], and the baby born with microcephaly is assumed by the fraction $(1-r)\pi_b q_R R_w$ in this paper.

Individuals with microcephaly experience a delay in development [33] and severe neurological disorders, so they are likely to have a short lifespan. Hence, we assumed that the adults with microcephaly do not reproduce or are involved in sexual transmission. The asymptomatically infected babies and adults are assumed to be noncontagious for both human and vector. Furthermore, the mortality rate from the disease is considered negligible. All parameter descriptions are summarized in Table 1.

Based on WHO [19] and Centers for Disease Control and Prevention (CDC) [20] suggestions, we considered vector, vertical, and sexual transmission the three control parameters selected for the model. The control variable $u_1(t)$ is the use of mosquito repellents to prevent or reduce the biting from mosquitoes. The control variable $u_2(t)$ is the procedure such as using protection for preventing pregnancy or unsafe sexual activity. The control variable $u_3(t)$ is the use of insecticide to control or eliminate the mosquito population. By applying control strategy, forces of infection in the adult and baby populations decrease by the fraction of $(1-u_1(t))$ and $(1-u_2(t))$. Correspondingly, the force of infection in vector population decreases by a fraction of $(1-u_1(t))$. Lastly, the recruitment rate of mosquito decreases by a fraction of $(1-u_3(t))$. It was also assumed that the mosquito death rate increases by $r_0 u_3(t)$, where $r_0 > 0$. Remark: If the value of each control parameters u_1, u_2, and u_3 is equal to zero for all given period, the model in the system in Equation (1) would be the Zika virus model without controls.

The controls u_1, u_2, and u_3 are determined to minimize the given objective function:

$$J(u_1,u_2,u_3) = \int_0^{t_f} \Big(A_1E_b + A_2I_b + A_3E_w + A_4I_w + A_5N_v + \frac{1}{2}(B_1u_1^2 + B_2u_2^2 + B_3u_3^2)\Big)dt, \tag{2}$$

subject to the state of system in Equation (1).

Table 1. Definition of variables and parameters.

Variable	Definition
$S_b(t), S_w(t)$	Susceptible newborn babies and adults
$E_b(t), E_w(t)$	Exposed newborn babies and adults
$A_b(t), A_w(t)$	Asymptomatically infected newborn babies and adults
$I_b(t), I_w(t)$	Symptomatically infected newborn babies without microcephaly and symptomatically infected adults
$I_{bm}(t), I_{wm}(t)$	Newborn babies and adults with microcephaly
$R_b(t), R_w(t)$	Recovered newborn babies and adults
$S_v(t)$	Susceptible vectors
$E_v(t)$	Exposed vectors
$I_v(t)$	Infected vectors
Parameter	**Definition**
π_b	Birth rate of newborn babies
α	Maturity rate of babies
p	Fraction of symptomatic infection
$1-p$	Remaining fraction of symptomatic infection
q_E, q_I, q_R	Fraction of newborn babies who are infected from pregnant adult of each class
$1-r$	Fraction of newborn babies who are infected with microcephaly
β_b, β_w	Transmission probability per contact from infected vectors to susceptible newborn babies and adults
β_v	Transmission probability per contact from infected humans to susceptible vectors
β_s	Transmission probability per sexual contact from infected humans to susceptible humans
η	Exposure modification parameter in babies
ρ_b, ρ_w	Infectivity modification parameters in exposed babies and adults
ρ_s	Sexual infectivity modification parameters in exposed adults
σ_b, σ_w	Progression rate of exposed newborn babies and adults
γ_b, γ_w	Recovery rate of newborn babies and adults
μ_b, μ_w	Natural death rate of newborn babies and adults
π_v	Recruitment rate of mosquitoes
b	Biting rate of mosquitoes
σ_v	Progression rate of exposed mosquitoes
μ_v	Mortality rate of mosquitoes

In Section 4, we find a set of controls that minimize the number of exposed humans, symptomatically infected humans, all mosquitoes and the costs related to the controlling implementation. Firstly, we introduce constants A_1, A_2, A_3, A_4, and A_5 as the weighted constants related to the exposed newborn babies, symptomatically infected newborn babies, exposed adults, symptomatically infected adults, and all mosquitoes, respectively. The constants B_1, B_2, and B_3 are the weighted constants of the control variables u_1, u_2, and u_3, respectively. The terms $\frac{1}{2}B_1u_1^2$, $\frac{1}{2}B_2u_2^2$, and $\frac{1}{2}B_3u_3^2$ are the implementation costs of the three controls. The cost included in the first control might be the prices of using insect repellent, mosquito net, and herbal spray for instance. The cost included in the second control might be the prices of providing protections or warning leaflets about the safe sexual activity. The cost included in the last control could come from the expenses of using mosquito pesticides and the process implementation.

3. Differential Evolution

The differential evolution algorithm is a population-based optimization algorithm, introduced by Storn and Price [27,28]. It is also one of the Evolutionary algorithms developed to find the parameter values and optimize real parameters or functions. DE is suitable for various practical problems which are nonlinear, non-differentiable, and multi-dimensional. The process consists of four parts, namely initialization, mutation, crossover, and selection.

Suppose optimizing a problem consisting of D predicted parameters with N_p number of population. Then, a $D-$dimensional vector is solved by the following main steps until the best solution vector reaches the termination criteria.

Step 1. Initialization

The DE algorithm randomly selected a population of the parameter vectors $\{x_{1,i}^G, x_{2,i}^G, \ldots, x_{D,i}^G\}$, $i = 1, 2, \ldots, N_p\}$ where G is the generation number. Define the initial population as

$$x_{i,j}^G = x_j^L + \phi_i(x_j^U - x_j^L), \qquad j = 1, 2, \ldots, D$$

that is each parameter $x_{i,j}^G \in [x_j^L, x_j^U]$, where x_j^L is the lower bound and x_j^U is the upper bound of the $j-$th parameter and ϕ_i is a uniformly distributed random number between 0 and 1. Then, each of the N_p parameter vectors will undergo mutation, crossover, and selection.

Step 2. Mutation

This process helps expand the search space. For each target vector x_i^G, randomly select three vectors x_{r1}^G, x_{r2}^G, and x_{r3}^G such that each index i, $r1$, $r2$, and $r3$ is different. Then, a mutant vector v_i^G is generated by

$$v_i^G = x_{r1}^G + F(x_{r2}^G - x_{r3}^G)$$

where $F \in [0, 2]$ is the mutation vector and v_i^G is called the donor vector.

Step 3. Crossover

Crossover or recombination process involves successful solutions from the previous generation. First, a trial vectors u_i^{G+1} is produced from v_i^G and x_i^G by

$$u_{i,j}^G = \begin{cases} v_{i,j}^G & \text{if } rand_j \leq CR, \text{ or } j = I_{rand} \\ x_{i,j}^G & \text{if } rand_j > CR \text{ and } j \neq I_{rand} \end{cases}$$

where $rand_j \sim U[0, 1]$ for a comparison to the crossover rate $CR \in [0, 1]$, and I_{rand} is a random index in $\{1, \ldots, D\}$ that u_i^G gets at least one component.

Step 4. Selection

The next generation vectors are selected by comparing the target vector x_i^G and trial vector u_i^G and selecting the one with the lowest objective function value. The objective function is given by comparing the value of E_b, I_b, E_h, I_h, and N_v in the system in Equation (1) by applying vector u_i^G and x_i^G. Thus, the next generation vector x_i^{G+1} is replaced by u_i^G or x_i^G under the following condition,

$$x_i^{G+1} = \begin{cases} u_i^G & \text{if } J(u_i^G) < J(x_i^G), \\ x_i^G & \text{otherwise,} \end{cases}$$

where J is the objective function (see Section 4).

Step 5. Repeating

After obtaining a new target vector, the mutation, crossover, and selection steps are repeated until the termination criteria are met.

4. Numerical Simulations for Sub-Optimal Control Problem

4.1. Sub-Optimal Control Problem

Given the objective function as defined in Equation (2), we assume there exists a solution (u_1^*, u_2^*, u_3^*) of the optimal control problem characterized by the Pontryagins Maximum Principle [21].

Since the implemented result of the optimal control problem is complicated for the real situation, a sub-optimal control problem is proposed. The idea of sub-optimal control problem is based on the works of Rodrigues et. al. [16] and Moulay et al. [17], and the Model Predictive Control [22]. Regarding MPC, its strategy is based on the model of the system, a cost function which penalizes the undesirable behaviors, and constraints which represent the system limits. Furthermore, the stability and suboptimality of MPC can be estimated from the control horizon and the cost function [23,24]. For our nonlinear unconstrained model with a finite horizon, the continuous time horizon is discretized into N time horizon as

$$t_0 = 0 < t_1 < t_2 < \ldots < t_{N-1} = t_f.$$

Then, each of the controls $u_1(t)$, $u_2(t)$, and $u_3(t)$, in each time horizon, is approximated by a piece-wise constant control such that

$$\begin{aligned} u_1(t) &= u_{1i}, \text{ for } t_{i-1} \le t \le t_i,\ i = 1, \ldots, N \\ u_2(t) &= u_{2i}, \text{ for } t_{i-1} \le t \le t_i,\ i = 1, \ldots, N \\ u_3(t) &= u_{3i}, \text{ for } t_{i-1} \le t \le t_i,\ i = 1, \ldots, N \end{aligned} \tag{3}$$

where u_{1i}, u_{2i}, and u_{3i} are constants and are referred to as the control parameters. Thus, for our model with now N control parameters for each u_1, u_2, and u_3, we have total $3N$ parameters to be minimized. Let

$$z = [u_{11}, u_{21}, u_{31}, \ldots, u_{1N}, u_{2N}, u_{3N}] \in R^{3N}.$$

Then, the sub-optimal control problem is to

$$\min_{z} J(z) \tag{4}$$

subject to the system in Equation (1) and J is as defined in Equation (2).

4.2. Numerical Simulations

The simulations in this work are constructed using the parameter values in Table 2. Since the Zika virus is still under investigation and discovery, some parameter values are obtained from the previous study of other mosquito-borne viruses. The initial conditions are also selected for the theoretical sense and illustration purpose. The initial conditions for the system in Equation (1) are given by $S_b(0) = 10,000$, $E_b(0) = 100$, $A_b(0) = 100$, $I_b(0) = 40$, $I_{bm}(0) = 10$, $R_b(0) = 50$, $S_w(0) = 200{,}000$, $E_w(0) = 1000$, $A_w(0) = 1000$, $I_w(0) = 100$, $I_{wm}(0) = 50$, $R_w(0) = 1000$, $S_v(0) = 10{,}000$, $E_v(0) = 1000$, and $I_v(0) = 100$. The weight constants are given as $A_1 = 0.025$, $A_2 = 0.025$, $A_3 = 0.025$, $A_4 = 0.025$, $A_5 = 0.025$, $E_1 = 20$, $E_2 = 20$, and $E_3 = 20$. The total time span is 120 days, which is approximately four months.

From [24], the performance bound, which depends on the time horizon N and the cost function, could be computed to estimate the distance between MPC cost and the optimal cost. It also suggests that the time horizon N should not be too small. Hence, in our simulation with the total time of 120 days, we consider two scenarios with the different time horizon N as follows: (i) with $N = 4$; and (ii) with $N = 8$. All controls must be non-negative and let the maximum value be 0.9 since it is almost impossible to fully implement each control strategy. The numerical simulation from each scenario is generated with the DE algorithm with the maximum iteration of 150, $F = 0.850$, and $CR = 1.000$. Number of population N_p is 120 and 240 for Scenarios $N = 4$ and $N = 8$, respectively.

Table 2. Parameter values.

Parameter	Value	References
π_b	$\frac{1}{15\times365}$	[34]
α	$\frac{1}{16\times365}$	[13]
p	0.5	[13]
q_E, q_I, q_R	0.5	[13]
r	0.5	[13]
β_b, β_w	0.33	[13,35]
β_v	0.5	[12]
β_s	0.05	[12]
η	0.5	Assumed
ρ_b, ρ_w	0.5	[13]
ρ_s	0.6	[12]
σ_b, σ_w	$\frac{1}{7.5}$	[36]
γ_b, γ_w	$\frac{1}{8.5}$	[36]
μ_b	$\frac{1}{18.60\times365}$	[3]
μ_w	$\frac{1}{70\times365}$	[35]
π_v	500	[36]
b	0.5	[35,37,38]
σ_v	$\frac{1}{10}$	[37]
μ_v	$\frac{1}{21}$	[38]
r_0	0.1	[18]

(i) Scenario $N = 4$

In this scenario, the period is 30 days or one month for each time horizon. Figure 2 represents the simulation of three controls over 120 days. From the graph, the levels of u_1 and u_2 are at the highest around 0.36 and 0.37 in the first month, respectively. After that, the levels of both controls u_1 and u_2 discretely drop in the second month and become approximately zero in the third month. Differently, the level of control u_3 stays at the maximum value 0.9 during the entire time.

The objective function value is 2.293810×10^5, where the control values are:

$$\begin{aligned} u_1 &= [0.379390 \quad 0.01558 \quad 0.00038 \quad 0.00007] \\ u_2 &= [0.365622 \quad 0.08256 \quad 0.01328 \quad 0.00136] \\ u_3 &= [0.900000 \quad 0.90000 \quad 0.90000 \quad 0.90000]. \end{aligned}$$

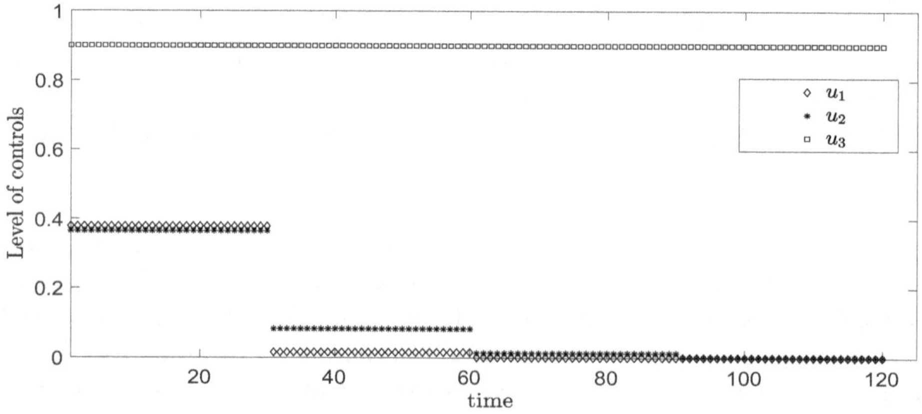

Figure 2. Level of controls from Scenario (i) $N = 4$.

(ii) Scenario $N = 8$

In this scenario, the period is 15 days or approximately two weeks for each time horizon. Figure 3 represents the simulation of three controls over 120 days. From the graph, u_1 is at the highest level around 0.6 in the first two weeks and discretely drops to approximately zero within one month. For u_2, it reaches around 0.45 for the first two weeks, then drops to nearly zero after the second month. Differently, the level of control u_3 stays at the maximum value 0.9 during the entire time.

The objective function value is 2.293640×10^5, where the control values are:

$$u_1 = [0.59757 \quad 0.17349 \quad 0.02167 \quad 0.01069 \quad 0.00873 \quad 0.00128 \quad 0.00201 \quad 0.00539]$$

$$u_2 = [0.45277 \quad 0.25743 \quad 0.12570 \quad 0.04461 \quad 0.01469 \quad 0.00008 \quad 0.01466 \quad 0.00918]$$

$$u_3 = [0.89999 \quad 0.89998 \quad 0.89999 \quad 0.89999 \quad 0.89997 \quad 0.89999 \quad 0.89999 \quad 0.89998].$$

The simulations in Figures 4–9 represent the number of exposed babies, infected babies, exposed adults, infected adults, exposed vectors and infected vectors, respectively. In each graph, the dotted line represents the model without control, the solid line represents the control model with Scenario $N = 4$, and the dashed line represents the control model with Scenario $N = 8$. In Figures 4–9, the numbers of exposed and infected individuals from the model with controls are less than those of the model without control. This implies that control strategy ceases the serve of the disease. The models from both scenarios yield very close results where $N = 8$ is slightly better. Notice in Figures 4–7 that the disease spreads heavily during the first month. Later, those patients decrease and become stable after two months, corresponding to the drop in control u_1 and u_2 after two months. Thus, both u_1 and u_2 contribute to the control of the disease. The control u_3 contributes the most and becomes the top priority in the strategy.

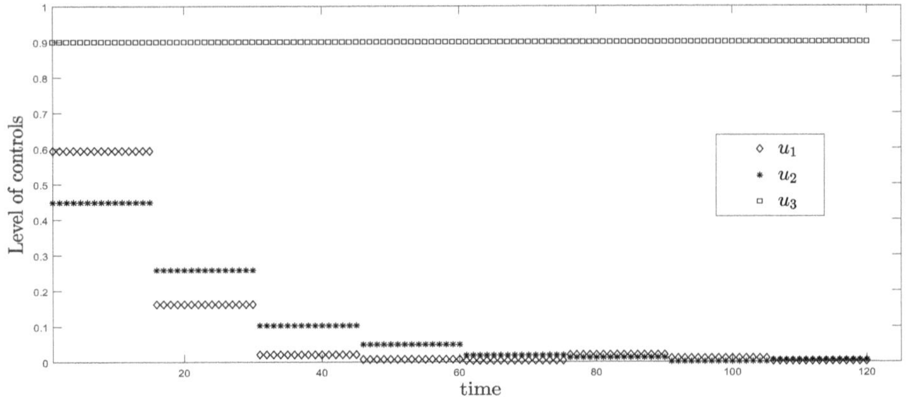

Figure 3. Level of controls from Scenario (ii) $N = 8$.

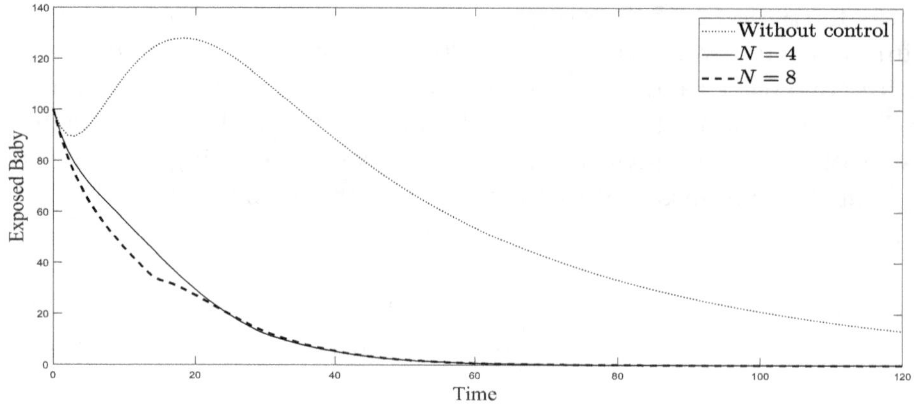

Figure 4. Simulation results of exposed newborn babies from: (i) the model without controls; (ii) the control model with $N = 4$; and (iii) the control model with $N = 8$.

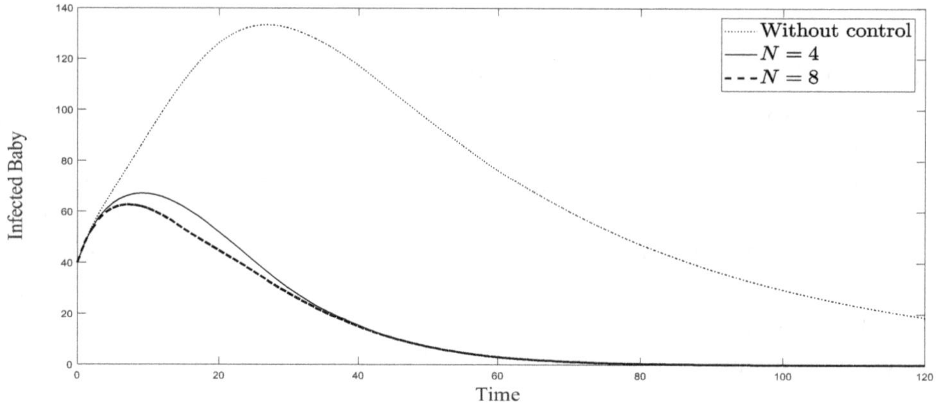

Figure 5. Simulation results of symptomatically infected newborn babies from: (i) the model without controls; (ii) the control model with $N = 4$; and (iii) the control model with $N = 8$.

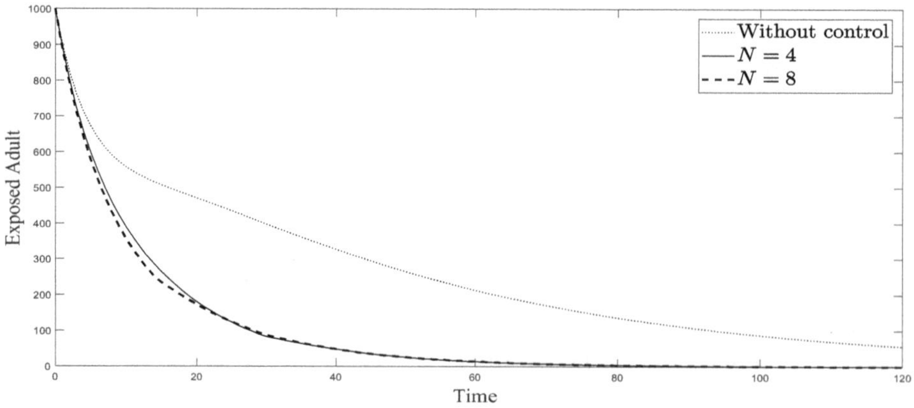

Figure 6. Simulation results of exposed adults from: (i) the model without controls; (ii) the control model with $N = 4$; and (iii) the control model with $N = 8$.

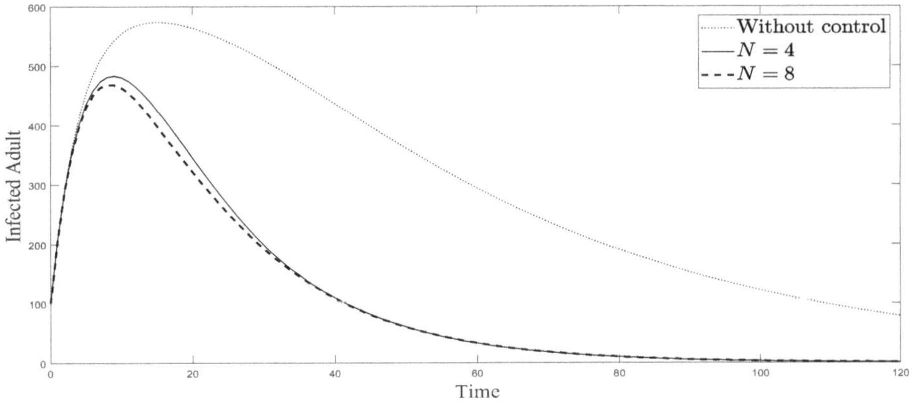

Figure 7. Simulation results of symptomatically infected adults from: (i) the model without controls; (ii) the control model with $N = 4$; and (iii) the control model with $N = 8$.

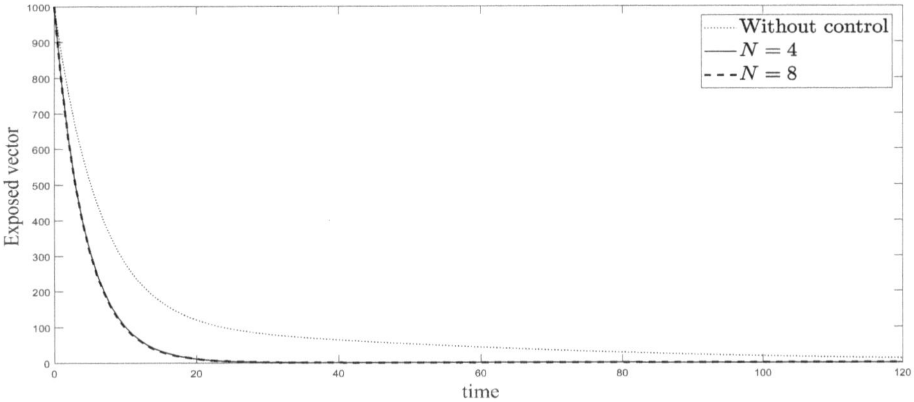

Figure 8. Simulation results of exposed vectors: (i) the model without controls; (ii) the control model with $N = 4$; and (iii) the control model with $N = 8$.

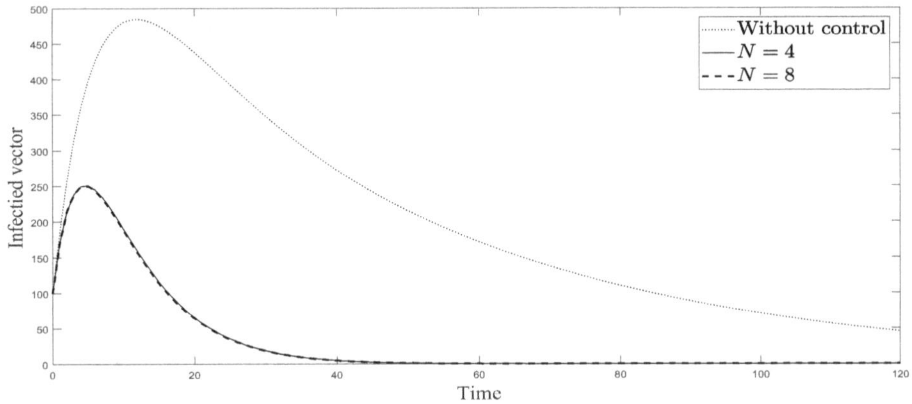

Figure 9. Simulation results of infected vectors from: (i) the model without controls; (ii) the control model with $N = 4$; and (iii) the control model with $N = 8$.

5. Conclusions

In this work, we developed the dynamical model of ZIKV disease with sexual transmission and vertical transmission. We proposed sub-optimal control problem with three control parameters: vector elimination, human–vector reduction, and human–human contact reduction. To find the control solution which is easy to implement, we partitioned the controls into discrete intervals. By using Differential Evolution, we solved and presented the numerical scenarios with four time horizons and eight time horizons. The simulations of both controls efficiently reduce the number of exposed and infected patients where the model with eight intervals yields a slightly better result. Controls u_1, using repellents to prevent mosquito biting, and u_2, avoiding pregnancy and unsafe sexual activity, should be applied mostly during the peak of the disease. After the disease dies out or becomes stable, the level of both controls will decrease. Mosquito elimination should be the first focus in controlling the disease, corresponding to the primary procedure announced by general health authorities. However, the cost associated with each control in this simulation is set to be equal. The consideration of adjusting the relative cost according to alternative scenarios might provide different results. Additionally, if more information is available in the future, the simulations should be conducted again using more precise data.

Author Contributions: Conceptualization, W.S.; Investigation, N.C. and W.S.; Methodology, W.S.; Visualization, N.C.; Writing—original draft, N.C.; and Writing—review and editing, W.S.

Acknowledgments: This research was supported by Centre of Excellence in Mathematics, PERDO, CHE, Thailand and Golden Jubilee Scholarship from Faculty of science, Mahidol university, Thailand.

Abbreviations

The following abbreviations are used in this manuscript:

WHO	World Health Organization
GA	Genetic Algorithm
DE	Differential Evolution

References

1. Dasti, J.I. Zika virus infections: An overview of current scenario. *Asian Pac. J. Trop. Med.* **2016**, *7* , 621–625. [CrossRef] [PubMed]
2. European Centre for Disease Prevention and Control. Zika Virus Epidemic in the Americas: Potential Association with Microcephaly and Guillain-Barre Syndrome. Available online: https://ecdc.europa.eu/sites/portal/files/media/en/publications/Publications/zika-virus-americas-association-with-microcephaly-rapid-risk-assessment.pdf (accessed on 12 February 2017) .
3. World Health Organization. Microcephaly-Brazil: Disease Outbreak News. Available online: http://www.who.int/csr/don/archive/disease/microcephaly/en/ (accessed on 12 February 2017).
4. World Health Organization. Zika Situation Report. Available online: http://www.who.int/emergencies/zika-virus/situation-report/12-february-2016/en/ (accessed on 12 February 2017).
5. Coelho, F.C.; Durovni, B.; Saraceni, V.; Lemos, C.; Codeco, C.T.; Camargo, S.; de Carvalho, L.M.; Bastos, L.; Arduini, D.; Villela, D.A.; et al. Higher incidence of Zika in adult women than adult men in Rio de Janeiro suggests a significant contribution of sexual transmission from men to women. *Int. J. Infect. Dis.* **2016**, *51*, 128–132. [CrossRef] [PubMed]
6. Atkinson, B.; Hearn, P.; Afrough, B.; Lumley, S.; Carter, D.; Aarons, E.J.; Simpson, A.J.; Brooks, T.J.; Hewson, R. Detection of Zika virus in semen. *Emerg. Infect. Dis.* **2016**, *22*, 940. [CrossRef] [PubMed]
7. Musso, D.; Roche, C.; Robin, E.; Nhan, T.; Teissier, A.; Cao-Lormeau, V.-M. Potential sexual transmission of Zika virus. *Emerg. Infect. Dis.* **2015**, *21*, 359–361. [CrossRef] [PubMed]
8. Foy, B.D.; Kobylinski, K.C.; Foy, J.L.; Blitvich, B.J.; da Rosa, A.T.; Haddow, A.D.; Lanciotti, R.S.; Tesh, R.B. Probable non-vector-borne transmission of Zika virus. *Emerg. Infect. Dis.* **2011**, *17*, 880–882. [CrossRef] [PubMed]
9. Chalishajar, D. Mathematical Analysis of Insulin-glucose feedback system of Diabetes. *J. Eng. Appl. Sci.* **2014**, *5*, 36–58.
10. Chalishajar, D.; Geary, D.; Cox, G. Review Study of Detection of Diabetes Models through Delay Differential Equations. *Appl. Math.* **2016**, *7*, 1087–1102. [CrossRef]

11. Kucharski, A.J.; Funk, S.; Eggo, R.M.; Mallet, H.-P.; Edmunds, W.J.; Nilles, E.J. Transmission dynamics of Zika virus in island populations: A modelling analysis of the 2013-14 French Polynesia outbreak. *PLoS Negl. Trop. Dis.* **2016**, *10*, 0004726. [CrossRef] [PubMed]
12. Gao, D.; Lou, Y.; He, D.; Porco, T.C.; Kuang, Y.; Chowell, G.; Ruan, S. Prevention and control of Zika as a mosquito-borne and sexually transmitted disease: a mathematical modeling analysis. *Sci. Rep.* **2016**, *6*, 28070. [CrossRef] [PubMed]
13. Agusto, F.B.; Bewick, S.; Fagan, W.F. Mathematical model of Zika virus with vertical transmission. *Infect. Dis. Model.* **2017**, *2*, 244–267. [CrossRef] [PubMed]
14. Lashari, A.A.; Zaman, G. Optimal control of a vector borne disease with horizontal transmission. *Nonlinear Anal.* **2012**, *13*, 203–212. [CrossRef]
15. Yan, X.; Zou, Y.; Li, J. Optimal quarantine and isolation strategies in epidemics control. *World J. Model. Simul.* **2007**, *3*, 202–211.
16. Rodrigues, H.S.; Monteiro, M.T.T.; Torres, D.F. Vaccination models and optimal control strategies to dengue. *Math. Biosci.* **2014**, *247*, 1–12. [CrossRef] [PubMed]
17. Moulay, D.; Aziz-Alaoui, M.; Kwon, H.-D. Optimal control of chikungunya disease: Larvae reduction, treatment and prevention. *Math. Biosci. Eng.* **2012**, *9*, 369–392. [CrossRef] [PubMed]
18. Chaikham, N.; Sawangtong, W. Optimal control of Zika virus infection by vector elimination, vector-to-human and human-to-human contact reduction. *Adv. Differ. Equ.* **2017**, *2017*, 177. [CrossRef]
19. World Health Organization. Mosquito Control: Can It Stop Zika at Source? Available online: http://www.who.int/emergencies/zika-virus/articles/mosquito-control/en/ (accessed on 12 Frebruary 2017).
20. Centers for Disease Control and Prevention (CDC), Update: Interim Guidelines for Prevention of Sexual Transmission of Zika Virus-United States. 2016. Available online: https://emergency.cdc.gov/han/han00388.asp (accessed on 5 July 2017) .
21. Lenhart, S.; Workman, J.T. Basic Optimal Control Problems. In *Optimal Control Applied to Biological Models*; CRC Press: Boca Raton, FL, USA, 2007; pp. 1–18.
22. Camacho, E.F.; Alba, C.B. Introduction to Model Predictive Control. In *Model Predictive Control*; Springer Science & Business Media: Berlin/Heidelberg, Germany, 2007; pp. 1–30.
23. Grüne, L. Analysis and design of unconstrained nonlinear MPC schemes for finite and infinite dimensional systems. *SIAM J. Control Optim.* **2009**,*48*, 1206–1228. [CrossRef]
24. Herty, M.; Zanella, M. Performance bounds for the mean-field limit of constrained dynamics. *arXiv* **2015**, arXiv:1511.08364.
25. Caetano, M.A.L.; Yoneyama, T. Optimal and sub-optimal control in Dengue epidemics. *Optim. Control Appl. Methods* **2011**, *22*, 63–73. [CrossRef]
26. Yan, X.; Zou, Y. Optimal and sub-optimal quarantine and isolation control in SARS epidemics. *Math. Comput. Model.* **2008**, *47*, 235–245. [CrossRef]
27. Price, K.; Storn, R.; Lampinen, J. The Differential Evolution Algorithm. In *Differential Evolution : A Practical Approach to Global Optimization*; Springer: New York, NY, USA, 2005; pp. 37–134, ISBN 978-3-540-31306-9.
28. Storn, R.; Price, K. Differential evolution—A simple and efficient heuristic for global optimization over continuous spaces. *J. Glob. Optim.* **1997**, *11*, 341–359. [CrossRef]
29. Hills, S.L. Transmission of Zika virus through sexual contact with travelers to areas of ongoing transmission-continental United States, 2016. *Morb. Mortal. Wkly. Rep.* **2016**, *65*, 215–216. [CrossRef] [PubMed]
30. Moore, C.A.; Staples, J.E.; Dobyns, W.B.; Pessoa, A.; Ventura, C.V.; Fonseca, E.B.D.; Ribeiro, E.M.; Ventura, L.O.; Neto, N.N.; Arena, J.E.; et al. Characterizing the Pattern of Anomalies in Congenital Zika Syndrome for Pediatric Clinicians. *JAMA Pediatr.* **2017**, *171*, 288–295. [CrossRef] [PubMed]
31. Schuler-Faccini, L.; Ribeiro, E.M.; Feitosa, I.M.; Horovitz, D.D.; Cavalcanti, D.P.; Pessoa, A.; Doriqui, M.J.; Neri, J.I.; Neto, J.M.; Wanderley, H.Y.; et al. Possible Association Between Zika Virus Infection and Microcephaly—Brazil, 2015. *Morb. Mortal. Wkly. Rep.* **2016**, *65*, 59–62. [CrossRef] [PubMed]
32. Besnard, M.; Lastere, S.; Teissier, A.; Cao-Lormeau, V.M.; Musso, D. Evidence of perinatal transmission of Zika virus, French Polynesia, December 2013 and February 2014. *Euro Surveill.* **2014**, *19*, 20751. [CrossRef] [PubMed]

33. Carter, M.T.; Mirzaa, G.; McDonell, L.M.; Boycott, K.M. Microcephaly-Capillary Malformation Syndrome. 2013 December 12. *GeneReviews*®, Seattle (WA): University of Washington, Seattle, 1993–2018. Available online: https://www.ncbi.nlm.nih.gov/books/NBK174452/ (accessed on 5 July 2017).
34. Wikipedia. List of Sovereign States and Dependent Territories by Birth Rate. Available online: https://en.wikipedia.org/wiki/List_of_sovereign_states_and_dependencies_by_total_fertility_rate (accessed on 1 February 2017).
35. Manore, C.; Hickmann, J.; Xu, S.; Wearing, H.; Hyman, J. Comparing dengue and chikungunya emergence and endemic transmission in a. aegypti and a. albopictus. *J. Theor. Biol.* **2014**, *356*, 174–191. [CrossRef] [PubMed]
36. Bewick, S.; Fagan, W.; Calabrese, J.; Agusto, F. Zika virus: Endemic versus epidemic dynamics and implications for disease spread in the Americas. *bioRxiv* **2016**, doi:10.1101/041897. [CrossRef]
37. Andraud, M.; Hens, N.; Marais, C.; Beutels, P. Dynamic epidemiological models for dengue transmission: A systematic review of structural approaches. *PLoS ONE* **2012**, *7*, 49085. [CrossRef] [PubMed]
38. Trpis, M.; Haussermann, W. Dispersal and other population parameters of aedes aegypti in an african village and their possible significance in epidemiology of vector-borne diseases. *Am. J. Trop. Med. Hyg.* **1986**, *35*, 1263–1279. [CrossRef] [PubMed]

17

On a New Class of Laplace-Type Integrals Involving Generalized Hypergeometric Functions

Wolfram Koepf [1], **Insuk Kim** [2,*] **and Arjun K. Rathie** [3]

1 Department of Mathematics, University of Kassel, Heinrich-Plett-Str. 40, D-34132 Kassel, Germany
2 Department of Mathematics Education, Wonkwang University, Iksan 570-749, Korea
3 Department of Mathematics, Vedant College of Engineering and Technology (Rajasthan Technical University), Bundi 323021, Rajasthan, India
* Correspondence: iki@wku.ac.kr

Abstract: In the theory of generalized hypergeometric functions, classical summation theorems for the series ${}_2F_1$, ${}_3F_2$, ${}_4F_3$, ${}_5F_4$ and ${}_7F_6$ play a key role. Very recently, Masjed-Jamei and Koepf established generalizations of the above-mentioned summation theorems. Inspired by their work, the main objective of the paper is to provide a new class of Laplace-type integrals involving generalized hypergeometric functions ${}_pF_p$ for $p = 2, 3, 4, 5$ and 7 in the most general forms. Several new and known cases have also been obtained as special cases of our main findings.

Keywords: generalized hypergeometric functions; classical summation theorems; generalization; laplace transforms

---

1. Introduction

The generalized hypergeometric function with p numerator and q denominator parameters is defined [1–4] as

$$ {}_pF_q\left[\begin{matrix} a_1, & \cdots, & a_p \\ b_1, & \cdots, & b_q \end{matrix}; z\right] = \sum_{n=0}^{\infty} \frac{(a_1)_n \cdots (a_p)_n}{(b_1)_n \cdots (b_q)_n} \frac{z^n}{n!}, \tag{1} $$

in which no denominator parameters b_j is allowed to be zero or a negative integer. If any numerator parameter a_j in Equation (1) is zero or a negative integer, the series terminates.

In addition, here, $(a)_n$ is the well known Pochhammer symbol [5] for any complex number a defined as

$$ \begin{aligned} (a)_n &= \frac{\Gamma(a+n)}{\Gamma(a)} \qquad (2) \\ &= \begin{cases} 1, & (n = 0,\ a \in \mathbb{C} \setminus \{0\}) \\ a(a+1)\cdots(a+n-1), & (n \in \mathbb{N},\ a \in \mathbb{C}), \end{cases} \end{aligned} $$

where $\Gamma(z)$ is the well known gamma function defined by

$$ \Gamma(z) = \int_0^{\infty} e^{-x} x^{z-1} dx \tag{3} $$

for $\mathrm{Re}(z) > 0$.

Further, application of the ratio test shows that the series in Equation (1):

(i) converges for all finite z if $p \leq q$;
(ii) converges for $|z| < 1$ if $p = q + 1$; and
(iii) diverges for all z $z \neq 0$ if $p > q + 1$.

In addition, following Bromwich [6] (p. 41 and 241), Knopp [7] (p. 401) or Luke [8], it can be shown that the ${}_{q+1}F_q$ series is

(i) absolutely convergent for $|z| = 1$ if $\text{Re}(\eta) < 0$;
(ii) conditionally convergent for $|z| = 1, z \neq 1$ if $0 \leq \text{Re}(\eta) < 1$; and
(iii) divergent for $|z| = 1$ if $1 \leq \text{Re}(\eta)$, where

$$\eta = \sum_{j=1}^{p} a_j - \sum_{j=1}^{q} b_j.$$

It is not out of place to mention here that, whenever a generalized hypergeometric function reduces to products and quotients of gamma functions, the results are very useful from the point of view of applications. For $p = 2, 3, 4, 5$ and 7 of the generalized hypergeometric function in Equation (1) with proper choice of parameters, the results in the form of summation theorems are available in the literature in terms of gamma function. However, for $p = 6$, we do not have any summation theorem available. Thus, in this paper, we do not consider the case for $p = 6$. Here, we mention the following classical summation theorems [1,2], so that the paper may be self contained.

- Gauss Theorem for $\text{Re}(c - a - b) > 0$

$${}_2F_1\left[\begin{matrix} a, & b \\ & c \end{matrix};1\right] = \frac{\Gamma(c)\Gamma(c-a-b)}{\Gamma(c-a)\Gamma(c-b)} \tag{4}$$

- Kummer's Theorem

$${}_2F_1\left[\begin{matrix} a, & b \\ 1+a-b & \end{matrix};-1\right] = \frac{\Gamma(1+a-b)\Gamma(1+\frac{1}{2}a)}{\Gamma(1-b+\frac{1}{2}a)\Gamma(1+a)} \tag{5}$$

- Second Gauss Theorem

$${}_2F_1\left[\begin{matrix} a, & b \\ \frac{1}{2}(a+b+1) & \end{matrix};\frac{1}{2}\right] = \frac{\sqrt{\pi}\,\Gamma(\frac{1}{2}(a+b+1))}{\Gamma(\frac{1}{2}(a+1))\Gamma(\frac{1}{2}(b+1))} \tag{6}$$

- Bailey's Theorem

$${}_2F_1\left[\begin{matrix} a, & 1-a \\ & b \end{matrix};\frac{1}{2}\right] = \frac{\Gamma(\frac{1}{2}b)\Gamma(\frac{1}{2}(b+1))}{\Gamma(\frac{1}{2}(a+b))\Gamma(\frac{1}{2}(b-a+1))} \tag{7}$$

- Dixon's Theorem for $\text{Re}(a - 2b - 2c) > -2$

$$\begin{aligned} &{}_3F_2\left[\begin{matrix} a, & b, & c \\ 1+a-b, & 1+a-c & \end{matrix};1\right] \\ &= \frac{\Gamma(1+\frac{1}{2}a)\Gamma(1+a-b)\Gamma(1+a-c)\Gamma(1-b-c+\frac{1}{2}a)}{\Gamma(1+a)\Gamma(1-b+\frac{1}{2}a)\Gamma(1-c+\frac{1}{2}a)\Gamma(1+a-b-c)} \end{aligned} \tag{8}$$

- Watson's Theorem for $\text{Re}(2c - a - b) > 1$

$$\begin{aligned} &{}_3F_2\left[\begin{matrix} a, & b, & c \\ \frac{1}{2}(a+b+1), & 2c & \end{matrix};1\right] \\ &= \frac{\sqrt{\pi}\,\Gamma(c+\frac{1}{2})\Gamma(\frac{1}{2}(a+b+1))\Gamma(c-\frac{1}{2}(a+b-1))}{\Gamma(\frac{1}{2}(a+1))\Gamma(\frac{1}{2}(b+1))\Gamma(c-\frac{1}{2}(a-1))\Gamma(c-\frac{1}{2}(b-1))} \end{aligned} \tag{9}$$

- Whipple's Theorem for $\mathrm{Re}(b) > 0$

$$
{}_3F_2\left[\begin{matrix} a, & 1-a, & b \\ c, & 2b-c+1 \end{matrix};1\right] = \frac{\pi\, 2^{1-2b}\,\Gamma(c)\Gamma(2b-c+1)}{\Gamma(\frac{1}{2}(a+c))\Gamma(b+\frac{1}{2}(a-c+1))\Gamma(\frac{1}{2}(1-a+c))\Gamma(b+1-\frac{1}{2}(a+c))} \tag{10}
$$

- Pfaff-Saalschütz Theorem

$$
{}_3F_2\left[\begin{matrix} a, & b, & -n \\ c, & 1+a+b-c-n \end{matrix};1\right] = \frac{(c-a)_n(c-b)_n}{(c)_n(c-a-b)_n} \tag{11}
$$

- Second Whipple's Theorem

$$
{}_4F_3\left[\begin{matrix} a, & 1+\frac{1}{2}a, & b, & c \\ \frac{1}{2}a, & a-b+1, & a-c+1 \end{matrix};-1\right] = \frac{\Gamma(a-b+1)\Gamma(a-c+1)}{\Gamma(a+1)\Gamma(a-b-c+1)} \tag{12}
$$

- Dougall's Theorem for $\mathrm{Re}(a-c-d-e) > -1$

$$
{}_5F_4\left[\begin{matrix} a, & 1+\frac{1}{2}a, & c, & d, & e \\ \frac{1}{2}a, & a-c+1, & a-d+1, & a-e+1 \end{matrix};1\right] = \frac{\Gamma(a-c+1)\Gamma(a-d+1)\Gamma(a-e+1)\Gamma(a-c-d-e+1)}{\Gamma(a+1)\Gamma(a-d-e+1)\Gamma(a-c-e+1)\Gamma(a-c-d+1)} \tag{13}
$$

- Second Dougall's Theorem

$$
{}_7F_6\left[\begin{matrix} a, & 1+\frac{1}{2}a, & b, & c, & d, & 1+2a-b-c-d+n, & -n \\ \frac{1}{2}a, & a-b+1, & a-c+1, & a-d+1, & b+c+d-a-n, & a+1+n \end{matrix};1\right] = \frac{(a+1)_n(a-b-c+1)_n(a-b-d+1)_n(a-c-d+1)_n}{(a+1-b)_n(a+1-c)_n(a+1-d)_n(a+1-b-c-d)_n} \tag{14}
$$

For very interesting applications of some of the above-mentioned classical summation theorems, we refer a very popular and useful paper by Bailey [9].

In addition, for finite sums of hypergeometric series, if we use the following symbol

$$
{}_pF_q^{(m)}\left[\begin{matrix} a_1, & \cdots, & a_p \\ b_1, & \cdots, & b_q \end{matrix};z\right] = \sum_{n=0}^{m} \frac{\prod_{i=1}^{p}(a_i)_n}{\prod_{i=1}^{q}(b_i)_n}\frac{z^n}{n!},
$$

where, for instance,

$$
{}_pF_q^{(-1)}(z) = 0, \quad {}_pF_q^{(0)}(z) = 1, \quad {}_pF_q^{(1)}(z) = 1 + \frac{a_1\cdots a_p}{b_1\cdots b_q}z,
$$

then, by using the following relation [10],

$$
\begin{aligned}
&{}_pF_q\left[\begin{matrix} a_1,\cdots, a_{p-1}, & 1 \\ b_1,\cdots, b_{q-1}, & m \end{matrix};z\right] \\
&= \frac{\Gamma(b_1)\cdots\Gamma(b_{q-1})}{\Gamma(a_1)\cdots\Gamma(a_{p-1})}\,\frac{\Gamma(a_1-m+1)\cdots\Gamma(a_{p-1}-m+1)}{\Gamma(b_1-m+1)\cdots\Gamma(b_{q-1}-m+1)}\,\frac{(m-1)!}{z^{m-1}} \\
&\times\left\{{}_{p-1}F_{q-1}\left[\begin{matrix} a_1-m+1,\cdots, a_{p-1}-m+1 \\ b_1-m+1,\cdots, b_{q-1}-m+1 \end{matrix};z\right]\right. \\
&\left.-\,{}_{p-1}F_{q-1}^{(m-2)}\left[\begin{matrix} a_1-m+1,\cdots, a_{p-1}-m+1 \\ b_1-m+1,\cdots, b_{q-1}-m+1 \end{matrix};z\right]\right\},
\end{aligned}
\tag{15}
$$

very recently Masjed-Jamei and Koepf [11] established generalizations of the classical summation theorems in Equations (4)–(14) in the following form:

$$
\begin{aligned}
&{}_3F_2\left[\begin{matrix} a, & b, & 1 \\ & c, & m \end{matrix};1\right] \\
&= \frac{\Gamma(m)\Gamma(c)\Gamma(a-m+1)\Gamma(b-m+1)}{\Gamma(a)\Gamma(b)\Gamma(c-m+1)} \\
&\times\left\{\frac{\Gamma(c-m+1)\Gamma(c-a-b+m-1)}{\Gamma(c-a)\Gamma(c-b)} - {}_2F_1^{(m-2)}\left[\begin{matrix} a-m+1, b-m+1 \\ c-m+1 \end{matrix};1\right]\right\} \\
&= \Omega_1
\end{aligned}
\tag{16}
$$

$$
\begin{aligned}
&{}_3F_2\left[\begin{matrix} a, & b, & 1 \\ a-b+m, & m & \end{matrix};-1\right] \\
&= (-1)^{m-1}\frac{\Gamma(m)\Gamma(a-b+m)\Gamma(a-m+1)\Gamma(b-m+1)}{\Gamma(a)\Gamma(b)\Gamma(a-b+1)} \\
&\times\left\{\frac{\Gamma(a-b+1)\Gamma(1+\frac{1}{2}(a-m+1))}{\Gamma(2+a-m)\Gamma(m-b+\frac{1}{2}(a-m+1))} - {}_2F_1^{(m-2)}\left[\begin{matrix} a-m+1, b-m+1 \\ a-b+1 \end{matrix};-1\right]\right\} \\
&= \Omega_2
\end{aligned}
\tag{17}
$$

$$
\begin{aligned}
&{}_3F_2\left[\begin{matrix} a, & b, & 1 \\ \frac{1}{2}(a+b+1), & m & \end{matrix};\frac{1}{2}\right] \\
&= 2^{m-1}\frac{\Gamma(m)\Gamma(\frac{1}{2}(a+b+1))\Gamma(a-m+1)\Gamma(b-m+1)}{\Gamma(a)\Gamma(b)\Gamma(-m+1+\frac{1}{2}(a+b+1))} \\
&\times\left\{\frac{\sqrt{\pi}\,\Gamma(-m+1+\frac{1}{2}(a+b+1))}{\Gamma(1+\frac{1}{2}(a-m))\Gamma(1+\frac{1}{2}(b-m))} - {}_2F_1^{(m-2)}\left[\begin{matrix} a-m+1, \; b-m+1 \\ -m+1+\frac{1}{2}(a+b+1) \end{matrix};\frac{1}{2}\right]\right\} \\
&= \Omega_3
\end{aligned}
\tag{18}
$$

$$
\begin{aligned}
&{}_3F_2\left[\begin{matrix} a,\ 2m-a-1,\ 1 \\ b,\quad m \end{matrix};\frac{1}{2}\right] \\
&= 2^{m-1}\frac{\Gamma(m)\Gamma(b)\Gamma(a-m+1)\Gamma(m-a)}{\Gamma(a)\Gamma(2m-a-1)\Gamma(b-m+1)} \\
&\times\left\{\frac{\Gamma(\frac{1}{2}(b-m+1))\Gamma(\frac{1}{2}(b-m+2))}{\Gamma(-m+1+\frac{1}{2}(a+b))\Gamma(\frac{1}{2}(b-a+1))} - {}^{(m-2)}_{\ \ 2}F_1\left[\begin{matrix} a-m+1,\ m-a \\ b-m+1 \end{matrix};\frac{1}{2}\right]\right\} \\
&= \Omega_4
\end{aligned}
\tag{19}
$$

$$
\begin{aligned}
&{}_4F_3\left[\begin{matrix} a,\quad b,\quad c,\quad 1 \\ a-b+m,\ a-c+m,\ m \end{matrix};1\right] \\
&= \frac{\Gamma(m)\Gamma(a-b+m)\Gamma(a-c+m)\Gamma(a+1-m)\Gamma(b+1-m)\Gamma(c+1-m)}{\Gamma(a)\Gamma(b)\Gamma(c)\Gamma(a-b+1)\Gamma(a-c+1)} \\
&\times\left\{\frac{\Gamma(\frac{1}{2}(a+3-m))\Gamma(a-b+1)\Gamma(a-c+1)\Gamma(-b-c+\frac{1}{2}(a+3m-1))}{\Gamma(a+2-m)\Gamma(-b+\frac{1}{2}(a+m+1))\Gamma(-c+\frac{1}{2}(a+m+1))\Gamma(a-b-c+m)}\right. \\
&\left. - {}^{(m-2)}_{\ \ 3}F_2\left[\begin{matrix} a-m+1,\ b-m+1,\ c-m+1 \\ a-b+1,\quad a-c+1 \end{matrix};1\right]\right\} \\
&= \Omega_5
\end{aligned}
\tag{20}
$$

$$
\begin{aligned}
&{}_4F_3\left[\begin{matrix} a,\quad b,\quad c,\quad 1 \\ \frac{1}{2}(a+b+1),\ 2c+1-m,\ m \end{matrix};1\right] \\
&= \frac{\Gamma(m)\Gamma(\frac{1}{2}(a+b+1))\Gamma(2c+1-m)\Gamma(a+1-m)\Gamma(b+1-m)\Gamma(c+1-m)}{\Gamma(a)\Gamma(b)\Gamma(c)\Gamma(-m+\frac{1}{2}(a+b+3))\Gamma(2c-2m+2)} \\
&\times\left\{\frac{\sqrt{\pi}\,\Gamma(c-m+\frac{3}{2})\Gamma(-m+\frac{1}{2}(a+b+3))\Gamma(c-\frac{1}{2}(a+b-1))}{\Gamma(1+\frac{1}{2}(a-m))\Gamma(1+\frac{1}{2}(b-m))\Gamma(c+1-\frac{1}{2}(a+m))\Gamma(c+1-\frac{1}{2}(b+m))}\right. \\
&\left. - {}^{(m-2)}_{\ \ 3}F_2\left[\begin{matrix} a-m+1,\ b-m+1,\ c-m+1 \\ -m+1+\frac{1}{2}(a+b+1),\ 2c-2m+2 \end{matrix};1\right]\right\} \\
&= \Omega_6
\end{aligned}
\tag{21}
$$

$$
\begin{aligned}
&{}_4F_3\left[\begin{matrix} a,\quad 2m-1-a,\quad b,\quad 1 \\ c,\quad 2b-c+1,\quad m \end{matrix};1\right] \\
&= \frac{\Gamma(m)\Gamma(c)\Gamma(2b-c+1)\Gamma(m-a)\Gamma(a+1-m)\Gamma(b+1-m)}{\Gamma(a)\Gamma(b)\Gamma(2m-1-a)\Gamma(c+1-m)\Gamma(2b-c-m+2)} \\
&\times\left\{\frac{\pi\,2^{2m-2b-1}\,\Gamma(c-m+1)}{\Gamma(-m+1+\frac{1}{2}(a+c))\Gamma(-m+1+b+\frac{1}{2}(a-c+1))\Gamma(\frac{1}{2}(1-a+c))}\right. \\
&\left.\times\frac{\Gamma(2b-c-m+2)}{\Gamma(b+1-\frac{1}{2}(a+c))} - {}^{(m-2)}_{\ \ 3}F_2\left[\begin{matrix} a-m+1,\ b-m+1,\ m-a \\ c-m+1,\ 2b-c-m+2 \end{matrix};1\right]\right\} \\
&= \Omega_7
\end{aligned}
\tag{22}
$$

$$
{}_4F_3\left[\begin{matrix} a,\ b,\ -n+m-1,\ 1 \\ c,\ 1+a+b-c-n,\ m \end{matrix};1\right] = \frac{(m-1)!\,(1-c)_{m-1}}{(1-a)_{m-1}(1-b)_{m-1}} \times \frac{(c-a-b+n)_{m-1}}{(n+2-m)_{m-1}} \times \left\{ \frac{(c-a)_n(c-b)_n}{(c+1-m)_n(c-a-b+m-1)_n} - {}_3F_2^{(m-2)}\left[\begin{matrix} a-m+1,\ b-m+1,\ -n \\ c-m+1,\ 2+a+b-c-m-n \end{matrix};1\right]\right\} = \Omega_8 \tag{23}
$$

$$
{}_5F_4\left[\begin{matrix} a,\ \frac{1}{2}(a+m+1),\ b,\ c,\ 1 \\ \frac{1}{2}(a+m-1),\ a-b+m,\ a-c+m,\ m \end{matrix};-1\right] = (-1)^{m-1}\Gamma(m) \times \frac{\Gamma(\frac{1}{2}(a+m-1))\Gamma(a-b+m)\Gamma(a-c+m)\Gamma(\frac{1}{2}(a-m+3))\Gamma(a-m+1)}{\Gamma(a)\Gamma(b)\Gamma(c)\Gamma(\frac{1}{2}(a+m+1))\Gamma(\frac{1}{2}(a-m+1))} \times \frac{\Gamma(b+1-m)\Gamma(c+1-m)}{\Gamma(a-b+1)\Gamma(a-c+1)} \times \left\{ \frac{\Gamma(1+a-b)\Gamma(1+a-c)}{\Gamma(2-m+a)\Gamma(m+a-b-c)} - {}_4F_3^{(m-2)}\left[\begin{matrix} a-m+1,\ b-m+1,\ \frac{1}{2}(a-m+3),\ c-m+1 \\ \frac{1}{2}(a-m+1),\ a-b+1,\ a-c+1 \end{matrix};-1\right]\right\} = \Omega_9 \tag{24}
$$

$$
{}_6F_5\left[\begin{matrix} a,\ \frac{1}{2}(a+m+1),\ c,\ d,\ e,\ 1 \\ \frac{1}{2}(a+m-1),\ a-c+m,\ a-d+m,\ a-e+m,\ m \end{matrix};1\right] = \frac{\Gamma(m)\Gamma(\frac{1}{2}(a+m-1))\Gamma(a-c+m)\Gamma(a-d+m)\Gamma(a-e+m)}{\Gamma(a-c+1)\Gamma(a-d+1)\Gamma(a-e+1)} \times \frac{\Gamma(a-m+1)\Gamma(\frac{1}{2}(a-m+3))\Gamma(c+1-m)\Gamma(d+1-m)\Gamma(e+1-m)}{\Gamma(a)\Gamma(c)\Gamma(d)\Gamma(e)\Gamma(\frac{1}{2}(a+m+1))\Gamma(\frac{1}{2}(a-m+1))} \times \left\{ \frac{\Gamma(a-c+1)\Gamma(a-d+1)\Gamma(a-e+1)\Gamma(a-c-d-e+2m-1)}{\Gamma(2-m+a)\Gamma(a-c-e+m)\Gamma(a-d-e+m)\Gamma(a-c-d+m)} - {}_5F_4^{(m-2)}\left[\begin{matrix} a-m+1,\ c-m+1,\ \frac{1}{2}(a-m+3),\ d-m+1,\ e-m+1 \\ \frac{1}{2}(a-m+1),\ a-c+1,\ a-d+1,\ a-e+1 \end{matrix};1\right]\right\} = \Omega_{10} \tag{25}
$$

$$
{}_8F_7\left[\begin{matrix} a,\ \frac{1}{2}(a+m+1),\ b,\ c,\ d,\ 2a-b-c-d+2m-1+n,\ m-n-1,\ 1 \\ \frac{1}{2}(a+m-1),\ a-b+m,\ a-c+m,\ a-d+m,\ b+c+d-a+1-m-n,\ a+n+1,\ m \end{matrix};1\right] = (-1)^{m-1}(m-1)! \times \frac{(\frac{1}{2}(3-a-m))_{m-1}(1-a+b-m)_{m-1}}{(\frac{1}{2}(1-a-m))_{m-1}(1-a)_{m-1}} \times \frac{(1-a+c-m)_{m-1}(1-a+d-m)_{m-1}(m+n+a-b-c-d)_{m-1}(-a-n)_{m-1}}{(1-b)_{m-1}(1-c)_{m-1}(1-d)_{m-1}(b+c+d-2a+2-2m-n)_{m-1}(n+2-m)_{m-1}} \times \left\{ \frac{(a-m+2)_n(a-b-c+m)_n(a-b-d+m)_n(a-c-d+m)_n}{(a-b+1)_n(a-c+1)_n(a-d+1)_n(a-b-c-d+2m-1)_n} - {}_7F_6^{(m-2)}\left[\begin{matrix} a-m+1,\ \frac{1}{2}(a-m+3),\ b-m+1,\ c-m+1,\ d-m+1,\ 2a-b-c-d+m+n,\ -n \\ \frac{1}{2}(a-m+1),\ a-b+1,\ a-c+1,\ a-d+1,\ b+c+d+-a+2-2m-n,\ a-m+n+2 \end{matrix};1\right]\right\} = \Omega_{11} \tag{26}
$$

Remark 1. *For other generalizations of the results in Equations* (5)–(10)*, we refer to [12–16].*

On the other hand, we define the (direct) Laplace transform of a function $f(t)$ of a real variable t as the integral $g(s)$ over a range of the complex parameter s as

$$g(s) = \mathcal{L}\{f(t); s\} = \int_0^\infty e^{-st} f(t) dt \tag{27}$$

provided the integral exists in the Lebesgue sense. For more details, see, for instance, the works of [17] or [18]. It is interesting to mention here that, in view of the formula

$$\int_0^\infty e^{-st} t^{\alpha-1} dt = \Gamma(\alpha) s^{-\alpha} \tag{28}$$

provided $\mathrm{Re}(s) > 0$ and $\mathrm{Re}(\alpha) > 0$, by using Equation (1), it is a simple exercise to show that the Laplace transform of a generalized hypergeometric function ${}_pF_q$ is obtained as [3,19,20]:

$$\begin{aligned}&\int_0^\infty e^{-st} t^{\nu-1} {}_pF_q \left[\begin{matrix} a_1, & \cdots, & a_p \\ b_1, & \cdots, & b_q \end{matrix}; wt\right] dt \\ &= \Gamma(\nu) s^{-\nu} {}_{p+1}F_q \left[\begin{matrix} \nu, a_1, & \cdots, & a_p \\ b_1, & \cdots, & b_q \end{matrix}; \frac{w}{s}\right]\end{aligned} \tag{29a}$$

provided that when $p < q$, $\mathrm{Re}(\nu) > 0$, $\mathrm{Re}(s) > 0$ for w arbitrary, or $p = q > 0$, $\mathrm{Re}(\nu) > 0$ and $\mathrm{Re}(s) > \mathrm{Re}(w)$.

Further, in Equation (29a), if we take $p = q = 1$, $\nu = b$, $a_1 = a$ and $b_1 = c$, we find that (see [20]),

$$\int_0^\infty e^{-st} t^{b-1} {}_1F_1 \left[\begin{matrix} a \\ c \end{matrix}; wt\right] dt = \Gamma(b) s^{-b} {}_2F_1 \left[\begin{matrix} a, b \\ c \end{matrix}; \frac{w}{s}\right], \tag{29b}$$

provided $\mathrm{Re}(b) > 0$, $\mathrm{Re}(s) > 0$, $\mathrm{Re}(s) > \mathrm{Re}(w)$ and $|s| > |w|$.

Finally, in Equation (29b), if we take $w = \frac{1}{2}s$ and either $c = \frac{1}{2}(a+b+1)$ or $b = 1-a$, then it is easy to see that the ${}_2F_1$ series appearing on the right-hand side of Equation (29b) are summable by known summation theorems in Equations (6) and (7), respectively, and we find that

$$\int_0^\infty e^{-st} t^{b-1} {}_1F_1 \left[\begin{matrix} a \\ \frac{1}{2}(a+b+1) \end{matrix}; \frac{1}{2}st\right] dt = s^{-b} \frac{\Gamma(\frac{1}{2})\Gamma(b)\Gamma(\frac{1}{2}(a+b+1))}{\Gamma(\frac{1}{2}(a+1))\Gamma(\frac{1}{2}(b+1))}, \tag{29c}$$

provided $\mathrm{Re}(b) > 0$ and $\mathrm{Re}(s) > 0$, and that

$$\int_0^\infty e^{-st} t^{-a} {}_1F_1 \left[\begin{matrix} a \\ c \end{matrix}; \frac{1}{2}st\right] dt = s^{a-1} \frac{\Gamma(1-a)\Gamma(\frac{1}{2}c)\Gamma(\frac{1}{2}(c+1))}{\Gamma(\frac{1}{2}(c+a))\Gamma(\frac{1}{2}(c-a+1))}, \tag{29d}$$

provided $\mathrm{Re}(1-a) > 0$ and $\mathrm{Re}(s) > 0$.

The results in Equations (29b) and (29c) are very well-known in the literature and are recorded, for example, in the work of [20].

In addition to Equation (29a), it is interesting to observe here that, when $w = \pm s$ and $q = p$, on similar lines, we can obtain the following result in view of the conditions of convergence of ${}_pF_q$ mentioned in Section 1.

$$\begin{aligned}&\int_0^\infty e^{-st} t^{\nu-1} {}_pF_p \left[\begin{matrix} a_1, & \cdots, & a_p \\ b_1, & \cdots, & b_p \end{matrix}; \pm st\right] dt \\ &= \Gamma(\nu) s^{-\nu} {}_{p+1}F_p \left[\begin{matrix} \nu, a_1, & \cdots, & a_p \\ b_1, & \cdots, & b_p \end{matrix}; \pm 1\right]\end{aligned} \tag{29e}$$

provided $\mathrm{Re}(\nu) > 0$, $\mathrm{Re}(s) > 0$ and $\mathrm{Re}(b_1 + \cdots + b_p - a_1 - \cdots - a_p - \nu) > 0$.

Remark 2. *1. Since there is no summation theorems for the series ${}_pF_p$ for argument $\frac{1}{2}$, 1 and -1 are available in the literature, at this moment, it is not possible to find the Laplace-type integrals for the generalized hypergeometric function ${}_pF_p$. We leave this open problem for the readers.*
2. Laplace-type integrals in the case $p = 2$ were given by Deepthi et al. [21] and connections with fractional integral operators were recently studied by Parmar and Purohit [22].

The aim of this paper is to provide a new class of Laplace-type integrals involving generalized hypergeometric functions by employing the summation theorems in Equations (16)–(26). Several new and known special cases have also been considered.

2. Laplace-Type Integrals Involving Generalized Hypergeometric Functions

In this section, we establish several new, interesting and elementary Laplace-type integrals in the most general form, involving generalized hypergeometric functions asserted in the following theorems that follow directly from Equations (29a) and (29e) and the known results in Equations (16)–(26). The results presented in this section would serve as key formulas from which, on specializing the parameters, lead to several results, some of which are known and others are believed to be new.

[A] Laplace-type integrals involving generalized hypergeometric function ${}_2F_2$

The results to be established are asserted in the following theorems.

Theorem 1. *For $m \in \mathbb{N}$, $Re(s) > 0$, $Re(a) > 0$ and $Re(c - a - b + m) > 1$, the following result holds true.*

$$\int_0^\infty e^{-st}\, t^{a-1}\, {}_2F_2\left[\begin{matrix} b, & 1 \\ c, & m \end{matrix}; st\right] dt = \Gamma(a)\, s^{-a}\, \Omega_1, \tag{30}$$

where Ω_1 is the same as given in Equation (16).

Theorem 2. *For $m \in \mathbb{N}$, $Re(s) > 0$ and $Re(c - a - b + m) > 1$, the following result holds true.*

$$\int_0^\infty e^{-st}\, {}_2F_2\left[\begin{matrix} a, & b \\ c, & m \end{matrix}; st\right] dt = s^{-1}\Omega_1, \tag{31}$$

where Ω_1 is the same as given in Equation (16).

Theorem 3. *For $m \in \mathbb{N}$, $Re(s) > 0$ and $Re(a) > 0$, the following result holds true.*

$$\int_0^\infty e^{-st}\, t^{a-1}\, {}_2F_2\left[\begin{matrix} b, & 1 \\ a-b+m, & m \end{matrix}; -st\right] dt = \Gamma(a)\, s^{-a}\, \Omega_2, \tag{32}$$

where Ω_2 is the same as given in Equation (17).

Theorem 4. *For $m \in \mathbb{N}$, $Re(s) > 0$ and $Re(b) > 0$, the following result holds true.*

$$\int_0^\infty e^{-st}\, t^{b-1}\, {}_2F_2\left[\begin{matrix} a, & 1 \\ a-b+m, & m \end{matrix}; -st\right] dt = \Gamma(b)\, s^{-b}\, \Omega_2, \tag{33}$$

where Ω_2 is the same as given in Equation (17).

Remark 3. *The reader should observe that the results given in Theorems 3 and 4 are different but obtained from the same known result in Equation* (17).

Theorem 5. *For $m \in \mathbb{N}$ and $Re(s) > 0$, the following result holds true.*

$$\int_0^\infty e^{-st}\,{}_2F_2\left[\begin{matrix} a, & b \\ a-b+m, & m \end{matrix}; -st\right] dt = s^{-1}\Omega_2, \tag{34}$$

where Ω_2 is the same as given in Equation (17).

Theorem 6. *For $m \in \mathbb{N}$, $Re(s) > 0$ and $Re(a) > 0$, the following result holds true.*

$$\int_0^\infty e^{-st}\,t^{a-1}\,{}_2F_2\left[\begin{matrix} b, & 1 \\ \frac{1}{2}(a+b+1), & m \end{matrix}; \frac{1}{2}st\right] dt = \Gamma(a)\,s^{-a}\,\Omega_3, \tag{35}$$

where Ω_3 is the same as given in Equation (18).

Theorem 7. *For $m \in \mathbb{N}$ and $Re(s) > 0$, the following result holds true.*

$$\int_0^\infty e^{-st}\,{}_2F_2\left[\begin{matrix} a, & b \\ \frac{1}{2}(a+b+1), & m \end{matrix}; \frac{1}{2}st\right] dt = s^{-1}\Omega_3, \tag{36}$$

where Ω_3 is the same as given in Equation (18).

Theorem 8. *For $m \in \mathbb{N}$, $Re(s) > 0$ and $Re(a) > 0$, the following result holds true.*

$$\int_0^\infty e^{-st}\,t^{a-1}\,{}_2F_2\left[\begin{matrix} 2m-a-1, & 1 \\ b, & m \end{matrix}; \frac{1}{2}st\right] dt = \Gamma(a)\,s^{-a}\,\Omega_4, \tag{37}$$

where Ω_4 is the same as given in Equation (19).

Theorem 9. *For $m \in \mathbb{N}$, $Re(s) > 0$ and $Re(2m-a-1) > 0$, the following result holds true.*

$$\int_0^\infty e^{-st}\,t^{2m-a-2}\,{}_2F_2\left[\begin{matrix} a, & 1 \\ b, & m \end{matrix}; \frac{1}{2}st\right] dt = \Gamma(2m-a-1)\,s^{a+1-2m}\,\Omega_4, \tag{38}$$

where Ω_4 is the same as given in Equation (19).

Theorem 10. *For $m \in \mathbb{N}$ and $Re(s) > 0$, the following result holds true.*

$$\int_0^\infty e^{-st}\,{}_2F_2\left[\begin{matrix} a, & 2m-a-1 \\ b, & m \end{matrix}; \frac{1}{2}st\right] dt = s^{-1}\Omega_4, \tag{39}$$

where Ω_4 is the same as given in Equation (19).

Proof. To establish the result in Equation (30) asserted in Theorem 1, we proceed as follows. In Equation (29e), if we take $p = q = 2$, $\nu = a$, $a_1 = b$, $a_2 = 1$, $b_1 = c$, and $b_2 = m$, considering positive sign, we get

$$\int_0^\infty e^{-st}\,t^{a-1}\,{}_2F_2\left[\begin{matrix} b, & 1 \\ c, & m \end{matrix}; st\right] dx = s^{-a}\Gamma(a)\,{}_3F_2\left[\begin{matrix} a, & b, & 1 \\ & c, & m \end{matrix}; 1\right]. \tag{40}$$

We now observe that the ${}_3F_2$ appearing on the right-hand side of Equation (40) can be evaluated with the help of the result in Equation (16) and we easily arrive at the right-hand side of Equation (30). This completes the proof of Equation (30) asserted in Theorem 1.

In exactly the same manner, the results in Equations (31)–(39) asserted in Theorems 2–10 can be evaluated. We however omit the details. □

Corollary 1. ***(a)*** *In Theorem 1, if we take $m = 1, 2, 3$, we get the following results.*

$$\int_0^\infty e^{-st}\, t^{a-1}\, {}_1F_1\left[\begin{matrix} b \\ c \end{matrix}; st\right] dt = \frac{\Gamma(a)\Gamma(c)\Gamma(c-a-b)}{s^a\,\Gamma(c-a)\Gamma(c-b)}, \tag{41}$$

$$\int_0^\infty e^{-st}\, t^{a-1}\, {}_2F_2\left[\begin{matrix} b, & 1 \\ c, & 2 \end{matrix}; st\right] dt = \frac{(c-1)\Gamma(a-1)}{s^a\,(b-1)} \left\{ \frac{\Gamma(c-1)\Gamma(c-a-b+1)}{\Gamma(c-a)\Gamma(c-b)} - 1 \right\} \tag{42}$$

and

$$\int_0^\infty e^{-st}\, t^{a-1}\, {}_2F_2\left[\begin{matrix} b, & 1 \\ c, & 3 \end{matrix}; st\right] dt = \frac{2\Gamma(a)(c-2)_2}{s^a\,(a-2)_2(b-2)_2} \left\{ \frac{\Gamma(c-2)\Gamma(c-a-b+2)}{\Gamma(c-a)\Gamma(c-b)} - \frac{ab+c-2a-2b+2}{c-2} \right\}. \tag{43}$$

(b) *In Theorem 4, if we take $m = 1, 2, 3$, we get the following results.*

$$\int_0^\infty e^{-st}\, t^{b-1}\, {}_1F_1\left[\begin{matrix} a \\ 1+a-b \end{matrix}; -st\right] dt = \frac{2^{-a}\Gamma(b)\Gamma(\frac{1}{2})\Gamma(1+a-b)}{s^b\,\Gamma(\frac{1}{2}a+\frac{1}{2})\Gamma(1+\frac{1}{2}a-b)}, \tag{44}$$

$$\int_0^\infty e^{-st}\, t^{b-1}\, {}_2F_2\left[\begin{matrix} a, & 1 \\ 2+a-b, & 2 \end{matrix}; -st\right] dt = \frac{(a-b+1)\Gamma(b)}{s^b\,(a-1)(b-1)} \left\{ 1 - \frac{\Gamma(1+a-b)\Gamma(\frac{1}{2}a+\frac{1}{2})}{\Gamma(a)\Gamma(\frac{1}{2}a-b+\frac{3}{2})} \right\} \tag{45}$$

and

$$\int_0^\infty e^{-st}\, t^{a-1}\, {}_2F_2\left[\begin{matrix} b, & 1 \\ 3+a-b, & 3 \end{matrix}; -st\right] dt = \frac{2(a-b+1)_2\,\Gamma(a)}{s^b\,(a-2)_2(b-2)_2} \left\{ \frac{\Gamma(\frac{1}{2}a)\Gamma(1+a-b)}{\Gamma(a-1)\Gamma(\frac{1}{2}a-b+2)} - \frac{3a+b-ab-3}{1+a-b} \right\}. \tag{46}$$

(c) *In Theorem 7, if we take $m = 1, 2, 3$, we get the following results.*

$$\int_0^\infty e^{-st}\, t^{a-1}\, {}_1F_1\left[\begin{matrix} b \\ \frac{1}{2}(a+b+1) \end{matrix}; \frac{1}{2}st\right] dt = \frac{\sqrt{\pi}\Gamma(a)\Gamma(\frac{1}{2}(a+b+1))}{s^a\,\Gamma(\frac{1}{2}(a+1))\Gamma(\frac{1}{2}(b+1))}, \tag{47}$$

$$\int_0^\infty e^{-st}\, t^{a-1}\, {}_2F_2\left[\begin{matrix} b, & 1 \\ \frac{1}{2}(a+b+1), & 2 \end{matrix}; \frac{1}{2}st\right] dt = \frac{(a+b-1)\Gamma(a-1)}{s^a\,(b-1)} \left\{ \frac{\sqrt{\pi}\,\Gamma(\frac{1}{2}(a+b-1))}{\Gamma(\frac{1}{2}a)\Gamma(\frac{1}{2}b)} - 1 \right\} \tag{48}$$

and

$$\int_0^\infty e^{-st}\, t^{a-1}\, {}_2F_2\left[\begin{matrix} b, & 1 \\ \frac{1}{2}(a+b+1), 3 \end{matrix}; \frac{1}{2}st\right] dt = \frac{2\Gamma(a)(a+b-1)(a+b-3)}{s^a\,(a-2)_2(b-2)_2} \times \left\{ \frac{\sqrt{\pi}\,\Gamma(\frac{1}{2}(a+b-3))}{\Gamma(\frac{1}{2}(a-1))\Gamma(\frac{1}{2}(b-1))} - \frac{ab-a-b+1}{a+b-3} \right\}. \tag{49}$$

(d) *In Theorem 10, if we take $m = 1, 2, 3$, we get the following results.*

$$\int_0^\infty e^{-st}\, t^{a-1}\, {}_1F_1\left[\begin{matrix} 1-a \\ b \end{matrix}; \frac{1}{2}st\right] dt = \frac{\Gamma(a)\Gamma(\frac{1}{2}b)\Gamma(\frac{1}{2}(b+1))}{s^a\,\Gamma(\frac{1}{2}(a+b))\Gamma(\frac{1}{2}(b-a+1))}, \tag{50}$$

$$\int_0^\infty e^{-st}\, t^{a-1}\, {}_2F_2\left[\begin{matrix} 3-a, & 1 \\ b, & 2 \end{matrix}; \frac{1}{2}st\right] dt = \frac{2(1-b)\Gamma(a)}{s^a\,(1-a)_2} \left\{ \frac{\Gamma(\frac{1}{2}(b-1))\Gamma(\frac{1}{2}b)}{\Gamma(\frac{1}{2}(a+b)-1)\Gamma(\frac{1}{2}(b-a+1))} - 1 \right\} \tag{51}$$

and

$$\int_0^\infty e^{-st}\, t^{a-1}\, {}_2F_2\left[\begin{matrix} 5-a, & 1 \\ b, & 3 \end{matrix}; \frac{1}{2}st\right] dt = \frac{8(b-2)_2\Gamma(a)}{s^a\,(a-4)_4} \times \left\{ \frac{\Gamma(\frac{1}{2}(b-1))\Gamma(\frac{1}{2}(b-2))}{\Gamma(\frac{1}{2}(a+b)-2)\Gamma(\frac{1}{2}(b-a+1))} - \frac{5a-a^2+2b-10}{2(b-2)} \right\}. \tag{52}$$

Similarly, other results can be obtained from Theorems 2, 3, 5, 6, 8 and 9.

Remark 4. *1. The results in Equations* (47) *and* (50) *were recorded by [23] as well as [20].*
2. The proofs of Theorems 11–35 given below are straight forward and can be proven with the help of the result in Equation (29e)*, thus they are given here without proof.*

[B] Laplace-type integrals involving generalized hypergeometric function ${}_3F_3$

The results to be established are asserted in the following theorems.

Theorem 11. *For $m \in \mathbb{N}$, $Re(s) > 0$, $Re(a) > 0$ and $Re(a-2b-2c+3m) > 1$, the following result holds true.*

$$\int_0^\infty e^{-st}\, t^{a-1}\, {}_3F_3\left[\begin{matrix} b, & c, & 1 \\ a-b+m, & a-c+m, & m \end{matrix}; st\right] dt = \Gamma(a)\, s^{-a}\, \Omega_5, \tag{53}$$

where Ω_5 is the same as given in Equation (20).

Theorem 12. *For $m \in \mathbb{N}$, $Re(s) > 0$, $Re(b) > 0$ and $Re(a-2b-2c+3m) > 1$, the following result holds true.*

$$\int_0^\infty e^{-st}\, t^{b-1}\, {}_3F_3\left[\begin{matrix} a, & c, & 1 \\ a-b+m, & a-c+m, & m \end{matrix}; st\right] dt = \Gamma(b)\, s^{-b}\, \Omega_5, \tag{54}$$

where Ω_5 is the same as given in Equation (20).

Theorem 13. *For $m \in \mathbb{N}$, $Re(s) > 0$ and $Re(a - 2b - 2c + 3m) > 1$, the following result holds true.*

$$\int_0^\infty e^{-st}\, {}_3F_3 \left[\begin{matrix} a, & b, & c \\ a-b+m, & a-c+m, & m \end{matrix} ; st \right] dt = s^{-1} \Omega_5, \tag{55}$$

where Ω_5 is the same as given in Equation (20).

Theorem 14. *For $m \in \mathbb{N}$, $Re(s) > 0$, $Re(a) > 0$ and $Re(2c - a - b) > -1$, the following result holds true.*

$$\int_0^\infty e^{-st}\, t^{a-1}\, {}_3F_3 \left[\begin{matrix} b, & c, & 1 \\ \frac{1}{2}(a+b+1), & 2c+1-m, & m \end{matrix} ; st \right] dt = \Gamma(a)\, s^{-a}\, \Omega_6, \tag{56}$$

where Ω_6 is the same as given in Equation (21).

Theorem 15. *For $m \in \mathbb{N}$, $Re(s) > 0$, $Re(c) > 0$ and $Re(2c - a - b) > -1$, the following result holds true.*

$$\int_0^\infty e^{-st}\, t^{c-1}\, {}_3F_3 \left[\begin{matrix} a, & b, & 1 \\ \frac{1}{2}(a+b+1), & 2c+1-m, & m \end{matrix} ; st \right] dt = \Gamma(c)\, s^{-c}\, \Omega_6, \tag{57}$$

where Ω_6 is the same as given in Equation (21).

Theorem 16. *For $m \in \mathbb{N}$, $Re(s) > 0$ and $Re(2c - a - b) > -1$, the following result holds true.*

$$\int_0^\infty e^{-st}\, {}_3F_3 \left[\begin{matrix} a, & b, & c \\ \frac{1}{2}(a+b+1), & 2c+1-m, & m \end{matrix} ; st \right] dt = s^{-1} \Omega_6, \tag{58}$$

where Ω_6 is the same as given in Equation (21).

Theorem 17. *For $m \in \mathbb{N}$, $Re(s) > 0$, $Re(a) > 0$ and $Re(b - m + 1) > 0$, the following result holds true.*

$$\int_0^\infty e^{-st}\, t^{a-1}\, {}_3F_3 \left[\begin{matrix} 2m-a-1, & b, & 1 \\ c, & 2b-c+1, & m \end{matrix} ; st \right] dt = \Gamma(a)\, s^{-a}\, \Omega_7, \tag{59}$$

where Ω_7 is the same as given in Equation (22).

Theorem 18. *For $m \in \mathbb{N}$, $Re(s) > 0$, $Re(b) > 0$ and $Re(b - m + 1) > 0$, the following result holds true.*

$$\int_0^\infty e^{-st}\, t^{b-1}\, {}_3F_3 \left[\begin{matrix} a, & 2m-a-1, & 1 \\ c, & 2b-c+1, & m \end{matrix} ; st \right] dt = \Gamma(b)\, s^{-b}\, \Omega_7, \tag{60}$$

where Ω_7 is the same as given in Equation (22).

Theorem 19. *For $m \in \mathbb{N}$, $Re(s) > 0$, $Re(2m - a - 1) > 0$ and $Re(b - m + 1) > 0$, the following result holds true.*

$$\int_0^\infty e^{-st}\, t^{2m-a-2}\, {}_3F_3 \left[\begin{matrix} a, & c, & 1 \\ c, & 2b-c+1, & m \end{matrix} ; st \right] dt = \Gamma(2m-a-1)\, s^{a+1-2m}\, \Omega_7, \tag{61}$$

where Ω_7 is the same as given in Equation (22).

Theorem 20. *For $m \in \mathbb{N}$, $Re(s) > 0$ and $Re(b - m + 1) > 0$, the following result holds true.*

$$\int_0^\infty e^{-st}\, {}_3F_3\left[\begin{matrix} a, & 2m-a-1, & b \\ c, & 2b-c+1, & m \end{matrix}; st\right] dt = s^{-1}\Omega_7, \tag{62}$$

where Ω_7 is the same as given in Equation (22).

Theorem 21. *For $m \in \mathbb{N}$, $n \in \mathbb{N}_0$, $Re(s) > 0$ and $Re(a) > 0$, the following result holds true.*

$$\int_0^\infty e^{-st}\, t^{a-1}\, {}_3F_3\left[\begin{matrix} b, & -n+m-1, & 1 \\ c, & 1+a+b-c-n, & m \end{matrix}; st\right] dt = \Gamma(a)\, s^{-a}\, \Omega_8, \tag{63}$$

where Ω_8 is the same as given in Equation (23).

Theorem 22. *For $m \in \mathbb{N}$, $n \in \mathbb{N}_0$ and $Re(s) > 0$, the following result holds true.*

$$\int_0^\infty e^{-st}\, {}_3F_3\left[\begin{matrix} a, & b, & -n+m-1 \\ c, & 1+a+b-c-n, & m \end{matrix}; st\right] dt = s^{-1}\Omega_8, \tag{64}$$

where Ω_8 is the same as given in Equation (23).

Corollary 2. *(e) In Theorem 11, if we take $m = 1, 2, 3$, we get the following results.*

$$\int_0^\infty e^{-st}\, t^{a-1}\, {}_2F_2\left[\begin{matrix} b, & c \\ 1+a-b, & 1+a-c \end{matrix}; st\right] dt = \frac{\Gamma(\frac{1}{2}a)\Gamma(1+a-b)\Gamma(1+a-c)\Gamma(1+\frac{1}{2}a-b-c)}{2\, s^a\, \Gamma(1+\frac{1}{2}a-b)\Gamma(1+\frac{1}{2}a-c)\Gamma(1+a-b-c)}, \tag{65}$$

$$\int_0^\infty e^{-st}\, t^{a-1}\, {}_3F_3\left[\begin{matrix} b, & c, & 1 \\ a-b+2, & a-c+2, & 2 \end{matrix}; st\right] dt = \frac{\Gamma(a-1)(1+a-b)(1+a-c)}{s^a\,(b-1)(c-1)} \times \left\{\frac{\Gamma(\frac{1}{2}(a+1))\Gamma(1+a-b)\Gamma(1+a-c)\Gamma(\frac{1}{2}a-b-c+\frac{5}{2})}{\Gamma(a)\Gamma(\frac{1}{2}a-b+\frac{3}{2})\Gamma(\frac{1}{2}a-c+\frac{3}{2})\Gamma(2+a-b-c)} - 1\right\} \tag{66}$$

and

$$\int_0^\infty e^{-st}\, t^{a-1}\, {}_3F_3\left[\begin{matrix} b, & c, & 1 \\ a-b+3, & a-c+3, & 3 \end{matrix}; st\right] dt = \frac{2(a-b+1)_2(a-c+1)_2\,\Gamma(a)}{s^a\,(a-2)_2(b-2)_2(c-2)_2} \times \left\{\frac{\Gamma(\frac{1}{2}a)\Gamma(1+a-b)\Gamma(1+a-c)\Gamma(\frac{1}{2}a-b-c+4)}{\Gamma(a-1)\Gamma(\frac{1}{2}a-b+2)\Gamma(\frac{1}{2}a-c+2)\Gamma(3+a-b-c)} - \frac{(a-2)(b-2)(c-2)}{(a-b+1)(a-c+1)} - 1\right\}. \tag{67}$$

(f) *In Theorem 14, if we take $m = 1, 2, 3$, we get the following results.*

$$\int_0^\infty e^{-st}\, t^{a-1}\, {}_2F_2\left[\begin{matrix} b, & c \\ \frac{1}{2}(a+b+1), & 2c \end{matrix}; st\right] dt \tag{68}$$
$$= \frac{\sqrt{\pi}\Gamma(a)\Gamma(c+\frac{1}{2})\Gamma(\frac{1}{2}(a+b+1))\Gamma(c-\frac{1}{2}(a+b-1))}{s^a\,\Gamma(\frac{1}{2}(a+1)\Gamma(\frac{1}{2}(b+1))\Gamma(c-\frac{1}{2}(a-1))\Gamma(c-\frac{1}{2}(b-1))},$$

$$\int_0^\infty e^{-st}\, t^{a-1}\, {}_3F_3\left[\begin{matrix} b, & c, & 1 \\ \frac{1}{2}(a+b+1), & 2c-1, & 2 \end{matrix}; st\right] dt \tag{69}$$
$$= \frac{(a+b-1)\Gamma(a-1)}{s^a\,(b-1)}\left\{\frac{\sqrt{\pi}\,\Gamma(c-\frac{1}{2})\Gamma(\frac{1}{2}(a+b-1))\Gamma(c-\frac{1}{2}(a+b-1))}{\Gamma(\frac{1}{2}a)\Gamma(\frac{1}{2}b)\Gamma(c-\frac{1}{2}a)\Gamma(c-\frac{1}{2}b)} - 1\right\}$$

and

$$\int_0^\infty e^{-st}\, t^{a-1}\, {}_3F_3\left[\begin{matrix} b, & c, & 1 \\ \frac{1}{2}(a+b+1), & 2c-2, & 3 \end{matrix}; st\right] dt \tag{70}$$
$$= \frac{(2c-3)(a+b-1)(a+b-3)\,\Gamma(a)}{s^a\,(c-1)(a-2)_2(b-2)_2}$$
$$\times\left\{\frac{\sqrt{\pi}\,\Gamma(c-\frac{3}{2})\Gamma(\frac{1}{2}(a+b-3))\Gamma(c-\frac{1}{2}(a+b-1))}{\Gamma(\frac{1}{2}(a-1))\Gamma(\frac{1}{2}(b-1))\Gamma(c-\frac{1}{2}(a+1))\Gamma(c-\frac{1}{2}(b+1))} - \frac{(a-2)(b-2)}{a+b-3} - 1\right\}.$$

(g) *In Theorem 17, if we take $m = 1, 2, 3$, we get the following results.*

$$\int_0^\infty e^{-st}\, t^{a-1}\, {}_2F_2\left[\begin{matrix} 1-a, & b \\ c, & 2b-c+1 \end{matrix}; st\right] dt \tag{71}$$
$$= \frac{\pi\, 2^{1-2b}\,\Gamma(a)\Gamma(c)\Gamma(2b-c+1)}{s^a\,\Gamma(\frac{1}{2}(a+c))\Gamma(b+\frac{1}{2}(a-c+1))\Gamma(\frac{1}{2}(1-a+c))\Gamma(b+1-\frac{1}{2}(a+c))},$$

$$\int_0^\infty e^{-st}\, t^{a-1}\, {}_3F_3\left[\begin{matrix} 3-a, & b, & 1 \\ c, & 2b-c+1, & 2 \end{matrix}; st\right] dt = \frac{(c-1)(c-2b)\Gamma(a)}{s^a\,(a-2)_2(b-1)} \tag{72}$$
$$\times\left\{\frac{\pi 2^{3-2b}\Gamma(c-1)\Gamma(2b-c)}{\Gamma(\frac{1}{2}(a+c)-1)\Gamma(b+\frac{1}{2}(a-c-1))\Gamma(\frac{1}{2}(1-a+c))\Gamma(b+1-\frac{1}{2}(a+c))} - 1\right\}$$

and

$$\int_0^\infty e^{-st}\, t^{a-1}\, {}_3F_3\left[\begin{matrix} 5-a, & b, & 1 \\ c, & 2b-c+1, & 3 \end{matrix}; st\right] dt = \frac{2(c-2)_2(2b-c-1)_2\Gamma(a)}{s^a\,(a-4)_4(b-2)_2} \tag{73}$$
$$\times\left\{\frac{\pi\, 2^{5-2b}\,\Gamma(c-2)\Gamma(2b-c+1)}{\Gamma(\frac{1}{2}(a+c)-2)\Gamma(b+\frac{1}{2}(a-c-3))\Gamma(\frac{1}{2}(1-a+c))\Gamma(b+1-\frac{1}{2}(a+c))} - \frac{(a-2)(3-a)(b-2)}{(c-2)(2b-c-1)} - 1\right\}.$$

(h) In Theorem 21, if we take $m = 1, 2, 3$, we get the following results.

$$\int_0^\infty e^{-st}\, t^{a-1}\, {}_2F_2\left[\begin{matrix} -n, & b \\ 1+a+b-c-n, & c \end{matrix}; st\right] dt = \frac{\Gamma(a)(c-a)_n(c-b)_n}{s^a\,(c)_n(c-a-b)_n}, \tag{74}$$

$$\begin{aligned}&\int_0^\infty e^{-st}\, t^{a-1}{}_3F_3\left[\begin{matrix} -n+1, & b, & 1 \\ 1+a+b-c-n, & c, & 2 \end{matrix}; st\right] dt \\ &= \frac{(1-c)(c-a-b+n)\Gamma(a-1)}{n(1-b)\,s^a}\left\{\frac{(c-a)_n(c-b)_n}{(c)_n(c-a-b+1)_n} - 1\right\}\end{aligned} \tag{75}$$

and

$$\begin{aligned}&\int_0^\infty e^{-st}\, t^{a-1}{}_3F_3\left[\begin{matrix} -n+2, & b, & 1 \\ 1+a+b-c-n, & c, & 3 \end{matrix}; st\right] dt \\ &= \frac{2(1-c)_2(c-a-b+n)_2\,\Gamma(a)}{s^a\,(1-a)_2(1-b)_2} \\ &\quad\times\left\{\frac{(c-a)_n(c-b)_n}{(c-2)_n(c-a-b+2)_n} + \frac{n(a-2)(b-2)}{(c-2)(a+b-c-n-1)} - 1\right\}.\end{aligned} \tag{76}$$

Remark 5. *The results in Equations* (65)*,* (68) *and* (71) *are known results due to Kim et al. [24].*

[C] Laplace-type integrals involving generalized hypergeometric function ${}_4F_4$

The results to be established are asserted in the following theorems.

Theorem 23. *For $m \in \mathbb{N}$, $Re(s) > 0$, $Re(a) > 0$ and $Re(a - 2b - 2c + 3m) > 2$, the following result holds true.*

$$\int_0^\infty e^{-st}\, t^{a-1}\, {}_4F_4\left[\begin{matrix} \frac{1}{2}(a+m+1), & b, & c, & 1 \\ \frac{1}{2}(a+m-1), & a-b+m, & a-c+m, & m \end{matrix}; -st\right] dt = \Gamma(a)\, s^{-a}\, \Omega_9, \tag{77}$$

where Ω_9 is the same as given in Equation (24)*.*

Theorem 24. *For $m \in \mathbb{N}$, $Re(s) > 0$, $Re(c) > 0$ and $Re(a - 2b - 2c + 3m) > 2$, the following result holds true.*

$$\int_0^\infty e^{-st}\, t^{c-1}\, {}_4F_4\left[\begin{matrix} a, & \frac{1}{2}(a+m+1), & b, & 1 \\ \frac{1}{2}(a+m-1), & a-b+m, & a-c+m, & m \end{matrix}; -st\right] dt = \Gamma(c)\, s^{-c}\, \Omega_9, \tag{78}$$

where Ω_9 is the same as given in Equation (24)*.*

Theorem 25. *For $m \in \mathbb{N}$, $Re(s) > 0$, $Re(a + m + 1) > 0$ and $Re(a - 2b - 2c + 3m) > 2$, the following result holds true.*

$$\begin{aligned}&\int_0^\infty e^{-st}\, t^{\frac{1}{2}(a+m-1)}\, {}_4F_4\left[\begin{matrix} a, & b, & c, & 1 \\ \frac{1}{2}(a+m-1), & a-b+m, & a-c+m, & m \end{matrix}; -st\right] dt \\ &= \Gamma(\frac{1}{2}(a+m+1))\, s^{-\frac{1}{2}(a+m+1)}\, \Omega_9,\end{aligned} \tag{79}$$

where Ω_9 is the same as given in Equation (24)*.*

Theorem 26. *For $m \in \mathbb{N}$, $Re(s) > 0$ and $Re(a - 2b - 2c + 3m) > 2$, the following result holds true.*

$$\int_0^\infty e^{-st}\, {}_4F_4\left[\begin{matrix} a, & \frac{1}{2}(a+m+1), & b, & c \\ \frac{1}{2}(a+m-1), & a-b+m, & a-c+m, & m \end{matrix}; -st\right] dt = s^{-1}\Omega_9, \tag{80}$$

where Ω_9 is the same as given in Equation (24).

Corollary 3. **(i)** *In Theorem 24, if we take $m = 1, 2, 3$, we get the following results.*

$$\begin{aligned} &\int_0^\infty e^{-st}\, t^{c-1}\, {}_3F_3\left[\begin{matrix} a, & \frac{1}{2}(a+2), & b \\ \frac{1}{2}a, & a-b+1, & a-c+1 \end{matrix}; -st\right] dt \\ &\quad = \frac{\Gamma(1+a-b)\Gamma(1+a-c)\Gamma(c)}{s^c\,\Gamma(1+a)\Gamma(1+a-b-c)}, \end{aligned} \tag{81}$$

$$\begin{aligned} &\int_0^\infty e^{-st}\, t^{c-1}\, {}_4F_4\left[\begin{matrix} a, & \frac{1}{2}(a+3), & b, & 1 \\ \frac{1}{2}(a+1), & a-b+2, & a-c+2, & 2 \end{matrix}; -st\right] dt \\ &= \frac{(1+a-b)(1+a-c)\Gamma(c)}{s^c\,(a+1)(b-1)(c-1)} \left\{1 - \frac{\Gamma(1+a-b)\Gamma(1+a-c)}{\Gamma(a)\Gamma(2+a-b-c)}\right\} \end{aligned} \tag{82}$$

and

$$\begin{aligned} &\int_0^\infty e^{-st}\, t^{c-1}\, {}_4F_4\left[\begin{matrix} a, & \frac{1}{2}(a+4), & b, & 1 \\ \frac{1}{2}(a+3), & a-b+3, & a-c+3, & 3 \end{matrix}; -st\right] dt \\ &= \frac{2(1+a-b)_2(1+a-c)_2\,\Gamma(c)}{s^c\,(a+2)(a-1)(b-2)_2(c-2)_2} \\ &\quad \times \left\{\frac{\Gamma(1+a-b)\Gamma(1+a-c)}{\Gamma(a-1)\Gamma(3+a-b-c)} + \frac{a(b-2)(c-2)}{(1+a-b)(1+a-c)} - 1\right\}. \end{aligned} \tag{83}$$

Similarly, other results can be obtained from Theorems 23, 25 and 26.

[D] Laplace-type integrals involving generalized hypergeometric function ${}_5F_5$

The results to be established are asserted in the following theorems.

Theorem 27. *For $m \in \mathbb{N}$, $Re(s) > 0$, $Re(a) > 0$ and $Re(2a - 2c - 2d - 2e + 3m) > 2$, the following result holds true.*

$$\begin{aligned} &\int_0^\infty e^{-st}\, t^{a-1}\, {}_5F_5\left[\begin{matrix} \frac{1}{2}(a+m+1), & c, & d, & e, & 1 \\ \frac{1}{2}(a+m-1), & a-c+m, & a-d+m, & a-e+m, & m \end{matrix}; st\right] dt \\ &= \Gamma(a)\, s^{-a}\, \Omega_{10}, \end{aligned} \tag{84}$$

where Ω_{10} is the same as given in Equation (25).

Theorem 28. *For $m \in \mathbb{N}$, $Re(s) > 0$, $Re(c) > 0$ and $Re(2a - 2c - 2d - 2e + 3m) > 2$, the following result holds true.*

$$\begin{aligned} &\int_0^\infty e^{-st}\, t^{c-1}\, {}_5F_5\left[\begin{matrix} a, & \frac{1}{2}(a+m+1), & d, & e, & 1 \\ \frac{1}{2}(a+m-1), & a-c+m, & a-d+m, & a-e+m, & m \end{matrix}; st\right] dt \\ &= \Gamma(c)\, s^{-c}\, \Omega_{10}, \end{aligned} \tag{85}$$

where Ω_{10} is the same as given in Equation (25).

Theorem 29. *For $m \in \mathbb{N}$, $Re(s) > 0$, $Re(a+m+1) > 0$ and $Re(2a-2c-2d-2e+3m) > 2$, the following result holds true.*

$$\int_0^\infty e^{-st}\, t^{\frac{1}{2}(a+m-1)}\, {}_5F_5\left[\begin{matrix} a, & c, & d, & e, & 1 \\ \frac{1}{2}(a+m-1), & a-c+m, & a-d+m, & a-e+m, & m \end{matrix}; st\right] dt \tag{86}$$
$$= \Gamma(\frac{1}{2}(a+m+1))\, s^{-\frac{1}{2}(a+m+1)}\, \Omega_{10},$$

where Ω_{10} is the same as given in Equation (25).

Theorem 30. *For $m \in \mathbb{N}$, $Re(s) > 0$ and $Re(2a-2c-2d-2e+3m) > 2$, the following result holds true.*

$$\int_0^\infty e^{-st}\, {}_5F_5\left[\begin{matrix} a, & \frac{1}{2}(a+m-1), & c, & d, & e \\ \frac{1}{2}(a+m-1), & a-c+m, & a-d+m, & a-e+m, & m \end{matrix}; st\right] dt \tag{87}$$
$$= s^{-1}\Omega_{10},$$

where Ω_{10} is the same as given in Equation (25).

Corollary 4. **(j)** *In Theorem 27, if we take $m = 1, 2, 3$, we get the following results.*

$$\int_0^\infty e^{-st}\, t^{a-1}\, {}_4F_4\left[\begin{matrix} c, & \frac{1}{2}(a+2), & d, & e \\ \frac{1}{2}a, & a-c+1, & a-d+1, & a-e+1 \end{matrix}; st\right] dt \tag{88}$$
$$= \frac{\Gamma(1+a-c)\Gamma(1+a-d)\Gamma(1+a-e)\Gamma(1+a-c-d-e)}{s^a\,\Gamma(1+a-d-e)\Gamma(1+a-c-e)\Gamma(1+a-c-d)},$$

$$\int_0^\infty e^{-st}\, t^{a-1}\, {}_5F_5\left[\begin{matrix} c, & \frac{1}{2}(a+3), & d, & e, & 1 \\ \frac{1}{2}(a+1), & a-c+2, & a-d+2, & a-e+2, & 2 \end{matrix}; st\right] dt \tag{89}$$
$$= \frac{(1+a-c)(1+a-d)(1+a-e)\,\Gamma(a)}{s^a\,(1+a)(c-1)(d-1)(e-1)}$$
$$\times \left\{\frac{\Gamma(1+a-c)\Gamma(1+a-d)\Gamma(1+a-e)\Gamma(3+a-c-d-e)}{\Gamma(a)\Gamma(2+a-d-e)\Gamma(2+a-c-e)\Gamma(2+a-c-d)} - 1\right\}$$

and

$$\int_0^\infty e^{-st}\, t^{a-1}\, {}_5F_5\left[\begin{matrix} c, & \frac{1}{2}(a+4), & d, & e, & 1 \\ \frac{1}{2}(a+2), & a-c+3, & a-d+3, & a-e+3, & 3 \end{matrix}; st\right] dt \tag{90}$$
$$= \frac{2(1+a-c)_2(1+a-d)_2(1+a-e)_2\,\Gamma(a)}{s^a\,(a-1)(a+2)(c-2)_2(d-2)_2(e-2)_2}$$
$$\times \left\{\frac{\Gamma(1+a-c)\Gamma(1+a-d)\Gamma(1+a-e)\Gamma(5+a-c-d-e)}{\Gamma(a-1)\Gamma(3+a-d-e)\Gamma(3+a-c-e)\Gamma(3+a-c-d)}\right.$$
$$\left. - \frac{a(c-2)(d-2)(e-2)}{(1+a-c)(1+a-d)(1+a-e)}\right\}.$$

Similarly, other results can be obtained from Theorems 28–30.

[E] Laplace-type integrals involving generalized hypergeometric function ${}_6F_6$

The results to be established are asserted in the following theorems.

Theorem 31. *For $m \in \mathbb{N}$, $Re(s) > 0$ and $Re(a) > 0$, the following result holds true.*

$$\int_0^\infty e^{-st}\, t^{a-1} \times \tag{91}$$
$${}_7F_7\left[\begin{matrix} \frac{1}{2}(a+m+1), & b, & c, & d, & 2a-b-c-d+2m-1+n, & m-n-1, & 1 \\ \frac{1}{2}(a+m-1), & a-b+m, & a-c+m, & a-d+m, & b+c+d-a+1-m-n, & a+n+1, & m \end{matrix}; st\right] dt$$
$$= \Gamma(a)\, s^{-a}\, \Omega_{11},$$

where Ω_{11} is the same as given in Equation (26).

Theorem 32. *For $m \in \mathbb{N}$, $Re(s) > 0$ and $Re(a+m+1) > 0$, the following result holds true.*

$$\int_0^\infty e^{-st}\, t^{\frac{1}{2}(a+m-1)} \times \tag{92}$$
$${}_7F_7\left[\begin{matrix} a, & b, & c, & d, & 2a-b-c-d+2m-1+n, & m-n-1, & 1 \\ \frac{1}{2}(a+m-1), & a-b+m, & a-c+m, & a-d+m, & b+c+d-a+1-m-n, & a+n+1, & m \end{matrix}; st\right] dt$$
$$= \Gamma\left(\frac{1}{2}(a+m+1)\right) s^{-\frac{1}{2}(a+m+1)}\, \Omega_{11},$$

where Ω_{11} is the same as given in Equation (26).

Theorem 33. *For $m \in \mathbb{N}$, $Re(s) > 0$ and $Re(b) > 0$, the following result holds true.*

$$\int_0^\infty e^{-st}\, t^{b-1} \times \tag{93}$$
$${}_7F_7\left[\begin{matrix} \frac{1}{2}(a+m+1), & b, & c, & d, & 2a-b-c-d+2m-1+n, & m-n-1, & 1 \\ \frac{1}{2}(a+m-1), & a-b+m, & a-c+m, & a-d+m, & b+c+d-a+1-m-n, & a+n+1, & m \end{matrix}; st\right] dt$$
$$= \Gamma(b)\, s^{-b}\, \Omega_{11},$$

where Ω_{11} is the same as given in Equation (26).

Theorem 34. *For $m \in \mathbb{N}$, $Re(s) > 0$ and $Re(2a-b-c-d+2m-1+n) > 0$, the following result holds true.*

$$\int_0^\infty e^{-st}\, t^{2a-b-c-d+2m+n-2} \times \tag{94}$$
$${}_7F_7\left[\begin{matrix} \frac{1}{2}(a+m+1), & a, & b, & c, & d, & m-n-1, & 1 \\ \frac{1}{2}(a+m-1), & a-b+m, & a-c+m, & a-d+m, & b+c+d-a+1-m-n, & a+n+1, & m \end{matrix}; st\right] dt$$
$$= \Gamma(2a-b-c-d+2m-1+n)\, s^{-(2a-b-c-d+2m-1+n)}\, \Omega_{11},$$

where Ω_{11} is the same as given in Equation (26).

Theorem 35. *For $m \in \mathbb{N}$ and $Re(s) > 0$, the following result holds true.*

$$\int_0^\infty e^{-st}\, {}_7F_7\left[\begin{matrix} \frac{1}{2}(a+m+1), & a, & b, & c, & d, & m-n-1, & 1 \\ \frac{1}{2}(a+m-1), & a-b+m, & a-c+m, & a-d+m, & b+c+d-a+1-m-n, & a+n+1, & m \end{matrix}; st\right] dt \tag{95}$$
$$= s^{-1}\Omega_{11},$$

where Ω_{11} is the same as given in Equation (26).

Corollary 5. (k) *In Theorem 31, if we take $m = 1, 2, 3$, we get the following results.*

$$\int_0^\infty e^{-st}\, t^{a-1} \tag{96}$$
$$\times\, {}_6F_6\left[\begin{matrix} b, & \frac{1}{2}(a+2), & c, & d, & 2a-b-c-d+n+1, & -n \\ \frac{1}{2}a, 1+a-b, 1+a-c, 1+a-d, b+c+d-a-n, a+n+1 \end{matrix} ; st\right] dt$$
$$= \frac{\Gamma(a)(1+a)_n(a-b-c+1)_n(a-b-d+1)_n(a-c-d+1)_n}{s^a\,(1+a-b)_n(1+a-c)_n(1+a-d)_n(1+a-b-c-d)_n},$$

$$\int_0^\infty e^{-st}\, t^{a-1} \tag{97}$$
$$\times\, {}_7F_7\left[\begin{matrix} b, & \frac{1}{2}(a+3), & c, & d, & 2a-b-c-d+n+3, & 1-n, & 1 \\ \frac{1}{2}(a+1), 1+a-b, 1+a-c, 1+a-d, b+c+d-a-n-1, a+n+1, 2 \end{matrix} ; st\right] dt$$
$$= \frac{(b-a-1)(c-a-1)(d-a-1)(n+2+a-b-c-d)(a+n)\Gamma(a)}{n\, s^a\,(1+a)(1-b)(1-c)(1-d)(b+c+d-2a-2-n)}$$
$$\times \left\{1 - \frac{(a)_n(a-b-c+2)_n(a-b-d+2)_n(a-c-d+2)_n}{(1+a-b)_n(1+a-c)_n(1+a-d)_n(3+a-b-c-d)_n}\right\}$$

and

$$\int_0^\infty e^{-st}\, t^{a-1} \tag{98}$$
$$\times\, {}_7F_7\left[\begin{matrix} b, & \frac{1}{2}(a+4), & c, & d, & 2a-b-c-d+n+5, & 2-n, & 1 \\ \frac{1}{2}(a+2), 3+a-b, 3+a-c, 3+a-d, b+c+d-a-n-2, a+n+1, 3 \end{matrix} ; st\right] dt$$
$$= \frac{(a-2)(b-a-2)_2(c-a-2)_2(d-a-2)_2}{s^a\,(a+2)(1-a)_2(1-b)_2(1-c)_2(1-d)_2}$$
$$\times \frac{(-a-n)_n(3+n+a-b-c-d)_2\,\Gamma(a)}{(n-1)_2(b+c+d-2a-4-n)_2}$$
$$\times \left\{\frac{(a-1)_n(a-b-c+3)_n(a-b-d+3)_n(a-c-d+3)_n}{(a-b+1)_n(a-c+1)_n(a-d+1)_n(a-b-c-d+5)_n}\right.$$
$$\left. + \frac{na(b-2)(c-2)(d-2)(2a-b-c+n+3)}{(a-b+1)(a-c+1)(a-d+1)(b+c+d-a-n-4)(n+a-1)} - 1\right\}.$$

Similarly, other results can be obtained from Theorems 32–35.

Remark 6. *For evaluation of Eulerian's type integrals involving generalized hypergeometric functions by employing the summation theorems, Equations* (16)–(26)*, we refer an interesting paper by Jun et al. [25].*

3. Concluding Remark

In the theory of generalized hypergeometric functions, classical summation theorems such as those of Gauss, Gauss second, Kummer, Bailey, Dixon, Watson, Whipple, Saalschütz and Dougall play a key role. Applications of the above-mentioned classical summation theorems are well-known. Very recently, Masjed-Jamei and Koepf established interesting and useful generalizations of the above-mentioned classical summation theorems in the most general form.

In this paper, an attempt has been made for providing a list of several Laplace-type integrals involving generalized hypergeometric functions ${}_pF_p$ for $p = 2, 3, 4, 5$ and 7 in the most general forms which would serve as key formulas from which, on specializing the parameters, lead to several results, some of which are known and others are believed to be new. The results established in this paper are simple, interesting, easily proven and may be potentially useful.

We conclude this section by remarking that other applications of the generalized summation theorems due to Masjed-Jamei and Koepf are under investigations and the same will form a part of the subsequent paper in this direction.

Author Contributions: All authors contributed equally to writing of this paper. All authors read and approved the final manuscript.

Acknowledgments: The authors are grateful to the referees for making certain very useful suggestions which led to a better presentation of the paper.

References

1. Andrews, G.E.; Askey, R.; Roy, R. Special Functions. In *Encyclopedia of Mathematics and Its Applications*; Cambridge University Press: Cambridge, UK, 1999; Volume 71.
2. Bailey, W.N. *Generalized Hypergeometric Series*; Cambridge University Press: Cambridge, UK, 1935; Reprinted by Stechert-Hafner, New York, NY, USA, 1964.
3. Oberhettinger, F.; Badi, L. *Tables of Laplace Transforms*; Springer: Berlin, Germany, 1973.
4. Rainville, E.D. *Special Functions*; The Macmillan Company: New York, NY, USA, 1960; Reprinted by Chelsea Publishing Company, Bronx, NY, USA, 1971.
5. Koepf, W. *Hypergeometric Summation: An Algorithmic Approach to Summation and Special Function Identities*, 2nd ed.; Springer: London, UK, 2014.
6. Bromwich, T.J. *An Introduction to the Theory of Infinite Series*; Macmillan: New York, NY, USA, 1948.
7. Knopp, K. *Theory and Applications of Infinite Series*: Hafner: New York, NY, USA, 1949; Reprinted by Dover, 1990.
8. Luke, Y.L. *The Special Functions and Their Approximations*; Academic Press: New York, NY, USA, 1969; Volume 1.
9. Bailey, W.N. Products of Generalized Hypergeometric Series. *Proc. Lond. Math. Soc.* **1928**, *28*, 242–254. [CrossRef]
10. Prudnikov, A.P.; Brychkov, Yu.A.; Marichev, O.I. *More Special Functions*; Integrals and Series; Gordon and Breach Science Publishers: Amsterdam, The Netherlands, 1990; Volume 3.
11. Masjed-Jamei, M.; Koepf, W. Some Summation Theorems for Generalized Hypergeometric Functions. *Axioms* **2018**, *7*, 38. [CrossRef]
12. Kim, Y.S.; Rakha, M.A.; Rathie, A.K. Extensions of Certain Classical Summation Theorems for the Series ${}_2F_1$, ${}_3F_2$ and ${}_4F_3$ with Applications in Ramanujan's Summations. *Int. J. Math. Math. Sci.* **2010**, *2010*, 309503. [CrossRef]
13. Lavoie, J.L.; Grondin, F.; Rathie, A.K. Generalizations of Watson's theorem on the sum of a ${}_3F_2$. *Indian J. Math.* **1992**, *34*, 23–32.
14. Lavoie, J.L.; Grondin, F.; Rathie, A.K.; Arora, K. Generalizations of Dixon's Theorem on the sum of a ${}_3F_2(1)$. *Math. Comp.* **1994**, *62*, 267–276.
15. Lavoie, J.L.; Grondin, F.; Rathie, A.K. Generalizations of Whipple's theorem on the sum of a ${}_3F_2$. *J. Comput. Appl. Math.* **1996**, *72*, 293–300. [CrossRef]
16. Rakha, M.A.; Rathie, A.K. Generalizations of classical summation theorems for the series ${}_2F_1$ and ${}_3F_2$ with applications. *Integral Transform. Spec. Funct.* **2011**, *22*, 823–840. [CrossRef]
17. Davis, B. *Integral Transforms and Their Applications*, 3rd ed.; Springer: New York, NY, USA, 2002.
18. Doetsch, G. *Introduction to the Theory and Applications of the Laplace Transformation*; Springer: New York, NY, USA, 1974.
19. Erdelyi, A.; Magnus, W.; Oberhettinger, F.; Tricomi, F.G. *Tables of Integral Transforms*; McGraw Hill: New York, NY, USA, 1954; Volume I–II.
20. Slater, L.J. *Confluent Hypergeometric Functions*; Cambridge University Press: Cambridge, UK, 1960.
21. Deepthi, P.; Prajapati, J.C.; Rathie, A.K. New Laplace transforms of the ${}_2F_2$ hypergeometric function. *J. Fract. Calc. Appl.* **2017**, *8*, 150–155.
22. Parmar, R.K.; Purohit, S.D. Certain integral transforms and fractional integral formulas for the extended hypergeometric functions. *TWMS J. Appl. Eng. Math.* **2017**, *7*, 74–81.

23. Kim, Y.S.; Rathie, A.K.; Civijovic, D. New Laplace transforms of Kummer's confluent hypergeometric functions. *Math. Comput. Model.* **2012**, *55*, 1068–1071. [CrossRef]
24. Kim, Y.S.; Rathie, A.K.; Lee, C.H. New Laplace transforms for the generalized hypergeometric functions ${}_2F_2$. *Honam Math. J.* **2015**, *37*, 245–252. [CrossRef]
25. Jun, S.; Kim, I.; Rathie, A.K. On a new class of Eulerian's type integrals involving generalized hypergeometric functions. *Aust. J. Math. Anal. Appl.* **2019**, *16*, 10.

Permissions

All chapters in this book were first published by MDPI; hereby published with permission under the Creative Commons Attribution License or equivalent. Every chapter published in this book has been scrutinized by our experts. Their significance has been extensively debated. The topics covered herein carry significant findings which will fuel the growth of the discipline. They may even be implemented as practical applications or may be referred to as a beginning point for another development.

The contributors of this book come from diverse backgrounds, making this book a truly international effort. This book will bring forth new frontiers with its revolutionizing research information and detailed analysis of the nascent developments around the world.

We would like to thank all the contributing authors for lending their expertise to make the book truly unique. They have played a crucial role in the development of this book. Without their invaluable contributions this book wouldn't have been possible. They have made vital efforts to compile up to date information on the varied aspects of this subject to make this book a valuable addition to the collection of many professionals and students.

This book was conceptualized with the vision of imparting up-to-date information and advanced data in this field. To ensure the same, a matchless editorial board was set up. Every individual on the board went through rigorous rounds of assessment to prove their worth. After which they invested a large part of their time researching and compiling the most relevant data for our readers.

The editorial board has been involved in producing this book since its inception. They have spent rigorous hours researching and exploring the diverse topics which have resulted in the successful publishing of this book. They have passed on their knowledge of decades through this book. To expedite this challenging task, the publisher supported the team at every step. A small team of assistant editors was also appointed to further simplify the editing procedure and attain best results for the readers.

Apart from the editorial board, the designing team has also invested a significant amount of their time in understanding the subject and creating the most relevant covers. They scrutinized every image to scout for the most suitable representation of the subject and create an appropriate cover for the book.

The publishing team has been an ardent support to the editorial, designing and production team. Their endless efforts to recruit the best for this project, has resulted in the accomplishment of this book. They are a veteran in the field of academics and their pool of knowledge is as vast as their experience in printing. Their expertise and guidance has proved useful at every step. Their uncompromising quality standards have made this book an exceptional effort. Their encouragement from time to time has been an inspiration for everyone.

The publisher and the editorial board hope that this book will prove to be a valuable piece of knowledge for researchers, students, practitioners and scholars across the globe.

List of Contributors

Ram K. Saxena
Department of Mathematics and Statistics, Jai Narain Vyas University, Jodhpur-342004, Rajasthan, India

Rakesh K. Parmar
Department of Mathematics, Government College of Engineering and Technology, Bikaner-334004, Rajasthan, India

Hari M. Srivastava
Department of Mathematics and Statistics, University of Victoria, Victoria, BC V8W 3R4, Canada
Department of Medical Research, China Medical University Hospital, China Medical University, Taichung 40402, Taiwan
Department of Mathematics and Informatics, Azerbaijan University, 71 Jeyhun Hajibeyli Street, Baku AZ1007, Azerbaijan

Jian Cao
Department of Mathematics, Hangzhou Normal University, Hangzhou City 311121, China

Sama Arjika
Department of Mathematics and Informatics, University of Agadez, Agadez 8000, Niger
International Chair of Mathematical Physics and Applications (ICMPA-UNESCO Chair), University of Abomey-Calavi, Cotonou 50, Benin

Nabil Mlaiki
Department of Mathematics and General Sciences, Prince Sultan University, Riyadh 11586, Saudi Arabia

Katarina Kukić
Faculty for Traffic and Transport Engineering, University of Belgrade, 11000 Belgrade, Serbia

Milanka Gardašević-Filipović
School of Computing, Union University, 11000 Belgrade, Serbia

Hassen Aydi
Institut Supérieur d'Informatique et des Techniques de Communication, Université de Sousse, H. Sousse 4000, Tunisia
China Medical University Hospital, China Medical University, Taichung 40402, Taiwan

Wolfram Koepf
Department of Mathematics, University of Kassel, Heinrich-Plett-Str. 40, D-34132 Kassel, Germany

Mohammad Masjed-Jamei
Department of Mathematics, University of Kassel, Heinrich-Plett-Str. 40, D-34132 Kassel, Germany
Department of Mathematics, K. N. Toosi University of Technology, 11369 Tehran, Iran

Hafize Gümüş
Department of Math Education, Faculty of Eregli Education, Necmettin Erbakan University, Konya 42060, Turkey

Nihal Demir
İkizli Middle School, Konya 42060, Turkey

Nihal Taş and Nihal Yılmaz Özgür
Department of Mathematics, Balıkesir University, 10145 Balıkesir, Turkey

Antti Rasila
Department of Mathematics and Systems Analysis, School of Science, Aalto University, FIN-00076 Aalto, Finland

Tommi Sottinen
Department of Mathematics and Statistics, Faculty of Technology, University of Vaasa, FIN-65101 Vaasa and Finland

Hari Mohan Srivastava
Department of Mathematics and Statistics, University of Victoria, Victoria, BC V8W 3R4, Canada
Department of Medical Research, China Medical University Hospital, China Medical University Taichung 40402, Taiwan
Department of Mathematics and Informatics, Azerbaijan University, 71 Jeyhun Hajibeyli Street, AZ1007 Baku, Azerbaijan

Gürhan İçöz and Bayram Çekim
Department of Mathematics, Gazi University, Ankara TR-06500, Turkey

Jean-Daniel Djida
Departamento de Análise Matemática, Universidade de Santiago de Compostela, 15782 Santiago de Compostela, Spain

Arran Fernandez
Department of Applied Mathematics & Theoretical Physics, University of Cambridge, Cambridge CB3 0WA, UK

Tariq Qawasmeh and Abdalla Tallafha
Department of mathematics, School of Science, University of Jordan, Amman 11942, Jordan

Wasfi Shatanawi
Department of Mathematics and general courses, Prince Sultan University, Riyadh 11586, Saudi Arabia
Department of Medical Research, China Medical University Hospital, China Medical University, Taichung 40402, Taiwan

Ömer Kişi
Department of Mathematics, Faculty of Science, Bartin University, Bartin 74100, Turkey

Ekrem Savas
Department of Mathematics, Istanbul Ticaret University, Üsküdar, Istanbul 34840, Turkey

Rekha Srivastava
Department of Mathematics and Statistics, University of Victoria, Victoria, BC V8W 3R4, Canada

Humera Naaz, Sabeena Kazi and Asifa Tassaddiq
College of Computer and Information Sciences, Majmaah University, Al Majmaah 11952, Saudi Arabia

Makrina Agaoglou
Mathematical Institute of Slovak Academy of Sciences, Štefánikova 49, 814 73 Bratislava, Slovakia
Department of Mechanical Engineering, Faculty of Engineering, Aristotle University of Thessaloniki, 54124 Thessaloniki, Greece

Vassilis M. Rothos
Department of Mechanical Engineering, Faculty of Engineering, Aristotle University of Thessaloniki, 54124 Thessaloniki, Greece

Michal Fečkan and Michal Pospíšil
Mathematical Institute of Slovak Academy of Sciences, Štefánikova 49, 814 73 Bratislava, Slovakia
Department of Mathematical Analysis and Numerical Mathematics, Comenius University in Bratislava, Mlynská dolina, 842 48 Bratislava, Slovakia

Alexander F. Vakakis
Department of Mechanical Science and Engineering, University of Illinois at Urbana-Champaign, Urbana, IL 61801, USA

Paolo Emilio Ricci
Section of Mathematics, International Telematic University UniNettuno, Corso Vittorio Emanuele II, 39, 00186 Roma, Italy

Nonthamon Chaikham and Wannika Sawangtong
Department of Mathematics, Faculty of Science, Mahidol University, Bangkok 10400, Thailand
Centre of Excellence in Mathematics, Commission on Higher Education, Ministry of Education, 328 Sri Ayuthaya Road, Bangkok 10400, Thailand

Insuk Kim
Department of Mathematics Education, Wonkwang University, Iksan 570-749, Korea

Arjun K. Rathie
Department of Mathematics, Vedant College of Engineering and Technology (Rajasthan Technical University), Bundi 323021, Rajasthan, India

Index

www.ingramcontent.com/pod-product-compliance
Lightning Source LLC
LaVergne TN
LVHW082204200926
848484LV00008B/2210